Ecological Studies, Vol. 203

Analysis and Synthesis

Edited by

M.M. Caldwell, Washington, USA
G. Heldmaier, Marburg, Germany
R.B. Jackson, Durham, USA
O.L. Lange, Würzburg, Germany
H.A. Mooney, Stanford, USA
E.-D. Schulze, Jena, Germany
U. Sommer, Kiel, Germany

Ecological Studies

Volumes published since 2006 are listed at the end of this book.

A. Johannes Dolman • Riccardo Valentini
Annette Freibauer

Editors

The Continental-Scale Greenhouse Gas Balance of Europe

 Springer

Editors
A. Johannes Dolman
VU University
Amsterdam
The Netherlands

Riccardo Valentini
University of Tuscia
Viterbo
Italy

Annette Freibauer
Max Planck Institute for Biogeochemistry
Jena
Germany

ISBN: 978-0-387-76568-6   e-ISBN: 978-0-387-76570-9
DOI: 10.1007/978-0-387-76570-9

Library of Congress Control Number: 2008920393

Printed on acid-free paper

9 8 7 6 5 4 3 2 1

springer.com

Preface

The human interference with the climate system, the perturbation of the carbon cycle through massive release of greenhouse gases, caused by fossil fuel burning and land-use change, is threatening society and represents a key challenge for research and policies in the twenty-first century. Growing evidence of human-induced climate change has raised public concern calling for urgent international policy actions. Initiatives culminated in the establishment of the United Nations Framework Convention for Climate Change (UNFCCC) and the Kyoto Protocol (1997), where Parties for the first time agreed on legally binding commitments to reduce greenhouse gas emissions. It is worth noting that the unfinished 'sink' business, the Articles in the Kyoto Protocol dealing with terrestrial biospheric carbon dioxide sources and sinks, gave carbon cycle research a real boost. In the 1990s, the regional carbon balance and how the different ecosystems contribute at different timescales under different environmental conditions were hardly known.

During the fourth Framework Programme (1994–1998), the European Union supported more than 20 research projects studying the components of the carbon cycle. These projects provided a solid basis for a more integrated attempt to tackle the research challenges and demands imposed by the Kyoto Protocol at European scale.

Both the European Commission and the scientific community felt that it was time to develop an integrated carbon cycle research programme taking the new challenges on board. Since an international science plan was lacking, the European Commission initiated an international workshop in Orvieto, 24 June 1998, in order to discuss the current status of research and find ways to overcome the European research fragmentation in this area. The CarboEurope idea was then presented for the first time.

The fifth Framework Programme (FP5, 1998–2002), Key Action: 'Global Change, Climate and Biodiversity', was focusing on applied research, with the specific aim to support the implementation of international conventions. Naturally, carbon cycle research received high priority. During the lifetime of FP5, the CarboEurope concept was implemented through a cluster of research projects, and also became a blueprint for other regional carbon programmes of the International Global Carbon Project (GCP).

During the sixth Framework Programme (2002–2006), and within the context of the European Research Area, a higher level of integration was achieved through the utilisation of specially designed funding instruments such as Integrated Projects (IP). This new funding instrument allowed the creation of large-scale consortia targeting research questions of strategic nature over periods of 5 years, the original concept now being implemented through the CarboEurope-IP.

The CARBOEUROPE-IP (2004–2008) brings together more than 60 research institutes and universities with the objective to assess the European Terrestrial Carbon Balance. The implementation of dual constrain concept (bottom–up, top–down) and the integration of long-term observations at different scales through modelling, followed over the past years, proved to be of great value for understanding and verifying sources and sinks at regional and continental scales. As a result, Europe has currently the densest and best integrated research network of in situ observations of ecosystem carbon and nitrogen fluxes and trace gas concentrations. These observations have discovered surprises in the ecosystem functioning and response to climate-related extremes and helped the further development of complex models. They also show that both scientific and societal questions about the carbon cycle can only be resolved by an integrated approach combining modelling with sustained long-term observations of key carbon variables. In the context of the seventh Framework Programme (2007–2013), integrated research on carbon cycle remains one of the key priorities under the Environment Programme.

We congratulate the authors for the present publication which provides an integrated assessment of our current capacity to observe the continental carbon cycle, understand the processes and quantify the uncertainties involved. It provides, with no doubt, a valuable contribution to the important ongoing scientific debate on carbon cycle and climate change.

Claus Bruening and **Anastasios Kentarchos**
Climate Change and Environmental Risks Unit
Environment Directorate,
Directorate General for Research
European Commission
Brussels
Belgium

Contents

Contributors

Gwenael Abril
Environnements et Paléoenvironnements OCéaniques (EPOC), Université de
Bordeaux 1. CNRS-UMR 5805, Avenue des Facultés, F-33405 Talence, France

Peter Bergamaschi
European Commission Joint Research Centre, Institute for Environment and
Sustainability, Ispra, Italy

Alberto V. Borges
University of Liège, Interfacultary Center for Marine Research (MARE),
Chemical Oceanography Unit, Institut de Physique (B5), Allée du 6 Août,
17, B-4000 Liège, Belgium

Philippe Bousquet
Laboratoire des Sciences du Climat et del'Environnement (LSCE), Gif sur Yvette,
France

Joseph G. Canadell
CSIRO Marine and Atmospheric Research, Canberra ACT 2601, Australia

Torben R. Christensen
GeoBiosphere Science Centre, Physical Geography and Ecosystems Analysis,
Lund University, Sweden

Galina Churkina
Max-Planck-Institute for Biogeochemistry P.O. Box 10 01 64, 07701 Jena and
Hans-Knoell-Strasse 10, 07745 Jena, Germany

Philippe Ciais
Laboratoire des Sciences du Climat et de l'Environnement, IPSL/LSCE
CEA-CNRS-UVSQ, F-91191 Gif sur Yvette, France

Mathias Disney
NERC Centre for Terrestrial Carbon Dynamics and University College London,
26 Bedford Way, London WC1H0AP, UK

A. Johannes Dolman
Department of Hydrology and Geo-Environmental Sciences, VU University
Amsterdam, Boelelaan 1085, 1081 HV Amsterdam, Netherlands

Matthias Drösler
Department of Vegetation Ecology, Technical University of Munich, Freising,
Germany

Gerd Folberth
School of Earth and Ocean Science (SEOS), University of Victoria, Victoria,
Canada

C. Mary R. Fowler
Department of Geology, Royal Holloway, Egham TW20 0EX, UK

Annette Freibauer
Max-Planck-Institute for Biogeochemistry, P.O. Box 10 01 64, 07701 Jena and
Hans-Knoell-Strasse 10, 07745 Jena, Germany

Thomas Friborg
Department of Geography and Geology, University of Copenhagen, Denmark,
tfj@geogr.ku.dk

Christoph Gerbig
Max-Planck-Institute for Biogeochemistry, P.O. Box 10 01 64, 07701 Jena and
Hans-Knoell-Strasse 10, 07745 Jena, Germany

Margriet Groenendijk
Department of Hydrology and Geo-Environmental Sciences, VU University
Amsterdam, Boelelaan 1085 1081 HV Amsterdam, Netherlands

Didier Hauglustaine
Laboratoire des Sciences du Climat et de l'Environnement, IPSL/LSCE
CEA-CNRS-UVSQ, F-91191 Gif sur Yvette, France

Martin Heimann
Max-Planck-Institute for Biogeochemistry, P.O. Box 10 01 64, 07701 Jena and
Hans-Knoell-Strasse 10, 07745 Jena, Germany

Dimmie Hendriks
Department of Hydrology and Geo-Environmental Sciences, Boelelaan 1085,
1081 HV, VU University Amsterdam, Netherlands

Ivan A. Janssens
Department of Biology, Universiteit Antwerpen, B-2160 Antwerpen, Belgium

Thomas Kaminski
FastOpt GmbH, Schanzenstr. 36, 20357 Hamburg, Germany

Ute Karstens
Max-Planck-Institute for Biogeochemistry, P.O. Box 10 01 64, 07701 Jena and
Hans-Knoell-Strasse 10, 07745 Jena, Germany

Thomas Lauvaux
Météo-France CNRM/GMME, Toulouse, France
LSCE, Paris, France

Aleksi Lehtonen
Finnish Forest Research Institute, Unioninkatu 40 A, FI-00170 Helsinki, Finland,
aleksi.lehtonen@metla.fi

Ingeborg Levin
Institut für Umweltphysik, University of Heidelberg, Heidelberg, Germany

Philip Lewis
NERC Centre for Terrestrial Carbon Dynamics and University College London,
26 Bedford Way, London WC1H0AP, UK, plewis@geog.ucl.ac.uk

Raisa Mäkipää
Finnish Forest Research Institute, Unioninkatu 40 A, FI-00170 Helsinki, Finland,
raisa.makipaa@metla.fi

Michel Meybeck
SISYPHE, Université Paris VI Jussieu, F-75005, Paris, France

Franco Miglietta
CNR IBIMET, Florence, Italy

Euan Nisbet
Department of Geology, Royal Holloway, University of London, Egham TW20
0EX, UK

Joel Noilhan
Météo-France CNRM/GMME, Toulouse, France

Mikko Peltoniemi
Finnish Forest Research Institute, Unioninkatu 40 A, FI-00170 Helsinki, Finland,
mikko.peltoniemi@metla.fi

Gorka Pérez-Landa
CEAM, Valencia, Spain

Heiko Pfeiffer
Institute of Energy Economics and the Rational Use of Energy, University of
Stuttgart, Hessbruehlstrasse 49a, 70565 Stuttgart, Germany

Phillip O'Brien
Department of Geology, Royal Holloway, Egham TW20 0EX, UK

Tristan Quaife
NERC Centre for Terrestrial Carbon Dynamics and University College London,
26 Bedford Way, London WC1H0AP, UK

Shaun Quegan
NERC Centre for Terrestrial Carbon Dynamics and the University of Sheffield,
Hicks Building, Hounsfield Road, Sheffield, UK

Michael R. Raupach
CSIRO Marine and Atmospheric Research, Canberra ACT 2601, Australia

Peter J. Rayner
LSCE/IPSL, Laboratoire CEA-CNRS-UVSQ, Bat. 701 LSCE - CEA de Saclay
Orme des Merisiers, 91191 Gif/Yvette, France

Stefan Reis
Centre for Ecology and Hydrology, Bush Estate, Penicuik, EH26 0QB, UK,
srei@ceh.ac.uk

Gareth Roberts
Department of Geography, Kings College London, Surrey Street Strand, London,
UK, gareth.j.roberts@kcl.ac.uk

Aodhagan Roddy
Department of Physics, Galway University, Galway, Ireland

Christian Rödenbeck
Max-Planck-Institute for Biogeochemistry, P.O. Box 10 01 64, 07701 Jena and
Hans-Knoell-Strasse 10, 07745 Jena, Germany

Yvonne Scholz
Deutsches Zentrum für Luft und Raumfahrt e.V. (DLR) in der
Helmholtzgemeinschaft, Institut für Technische Thermodynamik, Pfaffenwaldring
38-40, 70569 Stuttgart, Germany

Ingo Schöning
Max-Planck-Institute for Biogeochemistry, P.O. Box 10 01 64, 07701 Jena and
Hans-Knoell-Strasse 10, 07745 Jena, Germany

Marion Schrumpf
Max-Planck-Institute for Biogeochemistry, P.O. Box 10 01 64, 07701 Jena and
Hans-Knoell-Strasse 10, 07745 Jena, Germany

Ernst-Detlef Schulze
Max-Planck-Institute for Biogeochemistry, P.O. Box 10 01 64, 07701 Jena and
Hans-Knoell-Strasse 10, 07745 Jena, Germany

Jens Schumacher
Max-Planck-Institute for Biogeochemistry, P.O. Box 10 01 64, 07701 Jena and
Hans-Knoell-Strasse 10, 07745 Jena, Germany

Jean-François Soussana
INRA, UR874 Agronomie, Grassland Ecosystem Research, 234 Av. du Brézet,
Clermont-Ferrand F-63100, France

Jochen Theloke
Institute of Energy Economics and the Rational Use of Energy, University of
Stuttgart, Hessbruehlstrasse 49a, 70565 Stuttgart, Germany

Lieselotte Tolk
Department of Hydrology and Geo-Environmental Sciences, VU University
Amsterdam, Boelelaan 1085, 1081 HV Amsterdam, Netherlands

Riccardo Valentini
Department of Forest Science and Environment, University of Tuscia,
Via S. Camillo de Lellis 01100 Viterbo, Italy

Michel van der Molen
Department of Hydrology and Geo-Environmental Sciences, Boelelaan 1085,
1081 HV, VU University Amsterdam, Netherlands

Martin Wooster
Department of Geography, Kings College London, Surrey Street Strand, London,
UK, martin.wooster@kcl.ac.uk

Chapter 1
Introduction: Observing the Continental-Scale Greenhouse Gas Balance

A. Johannes Dolman, Riccardo Valentini, and Annette Freibauer

The concentrations of CO_2 and CH_4 in the atmosphere are at the highest level they have been in the past 25 million years. Current levels of CO_2 have increased by 30% from 280 ppm in pre-industrial times to 380 ppm today, and they continue to rise. These changes are caused by human activities. The primary inputs of carbon into the atmosphere arise from fossil fuel combustion and modifications of global vegetation cover through land-use change (e.g. land conversion to agriculture including pasture expansion, biomass burning). For the decade of the 1990s, an average of about 6.3 Pg C year^{-1} was released to the atmosphere from the burning of fossil fuels. For the recent period of 2000–2005, 7.2 Pg C year^{-1} was emitted from fossil fuel burning. It was estimated that an average of 1.5–2.5 Pg C year^{-1} was emitted because of deforestation and land-use change during the same intervals (Solomon et al. 2007).

The increasing CO_2 concentrations in the atmosphere raise concern regarding the heat balance of the atmosphere. In particular, the increasing concentration of these gases leads to an intensification of the Earth's natural greenhouse effect. This shift in the planetary heat balance may force the global climate system in ways which are currently not well understood. Complex interactions and feedbacks are involved, but there is a general consensus that global patterns of temperature and precipitation will change. The magnitude, distribution and timing of these changes, however, are far from certain. Recent Intergovernmental Panel on Climate Change (IPCC) assessments suggest a wide range of temperature and precipitation changes (Solomon et al. 2007).

Only half of the average of the CO_2 from anthropogenic emissions has remained in the atmosphere up till now. Analyses of the decreasing $^{13}C/^{12}C$ and O_2/N_2 ratios in the atmosphere have shown that the land and oceans have sequestered the other half, approximately in equal proportions. However, the partitioning of the carbon fluxes between ocean and land varies in time and space. The primary mechanism for current land carbon uptake is most likely the recovery from recent historical land use by afforestation and revegetation in North America (Pacala et al. 2001) and carbon-saving forestry practice in Europe (Janssens et al. 2003). There are several other terrestrial processes, such as possible enhanced plant growth due to the increase in atmospheric CO_2, nitrogen deposition and prolonged growing seasons in the temperate and boreal zones, whose effects on enhanced carbon uptake are not yet well understood.

The increases in greenhouse gases in the atmosphere are anthropogenically driven but are small compared to the large short-term natural fluxes of carbon in the global carbon cycle. Global photosynthesis for instance is estimated at $120 \, Pg \, C \, year^{-1}$ (Ciais et al. 1995). The current net sink of the terrestrial biosphere is however estimated to be only about 1% of this value, the difference being taken out by autotrophic and heterotrophic respiration of the biosphere and sudden disturbances such as forest fires, pest and harvests. Indeed, the carbon cycle has been qualified as 'Fast out Slow in' (Körner 2003). The perturbation of the global cycle thus takes place against large short-term background fluxes, and small long-term background fluxes.

The atmospheric growth rate of CO_2 exhibits large interannual fluctuations, on the order of the average long-term signal. This interannual variability signal cannot be explained only by the variability in fossil fuel use. Rather it appears to reflect primarily changes in terrestrial ecosystems induced by changing large-scale weather and climate patterns. These aspects in particular make the assessment of the global carbon balance so difficult, both in terms of magnitude and in terms of spatial and temporal variability.

Improved predictions of future CO_2 levels require better quantification and process-level understanding of the present state of the global carbon cycle, including both the natural components and the anthropogenic contributions. At present, limitations in our current understanding include an inability to locate well key sink or source regions. Accurate information on spatial and temporal patterns of CO_2 sources and sinks would however be of extraordinary value for challenging process-based terrestrial models, and thus for our ability to predict future CO_2 trajectories.

Observations document how climate change alters the Earth and help us to discover surprises and fill in knowledge gaps. Observations and their integration via monitoring are crucial for understanding and assessing the features and drivers of climate change and ecosystem responses to climate change. Quantifying present-day carbon sources and sinks and understanding the underlying carbon sequestration and emission mechanisms are prerequisites to informed policy decisions. Under the Kyoto Protocol, the majority of industrial countries have agreed to cut emissions and take other measures to reduce the amount of greenhouse gases in the atmosphere. The effectiveness of greenhouse gas emission reductions needs to be monitored through observations. The IPCC has issued guidelines for reporting carbon stock changes due to land use, forestry and land-use change (IPCC 2003). Observations and spatially and temporally explicit greenhouse gas budgets of anthropogenic and natural sources and sinks can also give guidance for further efforts to avoid dangerous anthropogenic interference with climate United Nations Framework Convention on Climate Change (UNFCCC, Article 2). Our current scientific and technological capacity is not yet fully up to the task of providing such regional and national budgets of greenhouse gas sources and sinks and of ecosystem response to climate change at sufficient accuracy. Approaches to establish greenhouse budgets so far have relied on combining a multitude of methods to assess the various budget components and this is likely to remain the case in the near future.

Establishing the CO_2 budget on a regional or national level can be achieved by integrating and modelling of measurements of the amount of carbon stored in different components of the ecosystems and soils, by direct measurements of CO_2 exchange between the surface and the air and 'top–down' CO_2 budgets inferred from measurement of CO_2 concentrations in higher parts of the atmosphere. This requires investing in both measurements of ecological, atmospheric and soil processes and attempting to provide realistic constraints on the spatial and temporal variability of fossil fuel emissions. Remotely sensed and other spatial information about land cover, land surface properties and radiation is often needed for integration, whereas remote sensing may also become available as a means of measuring atmospheric greenhouse gas concentrations in the near future. The combination of methods, while still subject to considerable uncertainties, is presently the most accurate manner to arrive at spatially explicit CO_2 budgets. It is referred to as the 'multiple constraint' method and has been pioneered by the CarboEurope community in Europe and several others in the USA and Australia (Fig. 1.1).

Fig. 1.1 Atmospheric sites, ecosystem flux towers, analytical facilities required for a greenhouse gas (GHG) observing system.

We focus in this book on the European continental-scale carbon balance. It is worth comparing the characteristics of the European versus the global carbon balance. In the global carbon balance, the terrestrial biosphere sink is roughly 30% of the anthropogenic emissions, and 2% of the net primary production (NPP). In the European carbon balance, the terrestrial biosphere sink is roughly less than 10% of the anthropogenic emissions and up to 10% of NPP, largely because of our carbon-saving forestry (Janssens et al. 2003). These numbers imply that in Europe, land management has potential for mitigation; some of which is already used. However, as the ratio of anthropogenic emission to biospheric uptake is rather different, it also suggests that a precise estimate of the fossil fuel emissions remains a critical issue in any monitoring system.

The aim of this book is to assess and review our current capability to observe and monitor the continental-scale carbon cycle, with an emphasis on quantifying the uncertainties involved. The book outlines the directions to and requirements for a continental-scale greenhouse gas monitoring network. The individual chapters provide a synthesis based on current research results of the European greenhouse gases budget, including both human-induced and biospheric sources and sinks. The book originated from discussions held at a CarboEurope workshop at The Royal Academy of Sciences and Arts in Amsterdam in 2005.

References

Ciais, P. et al. 1995. Partitioning of ocean and land uptake of CO_2 as inferred by $\delta^{13}C$ measurements from the NOAA Climate Monitoring and Diagnostics Laboratory Global Air Sampling Network. J. Geophys. Res., 100(D3): 5051–5070.

IPCC 2003. IPCC Special Report on Land Use, Land-Use Change and Forestry. Cambridge University Press.

Janssens, I.A. et al. 2003. Europe's terrestrial biosphere absorbs 7 to 12% of European anthropogenic CO_2 emissions. Science, 300(5625): 1538–1542.

Körner, C. 2003. Slow in, rapid out—carbon flux studies and Kyoto targets. Science, 300: 1242–1243.

Pacala, S.W. et al. 2001. Consistent land- and atmosphere-based U.S. carbon sink estimates. Science, 292(5525): 2316–2320.

Solomon, S., D. Qin, M. Manning, Z. Chen, M. Marquis, K.B. Averyt, M. Tignor and H.L. Miller (eds.) 2007. Climate Change 2007: The Physical Science Basis. Contribution of Working Group I to the Fourth Assessment Report of the Intergovernmental Panel on Climate Change. Cambridge University Press, Cambridge, United Kingdom and New York, NY, USA.

Chapter 2
Observing a Vulnerable Carbon Cycle

Michael R. Raupach and Joseph G. Canadell

2.1 Introduction

The carbon cycle and indeed the entire earth system are now inextricably linked with human activities (Global Carbon Project 2003; Steffen et al. 2004; Field and Raupach 2004), so that the 'carbon–climate–human system' constitutes a single, coupled entity in which interacting processes link all of its major components. Linking processes of primary significance include

1. The human drivers of energy consumption and land-use change, through increases in both population and per capita consumption
2. The role of human energy systems as sources of CO_2 and other greenhouse gases (GHGs)
3. Land-use change (deforestation, increases in agricultural and urban land use) and its consequences for both GHG emissions and resource (water, land, ecosystem) condition
4. Climate forcing by CO_2 and other GHGs, following from drivers 1, 2 and 3
5. The changing roles of the ocean and the terrestrial biosphere as sinks and sources of CO_2 and other GHGs, driven by the disequilibrium of the earth system through human activities
6. Impacts of climate change through declines in resource condition and human well-being
7. Attempts by human societies to reduce their impact on the global environment, for example, through reductions in GHG emissions to avoid 'dangerous climate change' (Schellnhuber et al. 2006)

Through the first six of these processes humankind is unintentionally influencing the earth system, while the seventh is an effort to manage global-scale human impacts on the earth system by mitigating their causes.

An integrated global carbon observation system (Ciais et al. 2004) is a contribution to monitoring the first six of the above processes, and bringing about the seventh. These underlying motivations lead to two broad goals for global carbon observations, respectively oriented towards understanding and management. The former goal is to provide increased understanding of the cycles of carbon and

5

related entities (water, energy, nutrients) in the earth system, contributing to our ability to diagnose trends and to predict future evolution of the carbon–climate system over timescales of decades to centuries. The latter is to provide the global-scale observations of carbon fluxes and GHG emissions needed to manage the carbon cycle, through emissions reduction programmes based on incentive, regulatory or trading mechanisms. Between them, these two goals largely determine the necessary broad attributes of a global carbon observing system. A recent analysis (Raupach et al. 2005) identified seven main attributes for terrestrial carbon observation, which (with slight extension) provide a broad specification of attributes for a complete global carbon observing system. These seven are (1) scientific rigour; (2) global scope and consistency; (3) spatial resolution sufficient to resolve and monitor all important processes, especially carbon fluxes associated with human land use and energy systems; (4) temporal resolution sufficient to monitor variability in fluxes from weather to climate timescales; (5) integrated monitoring of all relevant entities [CO_2, CH_4, CO, volatile organic compounds (VOCs), black carbon, together with fluxes of water, nutrients and other entities relevant in modulating carbon fluxes]; (6) process discrimination (for instance, between anthropogenic and non-anthropogenic fluxes, and between contributions to net fluxes such as assimilation, autotrophic and heterotrophic respiration); and (7) quantification of uncertainty.

Here, we discuss the implications of carbon–climate vulnerabilities for the attributes of an integrated carbon observation system. By 'carbon–climate vulnerability' we mean a positive, disturbance-amplifying feedback between an aspect of the carbon cycle (a pool or flux) and physical climate, including atmosphere, oceans and the hydrological cycle. In particular, carbon–climate vulnerabilities are processes causing global warming through the enhanced greenhouse effect to be larger than it otherwise would be in their absence.

Two ways have been used recently to quantify carbon–climate vulnerability in the above sense. The first is a risk-assessment methodology (Gruber et al. 2004, henceforth G2004) involving heuristic, judgement-based estimates of the releases of carbon to the atmosphere from several terrestrial and oceanic pools under projected changes (to 2100) in temperature, ocean circulation and other physical climate properties. G2004 expressed the results of the assessment as ellipses on a plane with axes defined by the mass of carbon released and a qualitatively judged probability of release (with small releases having high probability and vice versa). This approach is a valuable beginning, but cannot properly quantify carbon–climate feedbacks by estimating the extent to which a carbon release is modified by the extra climate change induced by the release itself.

The second, much more quantitative approach is through the use of fully coupled carbon–climate models. Eleven such models were compared in the recent Coupled Climate–Carbon Cycle Model Intercomparison Project (C^4MIP) (Friedlingstein et al. 2006). The models included full physical climate, ocean carbon biogeochemistry responsive to temperature and atmospheric CO_2 and terrestrial carbon dynamics responsive to light, water, temperature and CO_2. All models were run from 1850 to 2100 under a prescribed emissions scenario (the IPCC SRES[1] A2 scenario; see later

---

[1] IPCC, Intergovernmental Panel on Climate Change; SRES, Special Report on Emissions Scenarios.

for details). The results showed that coupling of the carbon cycle to climate through temperature-dependent processes led to increased atmospheric CO_2 in 2100 of 20–200 ppm (augmenting a CO_2 concentration of around 700 ppm, depending on the model) and an increase in predicted global temperature of 0.2–2 °C (augmenting an enhanced-greenhouse-induced warming of around 4 °C, likewise depending on the model). There were substantial differences among the ten models, stemming both from carbon cycle parameterisations (for instance, the temperature response of terrestrial heterotrophic respiration) and from the behaviour of modelled physical climate (for instance, the tendency of some of the models to dry out the Amazon as the model climate warms). As a means of studying carbon–climate vulnerabilities, fully coupled carbon–climate models are comprehensive, but they are laborious and difficult to parameterise because of model complexity. The extensive C^4MIP runs to date have focused on only a few of the potentially important feedback processes.

In this chapter, we analyse carbon–climate vulnerability and its implications for carbon cycle observations. The plan of the chapter (following this brief introductory section) is that Sect. 2.2 surveys the major feedbacks in the carbon–climate–human system at a general level, including forcing by and feedbacks on human actions. Section 2.3 focuses on carbon–climate feedbacks involving terrestrial processes, drawing from C^4MIP results and other sources. Attention is given to both CO_2 and CH_4. Section 2.4 proposes a perturbation-based approach using simple models for analysing carbon–climate vulnerabilities and illustrates the approach with a semi-quantitative evaluation of the response of permafrost carbon pools to global change. Finally, Sect. 2.5 discusses the implications of carbon vulnerability for integrated carbon observation.

2.2 Feedbacks and Vulnerabilities in the Carbon–Climate–Human System

The trajectories of climate and the carbon cycle are coupled by atmospheric composition. Of the linking groups of processes mentioned in Introduction, four are of central importance for feedbacks and vulnerabilities in the contemporary carbon–climate–human system. The first is enhanced radiative forcing by GHGs, CO_2 and CH_4 being the largest contributors. The other three correspond to the three major groups of fluxes in the atmospheric CO_2 and CH_4 budgets: emissions from human activities, ocean–atmosphere exchanges and land–atmosphere exchanges. Section 2.2.1 focuses on land–atmosphere exchanges in more detail, but before doing so, we examine all four groups of processes in general terms.

2.2.1 Radiative Forcing

The carbon cycle accounts for some, but not all, of the processes involved in the radiative forcing of climate. Total radiative forcing can be considered as the sum of three contributions: (1) from CO_2, (2) from non-CO_2 GHGs (mainly CH_4, halocarbons, N_2O, ozone) and (3) from non-gaseous mechanisms (mainly aerosols, albedo changes, solar variations). The first two are relatively well known (IPCC 2007): the current (2001–2005) radiative forcing from CO_2 is $+1.66 \pm 0.17$ W m^{-2}, and the forcing from non-CO_2 GHGs is about $+1.3$ W m^{-2} ($+0.48$ from CH_4, $+0.34$ from halocarbons, $+0.16$ from N_2O and $+0.30$ from ozone). The third contribution, from aerosols, albedo changes and solar variations, is highly uncertain but is considered to be negative, current estimates being around -1.3 ± 1 W m^{-2} (IPCC 2007). The current net radiative forcing ($+1.6$ W m^{-2}, range $+0.6$ to $+2.4$) drives global warming (at about 0.016 °C y^{-1}over the period 1980–2005). Thus, current net radiative forcing is approximately equal to the radiative forcing from CO_2 alone, with other contributions approximately cancelling. This does not imply that *future* forcing will behave this way, because all three contributions to radiative forcing are dependent on emissions scenarios and also on future climate through climate feedbacks, so the three contributions will evolve differently under these influences. In summary, a carbon budget and a radiative forcing budget are different entities, but they share a large common term associated with rising atmospheric CO_2.

2.2.2   Emissions from Human Activities

The global balance of atmospheric CO_2, shown in Fig. 2.1, demonstrates that human activities are the overwhelmingly dominant contribution to the current disequilibrium of the global carbon cycle. Fossil-fuel emissions were about 7.2 ± 0.3 Pg C y^{-1} for the period 2000–2005, and increased at over 3% y^{-1} for 2000–2005 compared with 1% y^{-1} for 1990–1999 (Canadell et al. 2007a; Raupach et al. 2007). Emissions from land clearing have changed more slowly, averaging about 1.5 ± 0.5 Pg C y^{-1} (Canadell et al. 2007a).

The atmospheric CH_4 balance involves ten major source terms and three sink terms, and is discussed in more detail later. Of the current total source, about 2/3 is anthropogenic.

The four main groups of IPCC SRES scenarios (Nakicenovic et al. 2000) all involve major increases in CO_2 emissions over the period to 2100 (Table 2.1), ranging from 12 to 17 Pg C y^{-1} in 2050 and from 7 to 30 Pg C y^{-1} in 2100. Methane emissions are also projected to increase in most scenarios. From the present standpoint of vulnerability analysis, the important question is: are these emissions scenarios significantly dependent on climate itself, so that different emissions scenarios would result from different climate change scenarios? While climate is just one among the many economic, social and environmental factors influencing the scenarios, there are potential mechanisms for feedbacks on emissions scenarios from climate change. Examples include increased energy use to buffer against adverse effects of climate change (e.g. air conditioning), increased energy use to augment resources threatened by climate change (e.g. desalination to supplement water supplies) and increased military energy use brought on by climate-induced geopolitical instability.

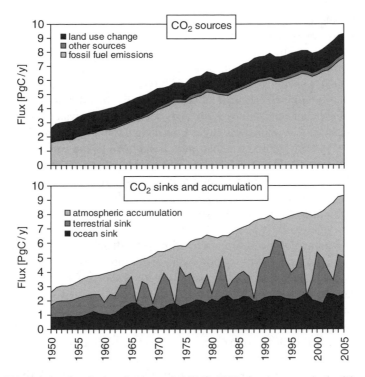

Fig. 2.1 The global carbon budget for the period 1950–2005, showing terms in the CO_2 mass balance: [atmospheric CO_2 accumulation] = [emission flux from fossil fuels] + [other industrial emissions] + [emissions from land-use change] − [flux to terrestrial sink] − [flux to ocean sink]. Data sources as in Canadell et al. (2007a), with atmospheric CO_2 change from ice core data (Law Dome, Antarctica; Etheridge et al. 1998b) before 1959 and direct measurements (Keeling and Whorf 2005) after 1959. Net terrestrial uptake is inferred by difference (total emissions less ocean uptake).

Table 2.1 Indicative fossil-fuel emissions of CO_2 under four major IPCC SRES emissions scenarios (Nakicenovic et al. 2000)

Scenario	Global-local orientation	Economic-environmental orientation	Fossil-fuel emission (Pg C y⁻¹) in 2050	Fossil-fuel emission (Pg C y⁻¹) in 2100
A1B	Global	Economic	17	15
A2	Regional	Economic	15	30
B1	Global	Environmental	11	15
B2	Regional	Environmental	12	7

There are three variants of the A1 (globalised, economically oriented) scenario: A1B (balanced), A1T (technologically innovative) and A1FI (fossil-fuel intensive), leading to very different emissions trajectories. Only the A1B scenario is used here

2.2.3 Ocean–Atmosphere Exchanges

The ocean is a major CO_2 sink, currently absorbing about 25% of fossil-fuel emissions (Fig. 2.1). This will continue over the next century according to C⁴MIP carbon–

climate models (Friedlingstein et al. 2006). There are several timescales for ocean–atmosphere exchanges related to different ocean carbon pools, but even the shortest is long enough for ocean uptake of CO_2 to be fairly smooth from year to year (Fig. 2.1). Several reviews (Jacobson et al. 2000; Steffen et al. 2004; Le Quere and Metzl 2004; Greenblatt and Sarmiento 2004) describe the processes involved, which include (1) *air–sea exchange* of CO_2, driven by the difference in CO_2 partial pressure (pCO_2) between atmosphere and ocean surface waters; (2) *buffering* between dissolved CO_2 and DIC (total inorganic carbon including CO_2, H_2CO_3, HCO_3^- and CO_3^{2-}), which means that only about 10% of the carbon crossing the air–sea interface contributes to aqueous pCO_2, with the rest appearing as other forms of DIC; (3) *the ocean circulation pump*, by which ocean circulations export carbon from surface to deep ocean waters and (4) *biological pumps*, by which soft-tissue and carbonate detritus from ocean biota in the surface layer export carbon to deep waters as they sink.

All of these processes are subject to climate feedbacks. G2004, in their risk-assessment-based analysis of carbon cycle vulnerabilities, identified six feedbacks. First, chemistry leads to a positive feedback because ocean pH falls as CO_2 is taken up, thereby altering the CO_2/DIC partition and reducing uptake. Second, temperature increases lead to a similar positive feedback through the CO_2/DIC partition. Third, changes in deep ocean circulation (mainly through increasing vertical stratification) increase ocean equilibration times, inducing a positive feedback through reduced uptake over timescales of 10–100 years since the mixing timescale becomes longer. Fourth, ocean circulation changes alter the equilibrium carbon distribution itself (in addition to changing the equilibration timescale), which when coupled with the biological pump leads to a negative feedback because upward transport of DIC from deep to surface waters is reduced but the downward biological pump is not. Fifth, the biological pump is subject to large influence by climate change, but uncertainties are so high that it is not yet possible to identify whether these influences add up to an overall positive or a negative feedback. Sixth, there is the possibility of release from the vast stores of methane hydrates in sediments under continental shelves (and in permafrost). Such a release would constitute a massive positive (heating) feedback on the climate system, but is rated by G2004 as a high-risk, low-probability scenario.

2.2.4 Land–Atmosphere Exchanges

Like the oceans, the terrestrial biosphere currently takes up about 25% of fossil-fuel emissions of CO_2, but unlike the ocean sink, the terrestrial CO_2 sink varies enormously from year to year (Fig. 2.1). Similarly, on longer (100 year) timescales, C^4MIP results suggest higher variability for the terrestrial than the ocean CO_2 sink, both in time and between individual models in C^4MIP. Land–atmosphere exchanges are also critical in the CH_4 budget. These issues are explored in more detail in Sect. 2.3.

2.3 Vulnerabilities in Terrestrial Carbon Pools and Fluxes

The terrestrial carbon balance equates the net change in terrestrial-biospheric carbon to the sum of carbon fluxes into the terrestrial carbon pool. These fluxes include land–air gaseous exchanges, waterborne and airborne particulate transport, and product removal by humans. The focus here is on the first, which is the most significant for the global carbon budget (though the others can be important particularly for regional carbon budgets). Gaseous carbon exchange between terrestrial systems and the atmosphere occurs through fluxes of CO_2 and several other species including CH_4, VOCs and CO. The CO_2 exchange dominates the mass flux, but exchanges of other species have significant effects on radiative forcing. Here, we summarise the main processes leading to vulnerabilities in land–air exchanges of carbon as CO_2 and CH_4, since these two entities are the most important from both mass-flux and radiative forcing standpoints.

2.3.1 Vulnerabilities Associated with CO_2 Exchanges

Table 2.2 summarises the main processes affecting the net land-atmosphere flux of CO_2 in terrestrial systems. The table identifies three classes of driver for changes in

Table 2.2 Processes contributing to net land–atmosphere exchange of CO_2

	Process	Driver	Sign of land-to-air flux (+,−) =(source, sink)
a1	CO_2 fertilisation	a	−
a2	Nutrient constraints on CO_2 fertilisation	a	+
a3	Fertilisation by nitrogen deposition	a	−
a4	Effects of pollution (e.g. acid rain, ozone, etc.)	a	+
b1	Response of respiration to warming and moisture	b	+ (warming), ± (moisture)
b2	Response of NPP to warming and moisture	b	− (warming), ± (moisture)
b3	Radiation effects (e.g. direct/diffuse partition)	b	−
b4	Biome shifts	b	±
b5	Permafrost thawing	b	+
b6	Changes in wildfire regime	b	+ (rapid), − (slow)
b7	Changes in herbivore (e.g. insect) ecology	b, c	+
c1	Changes in managed fire regime	c	+ (rapid), − (slow)
c2	Managed reforestation and afforestation	c	−
c3	Unmanaged forest regrowth (after cropland abandonment)	c	−
c4	Woody encroachment/woody thickening	c	−
c5	Deforestation and land clearing (e.g. forest to savannah)	c	+
c6	Peatland and wetland drainage	c	+
c7	Agricultural practices	c	±

Drivers are (*a*, shaded grey) changes in atmospheric composition and chemistry; (*b*, unshaded) physical climate changes; (*c*, shaded grey) changes in land use and land management. The sign of the land-to-air CO_2 flux is the same as the sign of the climate warming feedback (a positive land-to-air flux increases the CO_2 radiative forcing)
NPP net primary production

these processes: (*a*) changes in atmospheric composition (CO_2 fertilisation, nutrient effects, pollution effects); (*b*) changes in physical climate (temperature, precipitation, light) and (*c*) changes in land use and land management. Table 2.2 also identifies whether the process is a source or a sink, an attribute that has no simple correlation with the class of driver. All processes driven by changes in atmospheric composition (*a*) and physical climate (*b*) are directly involved in carbon–climate feedbacks. An additional dimension to terrestrial carbon vulnerability arises through processes driven by changes in land use and land management (*c*), which provide direct couplings between the terrestrial carbon cycle and human actions in the carbon–climate–human system.

Most of the processes listed in Table 2.1 are vulnerable to major changes over the next century, and many have already changed significantly over the last century. The most important feedbacks and vulnerabilities associated with these processes can be summarised as follows, drawing from a recent review (Canadell et al. 2007b; henceforth C2007). (Note that not all processes listed in Table 2.2 are discussed in the following.)

2.3.1.1 CO_2 Fertilisation and Its Limitation by Water and Nutrient Constraints *(a1, a2)*

It is expected on physiological grounds that plants respond to rising atmospheric CO_2 with increased assimilation, leading to increased biomass. The response saturates at around 1,000 ppm according to models and laboratory experiments (Farquhar et al. 1980; Farquhar and Sharkey 1982). This CO_2 fertilisation process is now incorporated in most terrestrial–biosphere and dynamic-vegetation models (Cramer et al. 2001). However, field results from free air CO_2 enrichment (FACE) and other elevated CO_2 studies have been variable (Oren et al. 2001; Nowak et al. 2004; Norby et al. 2005). Some actively growing vegetation types (such as temperate, well-watered young forests) show responses in net primary production (NPP) of up to 25% (Norby et al. 2005), while other, more stressed vegetation types show much lower responses (Nowak et al. 2004). Field observations also indicate a saturation of response to CO_2 fertilisation at 500–600 ppm, much lower than expected on physiological grounds. These results indicate that the full possible physiological response to CO_2 fertilisation is not manifested in most field environments because of constraints from factors other than CO_2, especially nitrogen limitations (Luo et al. 2004). There is also an interaction between water limitation and CO_2 fertilisation through the beneficial (to terrestrial carbon storage) effect of increasing CO_2 on water-use efficiency because of decreased stomatal conductance. This is most pronounced in dry ecosystems (Field et al. 1996; Owensby et al. 1997; Pataki et al. 2000).

2.3.1.2   Fertilisation by Nitrogen Deposition *(a3)*

Early studies (Townsend et al. 1996; Holland et al. 1997) suggested significant enhancement of the terrestrial CO_2 sink by N deposition, especially in mid-latitude

Northern Hemisphere forests where N limitation is common and deposition rates are high. Later work (Nadelhoffer et al. 1999) has suggested a lower contribution of about 0.25 Pg C y^{-1} to the current net terrestrial sink of 2–3 Pg C y^{-1}. C2007 concluded that it is unlikely that N deposition will create major new carbon sinks over the next century.

2.3.1.3 Response of Respiration to Warming and Moisture (b1)

Warming increases heterotrophic respiration of soil carbon (R), thus decreasing Net Ecosystem Exchange (NEE = NPP − R). Soil moisture is comparably important. Most models incorporate a strong temperature response of soil respiration, quantified by $Q_{10} = R(T + 10\,°K)/R(T)$, which is a dominant contributor to the source side of the terrestrial carbon balance and is a major reason for predictions that the current terrestrial sink will saturate or reverse in the future. Most of the 11 C⁴MIP models used $Q_{10} = 2$ (Friedlingstein et al. 2006). However, as usual, experimental evidence is less clear-cut than modelling assumptions. The review of C2007 is summarised here in three stages: first, many laboratory and some field studies support high Q_{10} values of 2 or more for soil respiration, and show that Q_{10} decreases with increasing temperature (Lloyd and Taylor 1994). Second, other field studies (Giardina and Ryan 2000; Valentini et al. 2000; Jarvis and Linder 2000; Luo et al. 2001) have suggested that Q_{10} declines after some time as labile soil carbon is respired, leaving more recalcitrant soil carbon for which turnover is slower and less sensitive to temperature. Third, recent studies (Knorr et al. 2005; Fang et al. 2005) have resolved the apparent paradox by analysing the data in terms of separate fast and slow carbon pools (in contrast with earlier studies which analysed all soil carbon in one pool). These newer studies find different temperature dependences in R for fast and slow soil carbon pools, with faster pools having more temperature sensitivity.

2.3.1.4 Response of NPP to Warming and Moisture (b2)

Global terrestrial NPP has increased by 6% (3.4 Pg C y^{-1}) over the two decades from 1981 to 2000, largely because of extension of the growing season in high–northern-latitude ecosystems because of global warming (Nemani et al. 2003). This is associated with an increase in the terrestrial CO_2 sink (excluding land-use change) from around 0.3 Pg C y^{-1} in the 1980s to 2–3 Pg C y^{-1} in the 1990s (IPCC 2001; Sabine et al. 2004). Most (though not all) C⁴MIP simulations predict a further increase in the terrestrial CO_2 sink through the twenty-first century, driven largely by CO_2 fertilisation (Friedlingstein et al. 2006). This would provide a negative feedback on further climate change. However, new observations of a reduced CO_2 sink due to increased climate variability are challenging the hypothesis of an increasing terrestrial CO_2 sink in the twenty-first century. An analysis of Northern Hemisphere terrestrial carbon fluxes (Angert et al. 2005) shows that since 1994, accelerated carbon uptake in early spring was cancelled by decreased uptake during summer, most likely due to hotter and drier summers in middle and high latitudes. The heatwave in 2003 alone reduced the gross primary productivity of European ecosystems by 30%, resulting in a net atmospheric CO_2 source of 0.5 Pg C y^{-1} or 4 years of carbon accumulation in these systems (Ciais et al. 2005).

2.3.1.5 Permafrost Thawing (b5)

The carbon store in frozen soils in the Northern Hemisphere has been estimated as over 400 Pg C, of which around 54% is in Eurasia, largely in Russia, and 46% in North America, largely in Canada (Tarnocai 1999). A recent estimate is substantially higher at around 900 Pg C (Zimov et al. 2006). Widespread observations already exist of permafrost thawing leading to the development of thermokarst and lake expansion, followed by lake drainage as permafrost further degrades (Camill 2005; Smith et al. 2005; Jorgenson et al. 2006). Preliminary estimates suggest that permafrost area could shrink by up to 25% with a mean global warming of 2 °C (Anisimov et al. 1999). Melting permafrost will increase both CO_2 and CH_4 emissions from frozen soils. It is estimated for the Canadian permafrost alone that up to 48 Pg C could be vulnerable to release under a 4 °C warming scenario (Tarnocai 1999). G2004 suggested that up to 5 Pg C could be released from melting permafrost over the next 20 years and up to 100 Pg C in the next 100 years, assuming a warming of 2 °C by 2100 and that this warming releases 25% of the carbon locked in frozen soils (following projected area reductions).

2.3.1.6 Fire (b6, c1)

The annual carbon flux to the atmosphere from savannah and forest fires (excluding biomass burning for fuel and land clearing) is estimated to be in the range of 1.7–4.1 Pg C y^{-1} (Mack et al. 1996), mostly as CO_2. A recent estimate (van der Werf et al. 2006) gives a fire emission of 2.5 Pg C y^{-1} 1997–2004. In high-intensity fire years such as during El Niño-Southern Oscillation (ENSO) events, emission from fires can be responsible for as much as 66% of the atmospheric CO_2 growth anomaly (van der Werf et al. 2003). In the long term or over large spatial regions, terrestrial carbon losses from fires may be compensated by the gains during vegetation regrowth. However, a terrestrial carbon imbalance is created during the transition from one disturbance regime to another: this leads to a CO_2 sink (or source) when the disturbance frequency is reduced (or increased). Furthermore, trends in fire disturbance frequency are not spatially uniform. For instance, fire exclusion during the twentieth century in many countries has resulted in an increase of biomass in forests and woodlands (Luger and Moll 1993; Houghton et al. 2000; Mouillot and Field 2005), and the potential exists for further accumulation especially in temperate and subtropical regions. By contrast, increases in annual burned area over the last two decades in boreal North America and some parts of Europe are shifting a long-term trend of terrestrial carbon accumulation into one of release to the atmosphere (Kurz and Apps 1999).

2.3.1.7   Managed and Unmanaged Forest Growth *(c2, c3, c4)*

Directly and indirectly human-induced forest regrowth (through cropland abandonment, vegetation thickening, afforestation and reforestation) accounts for a major contribution to the terrestrial carbon sink of 2–3 Pg C y^{-1} through the 1990s. Several processes are involved, all related to land use and land management. First, forest regrowth on abandoned agricultural land has been identified as one of the most significant mechanisms to explain the net CO_2 sink in the Northern Hemisphere, both in the USA (Houghton and Hackler 2000) and in Europe (Janssens et al. 2005). Crop abandonment results in the expansion of relatively young forests with fast growth rates, and therefore with high CO_2 sink capacity. Second, woody thickening and encroachment in semi-arid regions and savannahs, largely due to fire suppression policies and pasture management, accounts for 22–40% of the US terrestrial carbon sink (Pacala et al. 2001) and is a significant component of the sink in Australia (Gifford and Howden 2001; Burrows et al. 2002). However, these estimates have very large uncertainties. Third, managed afforestation and reforestation has a significant potential as a terrestrial CO_2 sink over the twenty-first century, though currently only China has seen a substantial increased terrestrial carbon storage (0.45 Pg C) by this mechanism, through large reforestation efforts over recent decades (Fang et al. 2001).

2.3.1.8   Deforestation and Land Clearing *(c5)*

Deforestation and land clearing, mainly to establish croplands, has released a total of 182–199 Pg C to the atmosphere over the period from 1800 to 2000 (DeFries et al. 1999) and is responsible for 33% of the increase in atmospheric CO_2 concentration observed over that period (Houghton 1998). Estimates of emissions from deforestation for the decades of the 1980s and 1990s range from 0.8 to 2.2 Pg C y^{-1} (DeFries et al. 2002; Houghton 2003; Achard et al. 2004) and are expected to continue being significant over the next decades to century.

2.3.1.9   Peatland and Wetland Drainage *(c6)*

Peatlands and wetlands in high, temperate and tropical latitudes contain a carbon store of over 450 Pg C, as soil organic matter. This carbon is largely isolated from decomposition by waterlogged environments and/or low temperatures, but is vulnerable to release to the atmosphere when water tables fall (through land-use change or precipitation change) or temperatures rise. When such carbon releases occur, a complex balance exists between CO_2 emissions (in oxidising conditions associated with falling water tables) and CH_4 emissions (in anoxic conditions associated with high water tables). A Siberian wetland illustrates this complexity (Friborg et al. 2003): despite being a net carbon sink, the wetland was a source of positive radiative forcing because of its emissions of CH_4 (which has a warming potential 21 times larger than that of CO_2 over 100 years). The case of tropical peatlands is particularly important because of high vulnerabilities driven by both land use and climate factors. The carbon store in tropical peatlands is about 70 Pg C, with deposits as deep as 20 m (Page et al. 2002) much of it in the south-east Asian archipelago (Indonesia, Malaysia, Papua New Guinea, Thailand, Philippines). These peatlands have been a net CO_2 sink since the late Pleistocene (Page et al. 2004). However, over the last decade, a combination of intense draining for agriculture and increasing climate variability (in the form of more intense El Niño-drought events) has resulted in a significant CO_2 source with discernible effects on atmospheric CO_2 growth. During El Niño 1997–1998 events in Indonesia, burning of peat and vegetation resulted in an estimated loss of carbon between 0.81 and 2.57 Pg C in 1997, equivalent to 13–40% of the mean annual global carbon emissions from fossil fuels (Page et al. 2002) and a 60% contribution to the atmospheric CO_2 growth anomaly due to fire activity (van der Werf et al. 2004).

2.3.2 Vulnerabilities Associated with CH_4 Exchanges

The atmospheric methane concentration is currently about 1,750 ppb (mole fraction), having risen from around 600 ppb in the pre-industrial era (Fig. 2.2). The growth rate of atmospheric CH_4 has declined over recent years to the point where it is now nearly zero (Allan et al. 2005; Bousquet et al. 2006). The CH_4 growth rate is controlled by an atmospheric methane budget which includes a number of different sources and sink terms, and can be written for the present purpose as:

$$r_{CH_4} \frac{d[CH_4]}{dt} = \underbrace{\left(\begin{array}{l} F_{Wetlands} + F_{Termites} + F_{Ocean} + F_{Geol} \\ F_{Fuels} + F_{Landfills} + F_{Ruminants} + F_{Rice} + F_{Fire} + F_{Other} \end{array} \right)}_{Sources} - \underbrace{(F_{Trop\,OH} + F_{Soils} + F_{Stat})}_{Sinks} \quad (2.1)$$

$$\frac{d[CH_4]}{dt} \approx r_{CH_4}^{-1} F_{Sources} - k_{CH_4}[CH_4]$$

where r_{CH4} is the mass of atmospheric methane per unit concentration and k_{CH4} is an overall decay rate. The sources can be categorised as non-anthropogenic and anthropogenic (respectively grouped on the first and second lines of the 'source' collection of fluxes above). Non-anthropogenic sources arise from wetlands, termites, ocean sources (including methane hydrates) and geological processes (geothermal and volcanic). Anthropogenic CH_4 sources, which are amenable to mitigation, result from burning and leakage of fossil fuels (natural gas, petroleum, coal), landfills, ruminant livestock, rice paddies and biomass burning associated with land clearing. The largest sink is from tropospheric OH oxidation, with smaller sinks

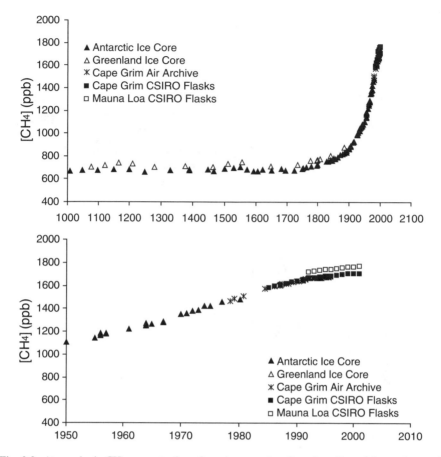

Fig. 2.2 Atmospheric CH$_4$ concentrations from ice core data from Law Dome, Antarctica, and Summit, Greenland (Etheridge et al. 1998a); the Cape Grim Air archive (Etheridge et al. 1998a) and flask measurements at Cape Grim, Tasmania, and Mauna Loa, Hawaii (CSIRO flask network, archived at http://cdiac.ornl.gov/trends/atm_meth/csiro/csiro_gaslabch4.html). Upper and lower panels show data on 1,000-year and 50-year time axes, respectively.

arising from soils, oceanic and stratospheric oxidation. The sinks can be parameterised as a first-order decay with a turnover time ($1/k_{CH4}$) of about 10 years. Provided k_{CH4} is steady, this implies that

$$[CH_4](t) \approx \int_0^\infty F_{Sources}(t-s)\exp(-k_{CH_4}s)\,ds \qquad (2.2)$$

so the CH$_4$ concentration is a lagged moving time average of the sources with an exponential weighting factor in time.

Figure 2.3 shows estimates from several studies of the CH$_4$ source and sink terms in Eq. (2.1). Approximately 2/3 of global emissions are currently from anthropogenic sources. Estimates of the sources are quite scattered, with each of the

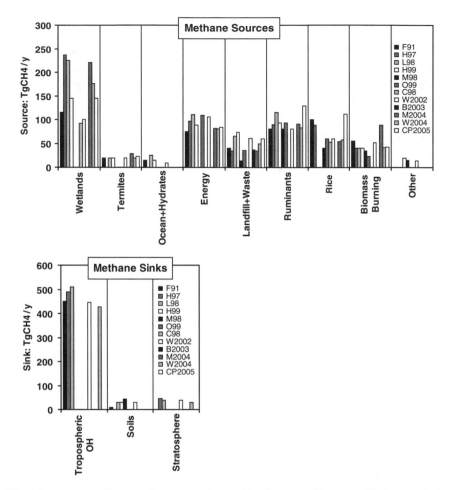

Fig. 2.3 Estimates of sources (upper panel) and sinks (lower panel) in the global atmospheric methane budget. Note different scales. References: Fung et al. 1991 (F91); Hein, Crutzen and Heimann 1997 (H97); Lelieveld, Crutzen and Dentener 1998 (L98); Houweling et al. 1999 (H99); Mosier et al. 1998 (M98); Olivier et al. 1999 (O99); Cao, Gregson and Marshall 1998 (C98); Wuebbles and Hayhoe 2002 (W2002); Bogner and Matthews 2003 (B2003); Mikaloff Fletcher et al. 2004a; Mikaloff Fletcher et al. 2004b (M2004); Wang et al. 2004 (W2004); Chen and Prinn 2005 (CP2005). A 4-level greyscale is used to distinguish studies, cycling 3 times through the 12 available studies.

major anthropogenic sources (fossil fuels, landfills, ruminants, rice, biomass burning) contributing significantly (in the order of 50–100 Tg CH_4 y^{-1}) to the total current source of about 500 Tg CH_4 y^{-1}. The recent reduction in the growth rate of atmospheric CH_4 to near zero (Fig. 2.2) results from decreased fossil-fuel-related anthropogenic emissions through the 1990s, and decreased emissions since 1999 from drying wetlands in temperate and tropical Asia and tropical South America, combined with a return to rising anthropogenic emissions (Bousquet et al. 2006).

Despite the recent slow down in CH_4 growth rate, most SRES scenarios predict that atmospheric CH_4 emissions will increase by a factor of typically 1.5–2 in the century to 2100 (with much scatter among scenarios). These predicted increased emissions result from all the anthropogenic sources listed above. In addition, future methane vulnerabilities can arise from mechanisms which cause either non-anthropogenic or anthropogenic CH_4 sources to increase in response to climate change. Increased temperatures and atmospheric CO_2 will increase CH_4 production from wetlands and rice paddies as more substrate is made available. More importantly, however, is the control that the water table exercises on CH_4 wetland sources. Overall increase in precipitation in a warmer climate will enhance this source by raising water tables, but regional drying trends will result in decreased CH_4 production and a simultaneous increase in CO_2 emissions (Christiansen et al. 2003; Cao et al. 1998). This issue is similar to that for peatland and wetland drainage, discussed above for CO_2 emissions.

2.4 Semi-Quantitative Evaluation of Carbon–Climate Vulnerabilities

Evaluation of future carbon–climate vulnerabilities is not easy. The first requirement is a quantitative definition, here constructed as follows: we imagine a model of the carbon–climate system which includes a coupling or feedback process P between carbon and climate (such as any of the processes in Table 2.2). Starting from a given initial state of the system, the model can be integrated forward in time to produce estimates of the state $\mathbf{x}(t)$ of the system at any future time t. This integration can be carried out for an 'uncoupled' version of the model in which the process P is omitted from the model equations, and for a 'coupled' version in which P is included. For a variable y in the carbon–climate system such as atmospheric CO_2 or temperature [or in general any function of the state vector $\mathbf{x}(t)$], the vulnerability of y to P at time t can be defined as

$$\delta y(t) = y_C(t) - y_U(t) \tag{2.3}$$

where $y_C(t)$ and $y_U(t)$ are respectively the values of $y(t)$ from the coupled and uncoupled versions of the model. Under this quantitative definition, the vulnerability is specific to a variable y (such as a carbon pool), a process P and a time t. It is a model-based, 'what-if' measure of the significance of process P. [This definition is not the same as the sensitivities used in C⁴MIP analyses, which are linearised relationships between $\delta y_1(t)$ and $\delta y_2(t)$ for different model variables y_1 and y_2, for example, the sensitivity of terrestrial or ocean carbon uptake to CO_2 or temperature.]

In Introduction to this chapter, we have already reviewed the two main ways that have been used recently to explore carbon–climate vulnerability: the judgement-based, risk-assessment approach of G2004, and fully coupled carbon–climate models as in the C\⁴MIP experiments (Friedlingstein et al. 2006). These approaches have complementary strengths: the risk-assessment approach is subjective, exploratory and largely non-quantitative. The fully coupled model approach is quantitative and rigorous (within the parameterisation choices in particular models) but is subject to the usual difficulties of complex, numerically intensive models: (1) model parameterisations (functional relationships between fluxes and model state variables) are scale dependent (Raupach et al. 2005) and hence not unique, so that their choice involves an element of subjective judgement; (2) the settings of parameter values (numbers in the functional relationships) are usually underdetermined by available information; (3) complex models are usually numerically intensive and expensive to integrate over long times (particularly in ensemble mode to assess the properties of model solutions which are chaotic attractors), which places a practical restriction on the range of processes (P) which can be assessed.

For these reasons, it is appropriate to explore an intermediate pathway for assessing carbon–climate vulnerability, based on very simple but still quantitative models which preserve the essential carbon–climate feedbacks leading to possible vulnerabilities. Criteria for a simple model to be used in this way are that the model broadly reproduces (1) observed features of past changes in the carbon–climate system over the last 200 years; (2) the future predictions of more complex models, such as carbon–climate trajectories under the four main classes of IPCC scenario (Table 2.1); (3) the model must include defensible (though necessarily simple) parameterisations of processes (P) to be assessed for vulnerability which are not currently included in more complex models. The basic assumption is that although a simple model cannot provide a stand-alone prediction of the trajectory of the carbon–climate system at the same level of sophistication as a fully coupled carbon–climate model, it can provide information about the perturbation resulting from the presence or absence of a process P.

To illustrate this approach we have estimated the vulnerability of the carbon–climate system to one significant feedback, the carbon release from permafrost soils under the influence of warming (Zimov et al. 2006). Here, we assess the vulnerability due to CO_2 only (recognising that this is an underestimate of the true vulnerability because of the additional effect of CH_4) by using a simple, globally averaged 'box' carbon–climate model with the following attributes:

1. The model state vector includes six variables: mean global temperature (T_A), atmospheric CO_2 concentration ($[CO_2]_A$), the terrestrial carbon biomass excluding frozen soils (C_B), carbon store in permafrost or frozen soils (C_F), marine carbon store as dissolved inorganic carbon (C_M) and partial pressure of CO_2 in the upper ocean (pCO_2). Model equations and parameters are given in Tables 2.3–2.5.

2. Atmospheric CO_2 is connected to climate (T_A) through an overall climate sensitivity to CO_2 which parameterises not only the direct radiative forcing by CO_2 increases but also all other associated climate feedbacks (water vapour, clouds, …). Two alternative formulations are used, in which changes in T_A are proportional either to changes in $[CO_2]_A$ itself or to changes in $\ln [CO_2]_A$ (Table 2.3). The latter is consistent with classical radiative transfer theory for infrared absorption in

Table 2.3 Governing equations for state variables in a simple box model of the carbon–climate system Carbon fluxes (F) are defined in Table 2.4; parameters are defined in Table 2.5

State variable		Governing equation
Mean global temperature (°C)	T_A	(1) Linear: $\dfrac{dT_A}{dt} = \alpha_1 \dfrac{d[CO_2]_A}{dt}$ (2) Logarithmic: $\dfrac{dT_A}{dt} = \dfrac{\alpha_2}{(\ln 2)[CO_2]_A} \dfrac{d[CO_2]_A}{dt}$
Atmospheric CO_2 concentration (ppm)	$[CO_2]_A$	$r_{CO_2} \dfrac{d[CO_2]_A}{dt} = F_{Foss} + F_{LUC} + F_{MA} + F_{FA} - \dfrac{dC_B}{dt}$
Terrestrial carbon store in biomass and soils (Pg C) excluding peatlands, frozen soils	C_B	$\dfrac{dC_B}{dt} = F_{NPP} - F_R - F_{LUC}$
Ocean (marine) carbon store (Pg C)	C_M	$\dfrac{dC_M}{dt} = -F_{MA}$
Ocean pCO_2 (Pa)	pCO_2	$\dfrac{dpCO_2}{dt} = \beta \dfrac{pCO_2}{C_M} \dfrac{dC_M}{dt}$
Carbon store in permafrost (frozen soils) (Pg C)	C_F	$\dfrac{dC_F}{dt} = -F_{FA}$

nearly saturated CO_2 bands (Arrhenius 1896; Goody 1964), while the former provides some account for positive feedbacks (e.g. from water vapour) which cause greenhouse warming to be greater than that for CO_2 alone. Each includes a single sensitivity parameter. Neither expression has a formal justification, but the differences between the two provide a first indication of the sensitivity of the carbon–climate system to assumptions about the CO_2–temperature coupling.

3. Frozen-soil carbon (C_F) is assumed to be released as CO_2 under the influence of global warming, from a pool initialised at a pre-industrial value $C_{F0} = 900$ Pg C (Zimov et al. 2006). The rate constant for this release (k_F) is zero in pre-industrial conditions ($T_A = T_{A0}$) and increases linearly with climate warming ($T_A - T_{A0}$) with a proportionality coefficient k_{FT}, so that $dC_F/dt = -k_{FT}(T_A - T_{A0})C_F$. To estimate k_{FT}, we use the estimate that a warming of 2 °C will lead to shrinking of the permafrost area by around 25% (Anisimov et al. 1999). Assuming that (1) this area decrease translates to a shrinkage of the C_F pool by the same factor and (2) the release occurs over 100 years, the implied value of k_{FT} is 0.25 per century per (2 °C) or 0.00125 °C^{-1} y^{-1}. A value of 0.001 °C^{-1} y^{-1} is used here. This is a conservative value, mainly because a warming of 2 °C is likely to occur over a shorter time than 100 years, which would increase the inferred value of k_{FT}.

Table 2.4 Phenomenological equations for fluxes (F) in a simple box model of the carbon–climate system

Carbon flux (Pg C y^{-1})		Phenomenological equation
CO_2 emissions from fossil fuels and other industry	F_{Foss}	1751–2005: data (Marland et al. 2006) future to 2100: scenarios (Table 2.1)
Emissions from land-use change	F_{LUC}	1751–2005: data (Houghton, 1999, C2007) future to 2100: linear decline to zero in 2100
Terrestrial NPP	F_{NPP}	$F_{NPP} = F_{NPP0} + (F_{NPP2} - F_{NPP0})f$ $(C_A - C_{A0}, C_{A1} - C_{A0})$ with $f(x,a) = x^2/(x^2 + a^2)$, $C_A = r_{CO_2}[CO_2]_A$
Terrestrial respiration	F_R	$F_R = k_{B0}C_B 2^{(T_A - T_{A0})/\delta_{2T}}$
Ocean-air CO_2 flux	F_{MA}	$F_{MA} = \dfrac{\{A_{Ocean}v_{Piston}([pCO_2]_M - [pCO_2]_A)\}}{k_{Henry}}$
Permafrost-air CO_2 flux	F_{FA}	$F_{FA} = k_{FT}(T_A - T_{A0})C_F$

NPP net primary production

Table 2.5 Parameters in a simple box model of the carbon–climate system

Parameter	Symbol	Value
Conversion for atmospheric CO_2 (ppm to Pg C)	r_{CO2}	2.181 Pg C ppm^{-1}
Linear climate sensitivity to C_A	α_1	0.008 °C ppm^{-1}
Logarithmic (CO_2 doubling) climate sensitivity to CO_2	α_2	2 °C
Revelle buffer factor for response of pCO_2 to C_M	B	12 (dimensionless)
Initial (pre-industrial) temperature	T_{A0}	15 °C
Initial (pre-industrial) CO_2 concentration	$[CO_2]_{A0}$	278 ppm
Initial (pre-industrial) terrestrial biomass C store	C_{B0}	Calculated (equilibrium)
Initial (pre-industrial) marine C store	C_{M0}	5,000 Pg C
Initial (pre-industrial) ocean pCO_2	$pCO_{2(0)}$	Calculated (equilibrium)
Initial (pre-industrial) C store in frozen soils	C_{F0}	500 Pg C
Initial (pre-industrial) NPP	F_{NPP0}	60 Pg C y^{-1}
NPP at saturation with respect to CO_2	F_{NPP2}	75 Pg C y^{-1}
CO_2 scale for NPP saturation with respect to CO_2	$[CO_2]_{A1}$	450 ppm
Turnover rate for terrestrial respiration at $T_A = T_{A0}$	k_{B0}	0.02 y^{-1}
Doubling temperature for F_R	δ_{2T}	15 °C
Ocean area	A_{Ocean}	3.6 × 10^{14} m^2
Ocean exchange (piston) velocity	v_{Piston}	1,500 m y^{-1}
Henry's law constant for CO_2	k_{Henry}	2,900 × 2$^{[(25 - T_A)/25]}$ Pa m^3 mol^{-1}
Rate per °C for C flux from frozen soils	k_{FT}	0 or 0.001 y^{-1} °C^{-1}

NPP net primary production

4. Parameters (particularly the climate sensitivities to CO_2) and initial conditions
   were chosen so that predictions approximately matched past (1750–2005) trends
   T_A and $[CO_2]_A$ (Table 2.5).

Figure 2.4 shows predictions for 1750–2100 under four IPCC SRES scenarios
(Table 2.1), with no frozen-carbon feedback ($k_{FT} = 0$). Agreement with past observations of T_A and $[CO_2]_A$ is satisfactory. Predictions with a linear climate sensitivity
to CO_2 (upper panels) yield a warming range in 2100 of 2 °C (scenario B1) to 4 °C
(scenario A2), compared with 1.5–2.5 °C from the logarithmic climate sensitivity
to CO_2. The former, larger warming is very close to the ensemble of climate model
results in the IPCC Fourth Assessment Report (IPCC 2007), while the latter is
lower. The $[CO_2]_A$ range is 560 ppm (scenario B1) to 820 ppm (scenario A2).

Figure 2.5 compares model predictions with the frozen-carbon pool disabled ($k_{FT} = 0$)
and enabled ($k_{FT} = 0.001$ °C^{-1} y^{-1}), using future emissions scenario A2 (the standard
scenario for C^{4}MIP). With linear climate sensitivity to CO_2 (which yields a model
behaviour broadly consistent with IPCC predictions as noted above), the positive

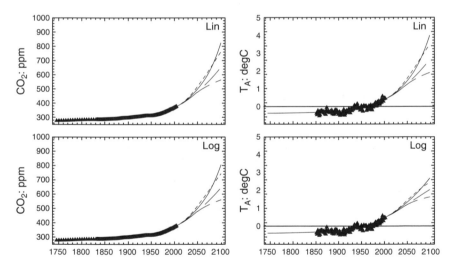

Fig. 2.4 Predictions from a simple box model of the carbon–climate system of atmospheric CO_2
concentration ($[CO_2]_A$, left panels) and global mean temperature (T_A, right panels). The zero line for
T_A is set to the average observed T_A over the period 1961–1990. Four emissions scenarios are shown,
comprising actual fossil-fuel and land-use-change emissions from 1751 to 2005 (data sources as in
Fig. 2.1), and the A2 (solid), A1B (short dashed), B1 (medium dashed) and B2 (long dashed) scenarios (Nakicenovic et al. 2000) from 2005 to 2100. Upper and lower panels respectively show predictions with linear and logarithmic climate sensitivities to $[CO_2]_A$. Data (points) for $[CO_2]_A$ are
composite observations from ice core data from Law Dome, Antarctica (Etheridge et al. 1998b)
(before 1959) and direct atmospheric measurements at Mauna Loa, Hawaii (Keeling and Whorf 2005)
(1959 onward). Upper and lower panels respectively show predic-
tions with linear and logarithmic climate sensitivities to $[CO_2]_A$. Data (points) for T_A are the global temperature series 1850–2004 (Jones et al. 2006).

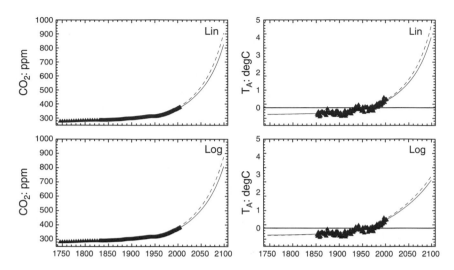

Fig. 2.5 Predictions from a simple box model of the carbon–climate system of atmospheric CO_2 concentration ($[CO_2]_A$, left panels) and global mean temperature (T_A, right panels) without (solid) and with (dashed) feedbacks between the frozen-carbon pool and global temperature. The emissions scenario comprises actual fossil-fuel and land-use-change emissions from 1751 to 2005 (data sources as in Fig. 2.1), and the A2 scenario from 2005 to 2100. Upper and lower panels respectively show predictions with linear and logarithmic climate sensitivities to $[CO_2]_A$. Data for $[CO_2]_A$ and T_A as in Fig. 2.4.

feedback on warming through CO_2 release from frozen soils yields an additional temperature increment in 2100 (δT_A) of about 0.7 °C and a CO_2 increment ($\delta[CO_2]_A$) of about 80 ppm, above the changes induced by other processes. The increments with the logarithmic climate sensitivity to CO_2 are lower, because vulnerability of frozen carbon increases with warming.

This simple calculation neglects the additional warming due to CH_4 release. From one estimate (Zimov et al. 2006), about 30% of the carbon released to the atmosphere from thawing permafrost is released as CH_4, with a present global warming potential (GWP) of about 21 relative to CO_2 (though the GWP of CH_4 will decline in the future). This methane has a residence time in the atmosphere of the order of 10 years, ultimately being oxidised to CO_2. Thus, the GWP of the released carbon decreases with time.

2.5 Implications of Carbon–Climate Vulnerabilities for Carbon Observation

Over several years, the global research and observation communities have been developing a concept for integrated global carbon observation. The carbon theme report of the Integrated Global Observing Strategy (Ciais et al. 2004) calls for integrated observations of fluxes, processes and pools, principally:

- Satellite measurements of atmospheric-column CO_2, CH_4 and other GHGs, together with in situ (flask and continuous) measurements
- An operational, optimised network of eddy-covariance flux towers, measuring the CO_2 flux as NEE, with water and energy fluxes, over all major ecosystems
- A global system for in situ (ship, drifter) ocean pCO_2 measurements
- Satellite observations of vegetation and other land surface properties
- Georeferenced fossil-fuel emission maps including temporal variability and uncertainties
- Regular inventories of forest above-ground biomass and soil carbon content
- Regular inventories of dissolved carbon and related biogeochemical quantities in the ocean
- Satellite and in situ observations of fire, land cover change and other disturbance processes
- Monitoring of land-to-ocean carbon fluxes in river run off
- Monitoring of carbon fluxes associated with goods (harvesting, trade, disposal)

Many of these measurements (particularly the atmospheric concentrations and ocean biogeochemistry) are to be integrated with near-real-time modelling frameworks, using data assimilation into climate and earth system models. This may involve collaboration with the atmosphere–ocean models run by operational weather forecasting agencies, together with longer timescale models of pool dynamics in the carbon–climate system.

In Introduction, we highlighted two broad goals for a global carbon observing system, oriented respectively towards understanding and management. Recognition of carbon–climate vulnerabilities gives added focus to both goals, through the need to provide early warning of feedbacks (both positive and negative), to understand the reasons for emerging trends, and to provide national and international policy communities with information to manage vulnerabilities. Considering the carbon–climate vulnerabilities associated with terrestrial processes (Sect. 2.3 and Table 2.2), we now analyse the modes of terrestrial observation which emerge as having particular importance. Some are already well covered by existing plans, others less so. The discussion is organised around (1) fluxes and processes and (2) pools.

2.5.1   Fluxes and Processes

Additional observations are needed to increase our capacity to attribute carbon fluxes to specific source and sink processes, and to detect changes in the statistical properties of vulnerabilities associated with episodic disturbances. The following are key areas:

- *Responses of NPP and respiration to extreme events*: NPP and respiration respond not only to long-term climate shifts but also to extreme events (sometimes through threshold-like transitions), as has already been observed (Sect. 2.3.1). Observational requirements to improve detection and understanding of these effects include:

○ *Satellite measurements of vegetation dynamics over multi-year periods:* Longevity of satellite records requires rehabilitation and harmonisation of data from old systems with records extending back to the 1980s (e.g. AVHRR), together with maintenance of modern systems (e.g. MODIS) and launching of well-equipped new systems on the timescales required to maintain continuity through overlap between systems.

○ *Process-based observations:* Flux towers and other process observations have a major role in the detection and attribution of the effects of drought and heat stress on production. Consistent long-term observations are critical for this purpose, to enable the extreme-event signal to emerge from the climatological background.

• *Multi-factorial of responses of NPP and respiration to temperature, water, nutrients and CO₂:* Climate-induced changes in the relationship between NPP and soil respiration have major implications for the terrestrial carbon balance, including NEE and pool dynamics. There is a wide scatter in existing process-based information on the responses of NPP, respiration and NEE to interactions among the drivers of temperature, water, nutrients and CO_2 (C2007; also see Sect. 2.3.1). It is important to attribute and hence reduce this scatter, which implies a continuing demand for ecosystem eddy flux measurements coupled with chamber-based soil measurements over diverse ecosystems (boreal, tundra, tropical, permafrost, temperate forests, semi-arid regions, etc.), together with ongoing multi-factorial manipulative experiments to disentangle the interactions.

• *Deforestation*: Carbon fluxes from deforestation remain the single most important direct human forcing of the terrestrial carbon balance, and the component of the global carbon budget with the largest uncertainty. To reduce this uncertainty, we need

○ Improved measurements of deforestation rates (including selective logging) from remote sensing

○ Improved biomass densities to calculate carbon emission factors

○ Improved knowledge of the time courses of carbon fluxes following deforestation (e.g. biomass cleared in 1 year may be burned in following years, and there are time lags associated with soil respiration). Yearly measurements deforestation would also help to attribute annual fluctuations in atmospheric CO_2 growth (current deforestation data is available only in increments of 5–10 years)

• *Fire*: This is another key process responsible for a large part of the perturbations in the annual CO_2 growth. Global improvement in estimates of both burned area and carbon emissions from fire will need to rely on atmospheric networks and remote sensing products. These observations include increased focus on (1) use of CO and CH_4 to quantify the contribution of fires to the atmospheric CO_2 growth (Langenfelds et al. 2002), (2) extraction of burned area (monthly) and hot spots (daily) from remote sensing, (3) ground and remote sensing observations of vegetation recovery after fire (when not linked to deforestation).

2.5.2 Pools

In addition to existing and currently planned inventory networks for monitoring terrestrial carbon pools, two kinds of observations emerge as critical from a vulnerability standpoint:

- *Carbon in frozen soils*: Section 2.4 has offered a semi-quantitative analysis of the significant carbon–climate vulnerability associated with carbon stored in frozen soils at high northern latitudes, which are susceptible to decomposition under global warming. To improve estimates such as this one, it is important to monitor thawing trends and the resulting carbon emission, particularly its distribution between CO_2 and CH_4. Key measurements are:

  ○ Monitoring of the permafrost southern limit and changes in depth of permafrost
  ○ Carbon content in frozen soils, to full profile depth (substantially deeper than 1 m)
  ○ Measurements of vertical fluxes of CO_2 and CH_4 with eddy-covariance (local) and atmospheric-inverse (large regional) methods
  ○ Measurement of lateral carbon fluxes to rivers and the coastal zone

- *Peatlands*: Both in high latitudes (as a continuum of permafrost) and in the tropics, peatlands require careful monitoring of changes in stocks and fluxes under the influences of warming and changes in the hydrological cycle. Particularly in the tropics, changes in hydrological regime result both from climate change and from human activities such as drainage. Key measurements (similar to those for frozen soils) include:

  ○ Monitoring of tropical peatland deforestation and drainage
  ○ Carbon content to full profile depth (substantially deeper than 1 m)
  ○ Monitoring of vertical and lateral fluxes in both cold and tropical peatlands
  ○ Monitoring of subsidence rates of drained peatland forests

2.6 Conclusion

In Sect. 2.3, we identified several major carbon–climate vulnerabilities: limitation of CO_2 fertilisation by water and nutrient constraints, the response of soil respiration and NPP to warming and moisture, permafrost thawing, fire and ecosystem responses to a variety of land-use changes. A semi-quantitative assessment in Sect. 2.4 of just one of these vulnerabilities (the CO_2 consequence of permafrost thawing) suggests a perturbation on global CO_2 in 2100 of about 80 ppm and a temperature perturbation of about 0.7 °C, relative to a prediction under the A2 emissions scenario in which this positive feedback is absent. The temperature perturbation is a conservative estimate because of the neglect of warming due to CH_4. It is reasonable to conclude that the sum of all the vulnerabilities discussed in Sect. 2.3 is not negligible relative to the primary forcing of the climate system by anthropogenic CO_2. The implication for carbon observation is to focus attention on the process measurements and monitoring programmes needed to track and better quantify the feedback processes leading to these vulnerabilities.

Acknowledgements The work described here is a contribution to Theme 2 (Processes and Interactions) of the Global Carbon Project (www.globalcarbonproject.org). We are grateful to Cathy Trudinger and Will Steffen for constructive comments on drafts of this work, and Peter Briggs for assistance with figures. MRR thanks the CarboEurope Integrated Program for support to attend the CarboEurope Greenhouse Gas Workshop, Amsterdam, 4–5 April 2005, which prompted the writing of this chapter. We thank the Australian Greenhouse Office and CSIRO for supporting the GCP International Project Office in Canberra.

References

Achard F., Eva H. D., Mayaux P., Stibig H. J., Belward A. 2004. Improved estimates of net carbon emissions from land cover change in the tropics for the 1990s. Global Biogeochem. Cycles **18**: doi:10.1029/2003GB002142.

Allan W., Lowe D. C., Gomez A. J. 2005. Interannual variations of ^{13}C in tropospheric methane: Implication for a possible atomic chlorine sink in the marine boundary layer. J. Geophys. Res. **110**: doi:10.1029/ 2004JD005650.

Angert A., Biraud S., Bonfils C. et al. 2005. Drier summers cancel out the CO_2 uptake enhancement induced by warmer springs. Proc. Natl. Acad. Sci. U.S.A. **102**: 10823–10827.

Anisimov O. A., Nelson F. E., Pavlov A. V. 1999. Predictive scenarios of permafrost development under conditions of global climate change in the XXI century. Earth Cryol. **3**: 15–25.

Arrhenius S. 1896. On the influence of carbonic acid in the air upon the temperature of the ground. Philos. Mag. J. Sci. **5**: 239–276.

Bogner J., Matthews E. 2003. Global methane emissions from landfills: New methodology and annual estimates 1980–1996. Global Biogeochem. Cycles **17**: doi:10.1029/2002GB001913.

Bousquet P., Ciais P., Miller J. B. et al. 2006. Contribution of anthropogenic and natural sources to atmospheric methane variability. Nature **443**: 439–443.

Burrows W. H., Henry B. K., Back P. V. et al. 2002. Growth and carbon stock change in eucalypt woodlands in northeast Australia: Ecological and greenhouse sink implications. Global Change Biol. **8**: 769–784.

Camill P. 2005. Permafrost thaw accelerates in boreal peatlands during late-20th century climate warming. Clim. Change **68**: 135–152.

Canadell J. G., Le Quéré C., Raupach M. R., Field C. B., Buitenhuis E. T., Ciais P, Conway T. J., Gillette N. P., Houghton R. A., Marland G. 2007a. Contributions to accelerating atmospheric CO_2 growth from economic activity, carbon intensity, and efficiency of natural sinks. Proceedings of the National Academy of Sciences, Early Edition 10.1073/pnas.0702737104.

Canadell J. G., Pataki D., Gifford R. M. et al. 2007b. Saturation of the terrestrial carbon sink. In Terrestrial Ecosystems in a Changing World, eds. J. G. Canadell, D. Pataki, L. Pitelka, pp. 59–78. Springer-Verlag, Berlin.

Cao M. K., Gregson K., Marshall S. 1998. Global methane emission from wetlands and its sensitivity to climate change. Atmos. Environ. **32**: 3293–3299.

Chen Y.-H., Prinn R. G. 2005. Atmospheric modeling of high- and low-frequency methane observations: Importance of interannually varying transport. J. Geophys. Res. **110**: D10303, doi:10.1029/ 2004JD005542.

Christiansen T. R., Ekberg A., Ström L., Mastepanov M. 2003. Factors controlling large scale variations in methane emission from wetlands. Geophys. Res. Lett. **30**: doi: 10.1029/ 2002GL016848.

Ciais P., Moore B. I., Steffen W. et al. 2004. *Integrated global carbon observation theme: A strategy to realise a coordinated system of integrated global carbon cycle observations.* Integrated Global Observing Strategy, Stockholm.

Ciais P., Reichstein M., Viovy N. et al. 2005. Europe-wide reduction in primary productivity caused by the heat and drought in 2003. Nature **437**: 529–533.

Cramer W., Bondeau A., Woodward F. I. et al. 2001. Global response of terrestrial ecosystem structure and function to CO_2 and climate change: Results from six dynamic global vegetation models. Global Change Biol. **7**: 357–373.

DeFries R. S., Field C. B., Fung I. Y., Collatz G. J., Bounoua L. 1999. Combining satellite data and biogeochemical models to estimate global effects of human-induced land cover change on carbon emissions and primary productivity. Global Biogeochem. Cycles **13**: 803–815.

DeFries R. S., Houghton R. A., Hansen M. C., Field C. B., Skole D., Townshend J. 2002. Carbon emissions from tropical deforestation and regrowth based on satellite observations for the 1980s and 1990s. Proc. Natl. Acad. Sci. U.S.A. **99**: 14256–14261.

Etheridge D. M., Steele L. P., Francey R. J., Langenfelds R. L. 1998a. Atmospheric methane between 1000 AD and present: Evidence of anthropogenic emissions and climatic variability. J. Geophys. Res. **103**: 15979–15993.

Etheridge D. M., Steele L. P., Langenfelds R. L., Francey R. J., Barnola J. M., Morgan V. I. 1998b. *Historical CO_2 records from the Law Dome DE08, DE08-2, and DSS ice cores.* Trends: A Compendium of Data on Global Change, Carbon Dioxide Information Analysis Center, Oak Ridge National Laboratory, U.S. Department of Energy, Oak Ridge, Tennessee, USA.

Fang J. Y., Chen A. P., Peng C. H., Zhao S. Q., Ci L. 2001. Changes in forest biomass carbon storage in China between 1949 and 1998. Science **292**: 2320–2322.

Fang C. M., Smith P., Moncrieff J. B., Smith J. U. 2005. Similar response of labile and resistant soil organic matter pools to changes in temperature. Nature **433**: 57–59.

Farquhar G. D., Sharkey T. D. 1982. Stomatal conductance and photosynthesis. Annu. Rev. Plant Physiol. **33**: 317–345.

Farquhar G. D., Caemmerer von S., Berry J. A. 1980. A biochemical model of photosynthetic CO_2 assimilation in leaves of C_3 species. Planta **149**: 78–90.

Field C. B., Raupach M. R. 2004. The Global Carbon Cycle: Integrating Humans, Climate, and the Natural World. Island Press, Washington, p. 526.

Field C. B., Chapin III F. S., Chiariello N. R., Holland E. A., Mooney H. A. 1996. The Jasper Ridge CO_2 experiment: Design and motivation. In Carbon Dioxide and Terrestrial Ecosystems, eds. G. W. Koch, H. A. Mooney, pp. 121–145. Academic Press, San Diego.

Friborg T., Soegaard H., Christensen T. R., Lloyd C. R., Panikov N. S. 2003. Siberian wetlands: Where a sink is a source. Geophys. Res. Lett. **30**.

Friedlingstein P., Cox P., Betts R. et al. 2006. Climate-carbon cycle feedback analysis: Results from the C4MIP model intercomparison. J. Clim. **19**: 3337–3353.

Fung I. Y., John J., Lerner J., Mathews E., Prather M., Steele L. P., Fraser P. J. 1991. Three-dimensional model synthesis of the global methane cycle. J. Geophys. Res. **96**: 13033–13065.

Giardina C. P., Ryan M. G. 2000. Biogeochemistry: Soil warming and organic carbon content - Reply. Nature **408**: 790.

Gifford R. M., Howden M. 2001. Vegetation thickening in an ecological perspective: Significance to national greenhouse gas inventories. Environ. Sci. Policy **4**: 59–72.

Global Carbon Project 2003. *Science Framework and Implementation.* Earth System Science Partnership (IGBP, IHDP, WCRP, Diversitas) Report No. 1; GCP Report No. 1, Global Carbon Project, Canberra.

Goody R. M. 1964. Atmospheric Radiation. I. Theoretical Basis. Clarendon Press, Oxford, 436 pp.

Greenblatt J. B., Sarmiento J. L. 2004. Variability and climate feedback mechanisms in ocean uptake of CO_2. In The Global Carbon Cycle: Integrating Humans, Climate, and the Natural World, eds. C. B. Field, M. R. Raupach, pp. 257–275. Island Press, Washington.

Gruber N., Friedlingstein P., Field C. B. et al. 2004. The vulnerability of the carbon cycle in the 21st century: An assessment of carbon-climate-human interactions. In The Global Carbon Cycle: Integrating Humans, Climate, and the Natural World, eds. C. B. Field, M. R. Raupach, pp. 45–76. Island Press, Washington.

Hein R., Crutzen P. J., Heimann M. 1997. An inverse modelling approach to investigate the global atmospheric methane cycle. Global Biogeochem. Cycles **11**: 43–76.

Holland E. A., Braswell B. H., Lamarque J. F. et al. 1997. Variations in the predicted spatial distribution of atmospheric nitrogen deposition and their impact on carbon uptake by terrestrial ecosystems. J. Geophys. Res. Atmos. **102**: 15849–15866.

Houghton R. A. 1998. Historic role of forests in the global carbon cycle. In Carbon Dioxide Mitigation in Forestry and Wood Industry, eds. G. H. Kohlmaier, M. Weber, R. A. Houghton, pp. 1–24. Springer-Verlag, Berlin.

Houghton R. A. 1999. The annual net flux of carbon to the atmosphere from changes in land use 1850–1990. Tellus Ser. B **51**: 298–313.

Houghton R. A. 2003. Why are estimates of the terrestrial carbon balance so different? Global Change Biol. **9**: 500–509.

Houghton R. A., Hackler J. L. 2000. Changes in terrestrial carbon storage in the United States. 1: The roles of agriculture and forestry. Global Ecol. Biogeog. **9**: 125–144.

Houghton R. A., Hackler J. L., Lawrence K. T. 2000. Changes in terrestrial carbon storage in the United States. 2: The role of fire and fire management. Global Ecol. Biogeog. **9**: 145–170.

Houweling S., Kaminski T., Dentener F. J., Lelieveld J., Heimann M. 1999. Inverse modeling of methane sources and sinks using the adjoint of a global transport model. J. Geophys. Res. **104**: 26137–26160.

IPCC 2001. *Climate Change 2001: The Scientific Basis*. Contribution of Working Group I to the Third Assessment Report of the Intergovernmental Panel on Climate Change. Cambridge University Press, Cambridge, United Kingdom and New York.

IPCC 2007. *Climate change 2007: The physical science basis. Summary for policymakers*. IPCC Secretariat, Geneva.

Jacobson M., Charleson R. J., Rodhe H., Orians G. H. 2000. Earth System Science: From Biogeochemical Cycles to Global Change. Academic Press, New York, p. 527.

Janssens I. A., Freibauer A., Schlamadinger B. et al. 2005. The carbon budget of terrestrial ecosystems at country-scale - a European case study. Biogeosciences **2**: 15–26.

Jarvis P., Linder S. 2000. Botany: Constraints to growth of boreal forests. Nature **405**: 904–905.

Jones P. D., Parker D. E., Osborn T. J., Briffa K. R. 2006. *Global and hemispheric temperature anomalies - land and marine instrumental records*. Trends: A Compendium of Data on Global Change, Carbon Dioxide Information Analysis Center, Oak Ridge National Laboratory, U.S. Department of Energy, Oak Ridge, Tennessee, USA.

Jorgenson T. M., Shur Y. L., Pullman E. R. 2006. Abrupt increase in permafrost degradation in Arctic Alaska. Geophys. Res. Lett. **33**: doi:10. 1029/2005GL024960.

Keeling C. D. and Whorf T. P. 2005. *Atmospheric CO₂ records from sites in the SIO air sampling network*. Trends: A Compendium of Data on Global Change, Carbon Dioxide Information Analysis Center, Oak Ridge National Laboratory, U.S. Department of Energy, Oak Ridge, Tennessee, USA.

Knorr W., Prentice I. C., House J. I., Holland E. A. 2005. Long-term sensitivity of soil carbon turnover to warming. Nature **433**: 298–301.

Kurz W. A., Apps M. J. 1999. A 70-year retrospective analysis of carbon fluxes in the Canadian forest sector. Ecol. Appl. **9**: 526–547.

Langenfelds R. L., Francey R. J., Pak B. C., Steele L. P., Lloyd J., Trudinger C. M., Allison C. E. 2002. Interannual growth rate variations of atmospheric CO_2 and its delta C-13, H-2, CH_4, and CO between 1992 and 1999 linked to biomass burning. Global Biogeochem. Cycles **16**.

Le Quere C., Metzl N. 2004. Natural processes regulating the ocean uptake of CO_2. In The Global Carbon Cycle: Integrating Humans, Climate, and the Natural World, eds. C. B. Field, M. R. Raupach, pp. 243–255. Island Press, Washington.

Lelieveld J., Crutzen P. J., Dentener F. J. 1998. Changing concentration, lifetime and climate forcing of atmospheric methane. Tellus Ser. B **50**: 128–150.

Lloyd J., Taylor J. A. 1994. On the temperature dependence of soil respiration. Functional Ecology **8**: 315–323.

Luger A. D., Moll E. J. 1993. Fire protection and afromontane forest expansion in Cape Fynbos. Biol. Conserv. **64**: 51–56.

Luo Y. Q., Wan S. Q., Hui D. F., Wallace L. L. 2001. Acclimatization of soil respiration to warming in a tall grass prairie. Nature **413**: 622–625.

Luo Y., Su B., Currie W. S. et al. 2004. Progressive nitrogen limitation of ecosystem responses to rising atmospheric carbon dioxide. BioScience **54**: 731–739.

Mack F., Hoffstadt J., Esser G., Goldammer J. G. 1996. Modeling the influence of vegetation fires on the global carbon cycle. In Biomass Burning and Global Change, ed. J. S. Levine. MIT Press, Cambridge, MA.

Marland G., Boden T. A., Andres R. J. 2006. *Global, regional, and national CO_2 emissions*. Trends: A Compendium of Data on Global Change, Carbon Dioxide Information Analysis Center, Oak Ridge National Laboratory, U.S. Department of Energy, Oak Ridge, Tennessee, USA.

Mikaloff Fletcher S. E., Tans P. P., Bruhwiler L. M., Miller J. B., Heimann M. 2004a. CH_4 sources estimated from atmospheric observations of CH_4 and its $^{13}C/^{12}C$ isotopic ratios: 1. Inverse modelling of source processes. Global Biogeochem. Cycles **18**: doi:10.1029/2004GB002223.

Mikaloff Fletcher S. E., Tans P. P., Bruhwiler L. M., Miller J. B., Heimann M. 2004b. CH_4 sources estimated from atmospheric observations of CH_4 and its $^{13}C/^{12}C$ isotopic ratios: 2. Inverse modelling of CH_4 fluxes from geographical regions. Global Biogeochem. Cycles **18**: doi:10.1029/2004GB002224.

Mosier A. R., Duxbury J. M., Freney J. R., Heinemeyer O., Minami K., Johnson D. E. 1998. Mitigating agricultural emissions of methane. Clim. Change **40**: 39–80.

Mouillot F., Field C. B. 2005. Fire history and the global carbon budget: A 1 degrees x 1 degrees fire history reconstruction for the 20th century. Global Change Biol. **11**: 398–420.

Nadelhoffer K. J., Emmett B. A., Gundersen P. et al. 1999. Nitrogen deposition makes a minor contribution to carbon sequestration in temperate forests. Nature **398**: 145–148.

Nakicenovic N., Alcamo J., Davis G. et al. 2000. *IPCC Special Report on Emissions Scenarios*. Cambridge University Press, Cambridge, U.K. and New York.

Nemani R. R., Keeling C. D., Hashimoto H. et al. 2003. Climate-driven increases in global terrestrial net primary production from 1982 to 1999. Science **300**: 1560–1563.

Norby R. J., DeLucia E. H., Gielen B. et al. 2005. Forest response to elevated CO_2 is conserved across a broad range of productivity. Proc. Natl. Acad. Sci. U.S.A. **102**: 18052–18056.

Nowak R. S., Ellsworth D. S., Smith S. D. 2004. Functional responses of plants to elevated atmospheric CO_2: Do photosynthetic and productivity data from FACE experiments support early predictions? New Phytol. **162**: 253–280.

Olivier J. G. J., Bouwman A. F., Berdowski J. J. M. et al. 1999. Sectoral emission inventories of greenhouse gases for 1990 on a per country basis as well as on 1x1. Environ. Sci. Policy **2**: 241–263.

Oren R., Ellsworth D. S., Johnsen K. H. et al. 2001. Soil fertility limits carbon sequestration by forest ecosystems in a CO_2-enriched atmosphere. Nature **411**: 469–472.

Owensby C. E., Ham J. M., Knapp A. K., Bremer D., Auen L. M. 1997. Water vapour fluxes and their impact under elevated CO_2 in a C4-tallgrass prairie. Global Change Biol. **3**: 189–195.

Pacala S. W., Hurtt G. C., Baker D. et al. 2001. Consistent land- and atmosphere-based US carbon sink estimates. Science **292**: 2316–2320.

Page S. E., Siegert F., Rieley J. O., Boehm H. D. V., Jaya A., Limin S. 2002. The amount of carbon released from peat and forest fires in Indonesia during 1997. Nature **420**: 61–65.

Page S. E., Wust R. A. J., Weiss D., Rieley J. O., Shotyk W., Limin S. H. 2004. A record of Late Pleistocene and Holocene carbon accumulation and climate change from an equatorial peat bog (Kalimantan, Indonesia): Implications for past, present and future carbon dynamics. J. Quat. Sci. **19**: 625–635.

Pataki D. E., Huxman T. E., Jordan D. N. et al. 2000. Water use of two Mojave Desert shrubs under elevated CO_2. Global Change Biol. **6**: 889–897.

Raupach M. R., Marland G., Ciais P., LeQuere C., Canadell J. G., Field C. B. 2007. Global and regional drivers of accelerating CO_2 emissions. Proceedings of the National Academy of Sciences **14**: 10288–10293.

Raupach M. R., Rayner P. J., Barrett D. J. et al. 2005. Model-data synthesis in terrestrial carbon observation: Methods, data requirements and data uncertainty specifications. Global Change Biol. **11**: 10.1111/j.1365–2486.2005.00917.x.

Sabine C. L., Heimann M., Artaxo P. et al. 2004. Current status and past trends of the global carbon cycle. In The Global Carbon Cycle: Integrating Humans, Climate, and the Natural World, eds. C. B. Field, M. R. Raupach, pp. 17–44. Island Press, Washington.

Schellnhuber H. J., Cramer W., Nakicenovic N., Wigley T. M. L., Yohe G. 2006. Avoiding Dangerous Climate Change. Cambridge University Press, Cambridge, 392 pp.

Smith L. C., Sheng Y., MacDonald G. M., Hinzman L. D. 2005. Disappearing Arctic lakes. Science **308**: 1429.

Steffen W. L., Sanderson A., Tyson P. D. et al. 2004. Global Change and the Earth System: A Planet Under Pressure. Springer, Berlin, 336 pp.

Tarnocai C. 1999. The effect of climate warming on the carbon balance of cryosols in Canada. Permafrost Periglac. **10**: 251–263.

Townsend A. R., Braswell B. H., Holland E. A., Penner J. E. 1996. Spatial and temporal patterns in terrestrial carbon storage due to deposition of fossil fuel nitrogen. Ecol. Appl. **6**: 806–814.

Valentini R., Matteucci G., Dolman A. J. et al. 2000. Respiration as the main determinant of carbon balance in European forests. Nature **404**: 861–865.

van der Werf G. R., Randerson J. T., Collatz G. J., Giglio L. 2003. Carbon emissions from fires in tropical and subtropical ecosystems. Global Change Biol. **9**: 547–562.

van der Werf G. R., Randerson J. T., Collatz G. J. et al. 2004. Continental-scale partitioning of fire emissions during the 1997 to 2001 El Nino/La Nina period. Science **303**: 73–76.

van der Werf G. R., Randerson J. T., Giglio L., Collatz G. J., Kasibhatla P. S., Arellano A. F. 2006. Interannual variability in global biomass burning emissions from 1997 to 2004. Atmos. Chem. Phys. **6**: 3423–3441.

Wang J. S., Logan J. A., McElroy M. B., Duncan B. N., Megretskaia I. A., Yantosca R. M. 2004. A 3-D model analysis of the slowdown and interannual variability in the methane growth rate from 1988 to 1997. Global Biogeochem. Cycles **18**: GB3011, doi:10.1029/2003GB002180.

Wuebbles D. J., Hayhoe K. 2002. Atmospheric methane and global change. Earth-Sci. Rev. **57**: 177–210.

Zimov S. A., Schuur E. A. G., Chapin F. S. 2006. Permafrost and the global carbon budget. Science **312**: 1612–1613.

Chapter 3
Assimilation and Network Design

Thomas Kaminski and Peter J. Rayner

3.1 Introduction

Information on the carbon cycle comes from a variety of sources. The methods described in this chapter provide a formalism for combining this information. Without such a formalism we are left making ad hoc choices about how to improve our understanding in the light of disagreements among various streams of information. The introduction of such methods into carbon cycle research, principally via the atmospheric studies of Enting et al. (1993, 1995), revolutionised the field and laid the groundwork for most of the subsequent investigations.

The methods in question are fundamentally statistical. They hence provide estimates of the confidence we should have in quantitative statements about the carbon cycle. These statements are usually couched as spreads of probability distributions or as confidence intervals. We refer to them generally as posterior uncertainties. These posterior uncertainties depend on the prior uncertainties of the various data streams that feed the estimation process, the method for combining these data streams (usually some kind of model) and the particular state of the system. Of course, an important aim of measurements is to reduce the posterior uncertainty.

The present chapter is concerned with quantitative network design, by which we understand the optimisation of a measurement strategy via minimisation of this posterior uncertainty for target quantities of particular interest. Examples of such target quantities are the long-term global mean terrestrial flux to the atmosphere over a period in the past or in the future. The computational tool that transforms the information provided by an observational network of the carbon cycle into an estimate of posterior uncertainty is a Carbon Cycle Data Assimilation System (CCDAS). Hence, network design is closely linked to assimilation both conceptually and computationally. Much of the work reviewed in this chapter lies in a small subset of possible network design applications for the carbon cycle. In particular, it uses a limited set of types of observations. This is not an inherent limitation of the approach but rather a limitation in modelling approaches that can combine many streams of measurements. This is changing now. Hence, much of the chapter looks forward to applications that combine different measurement approaches. It is useful, therefore, to describe the problem in general even if most cited examples are from simpler cases.

The first part of the chapter presents the formalism of carbon cycle data assimilation. It will describe the generation of posterior uncertainties for both simple and more complex cases. Next, we review applications of that methodology to the simpler case of atmospheric transport inversion along with the presentation of important caveats. This is followed by a sketch of how network design might look in a more comprehensive CCDAS. Finally, we give perspectives and recommendations.

3.2   Methodology

It is useful to look at a data assimilation system as a tool that combines various sources of information to form a consistent picture of the underlying system, which in our case is the global or regional carbon cycle. Among the pieces of available information are the observational data we would like to assimilate, estimates for various rate constants of the system and dynamical equations describing the system's evolution.

One usually has one or more target quantities that one is interested in, for instance the terrestrial uptake of a continent, such as Europe, over a particular time interval. As we typically cannot observe the target quantity itself, we use a numerical model to link the target quantity to the observations. Often this is most conveniently achieved in the two-step procedure sketched in Fig. 3.1. First, a set of control variables (a combination of initial and boundary conditions and tuning parameters in the model equations) is nominated, and the underlying numerical model is run in inverse mode. This means prior information on the control variables, observational information and, if available, information from other sources are combined with the model to form posterior information on the control variables. If the set of control

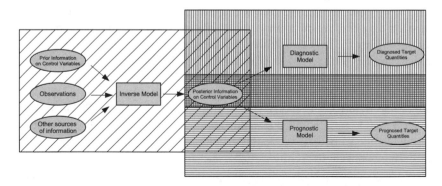

Fig. 3.1 Schematic overview on two-step procedure for inferring diagnosed and prognosed target quantities from data. Rectangular boxes denote processes and oval boxes denote data. The diagonally hatched box includes the inversion (calibration) step, the vertically hatched box the diagnostic step and the horizontally hatched box the prognostic step.

variables includes model parameters, this inversion process is also called calibration of the model. In Fig. 3.1, the inversion step corresponds to the diagonally hatched box.

In a second step, the posterior information on the control variables is used to infer information on the target quantities. Figure 3.1 distinguishes between diagnosed and prognosed target quantities, that is between quantities computed in a diagnostic/prognostic model run. A diagnostic run occurs within the same domain (spatial or temporal) as the inversion step, whereas a prognostic run is (at least in part) outside this domain.

When nominating the control variables, it is not important that they are interesting in themselves. In the two-step approach, they take the role of intermediate quantities on the way to the target quantities. The nomination of the control variables must rather attempt the minimisation of biases and other errors in the inversion step.

It is convenient to quantify the state of information on a specific physical quantity by a probability density function (PDF): the prior information is quantified by a PDF in the space of control variables, the observational information by a PDF in the space of observations and so on. Tarantola (1987) describes the probabilistic framework in detail and provides examples. When the input to the inverse problem can be characterised by Gaussian PDFs, the model is linear, and the model error is Gaussian as well, Tarantola (1987) shows that the posterior information is also quantified by a Gaussian PDF. The mean of that PDF is given by

$$x = x_0 + [M^T C(d)^{-1} M + C(x_0)^{-1}]^{-1} M^T C(d)^{-1} (d - Mx_0),\qquad(3.1)$$

and the covariance of its uncertainty is given by

$$C(x)^{-1} = M^T C(d)^{-1} M + C(x_0)^{-1}.\qquad(3.2)$$

where M denotes (the Jacobian matrix of) the model (that links control variables to observations) and x_0 and $C(x_0)$ the mean and the covariance of the prior information's PDF. On the observations' side, d and $C(d)$ denote the mean and the covariance of uncertainty. In the inversion procedure, the corresponding PDF has to reflect errors in both the observational process and our ability to correctly model the observations. We achieve this via

$$C(d) = C(d_{obs}) + C(d_{mod})\qquad(3.3)$$

and by subtracting the mean model and observational errors from Mx_0 and d, respectively.

We also note that with a little linear algebra, (3.2) can be reformulated to

$$C(x) = C(x_0) - C(x_0) M^T [C(d) + MC(x_0) M^T]^{-1} M^T C(x_0)\qquad(3.4)$$

One can easily verify that x (from 3.1) also minimises the cost function (the exponent of the Gaussian posterior PDF)

$$J(\tilde{x}) = \frac{1}{2}[(M\tilde{x} - d)^T C(d)^{-1}(M\tilde{x} - d) \tag{3.5}$$
$$+ (\tilde{x} - x_0)^T C(X_0)^{-1}(\tilde{x} - x_0)]$$

and that the Hessian matrix $H(\tilde{x})$ of j, that is the matrix composed of its second partial derivatives $\frac{\partial^2 J}{\partial x_i x_j}$, is constant and given by

$$C(x)^{-1} = H(\tilde{x}) \tag{3.6}$$

If the model is non-linear or any of the PDFs of the inputs are non-Gaussian, the Gaussian PDF with the minimum of (3.5) as mean and covariance given by (3.6) is an approximation of the posterior PDF.

For the second step, that is the estimation of a diagnostic or prognostic target quantity y, its PDF can be approximated by a Gaussian with mean

$$y = N(x) \tag{3.7}$$

and the covariance

$$C(y) = D(N)C(x)D(N)^T + C(y_{mod}), \tag{3.8}$$

where N is the model (in Fig. 3.1 denoted as diagnostic/prognostic model) that maps the control variables onto the target quantity, D(N) is its linearisation around the mean of the posterior PDF of the control variables, also denoted as the Jacobian matrix of N, and $C(y_{mod})$ is the uncertainty in the model result from errors in the model. Only if y coincides with one of the observations used in the inversion step, this uncertainty is already accounted for in $C(x)$, and we omit the $C(y_{mod})$ contribution. If N is linear and the posterior PDF of the control variables Gaussian, then the PDF of the target quantity is Gaussian as well, and completely described by (3.7) and (3.8).

3.3 Transport inversion

This section demonstrates network design for atmospheric transport inversion before addressing network design in an entire CCDAS. The set-up for atmospheric transport inversion as introduced by Enting et al. (1995) uses a set of atmospheric CO_2 observations provided by a regional or global sampling network to constrain surface fluxes. Adding this intermediate step before addressing network design in an entire CCDAS is useful for a number of reasons:

- Transport inversions allow us to introduce the main network design concepts for a single component of the carbon cycle.

- From the methodological point of view, it is convenient that, at least for the typical time scales of interest (from diurnal to decadal), the atmospheric transport of CO_2 is modelled as linear.
- Transport inversions often use only a single data stream.
- There have been successful applications of network design concepts.
- Finally, transport inversion has been a typical and, hence, familiar starting point into CCDAS for many colleagues (and the authors).

In transport inversions, the PDFs for priors and data are usually assumed Gaussian, that is the posterior PDF is given by (3.1) and (3.2), where M denotes the transport model and x and d the surface fluxes and data, respectively. Enting (2002) provides details on transport inversion and an exhaustive list of examples and references.

As soon as the pioneering paper of Enting et al. (1995) introduced the calculation of posterior uncertainties into the atmospheric inversion problem, it became clear that they were disturbingly large. As a more recent example, the posterior estimates in Fig. 2 of Gurney et al. (2002) show that even for Europe, a relatively well-sampled region, the posterior uncertainties are very large. Indeed, there were large areas of the world where, according to that study, the atmospheric data added little information. In the optimistic case of the ocean, this partly indicated the tight prior constraint. Here the atmosphere is really a consistency check. For many terrestrial regions the position is reversed. Even with large prior uncertainties, which imply enhanced sensitivity of the posterior flux to data, there is little reduction in uncertainty between the prior and posterior fluxes. Clearly more measurements are needed.

Such measurements are expensive and difficult. Either they require development and deployment of expensive instruments or the painstaking analysis of flasks returned to central measurement facilities. It would be helpful if augmentation of the network was guided so as to produce maximal return. We exploit the important property of (3.2) that, for linear systems and a given $C(d)$, the posterior uncertainty depends only on prior uncertainty and the Jacobian. Thus we can predict the information content of a measurement without actually making it. We can use this property to study optimal deployment of measurements.

Like much else in the basic theory of atmospheric inversions, this idea was imported from solid geophysics. Hardt and Scherbaum (1992) (later published as Hardt and Scherbaum 1994) optimised various quality measures of a seismographic network using station locations as parameters. This is a conventional minimisation problem but of a highly non-linear function. They used the procedure of simulated annealing for the optimisation. The algorithm mimics the 'shaking down' of ions into a crystal on a sufficiently gradual phase transition. Put briefly, a function is calculated for a given vector of parameter values. The parameters are perturbed and the function recalculated. If the new function value is lower, the new vector will be retained. There is also a finite probability that a higher value will be retained, avoiding the algorithm being stuck in a local minimum. The perturbations are proportional to a 'temperature' and the probability of retension of higher values is

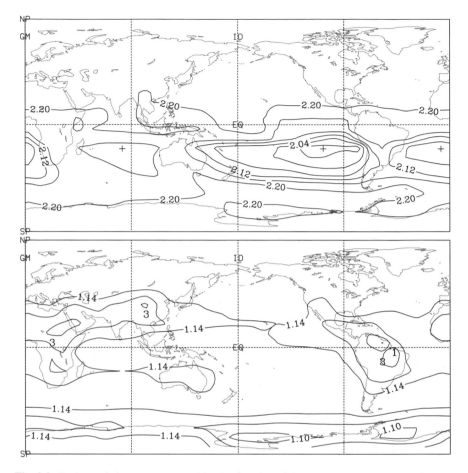

Fig. 3.2 Each panel shows a separate objective function of the location of one additional station to the current network. The station measures the same species with the same uncertainties as the existing station at Cape Grim, Tasmania. The upper panel's objective function is the standard deviation (in GTC) of global ocean uptake, and the lower panel's objective function is the square root of the sum of the variance of individual oceanic source components (in GTC), which provides a separate constraint for each ocean region. The upper panel has a contour interval of 0.02 and the lower panel of 0.04. Contouring is curtailed at 80° to avoid problems arising from multi-valued responses at the poles. The upper panel indicates by crosses the optimal locations of three additional stations, and the lower panel uses the numbers 1 and 3 to indicate optimal locations of one or three more stations. Both panels taken from Rayner et al. (1996).

based on a Boltzmann factor (a negative exponent of 'temperature'). The 'temperature' is reduced sufficiently gradually that the algorithm may find a global minimum. As the 'temperature' is reduced, the probability of retaining suboptimal values is also reduced so that the iterations of the system converge.

Rayner et al. (1996) applied the same technique to atmospheric inversions using CO_2 and $\delta^{13}C$ measurements. The parameters were the longitude and latitude of the observing sites. They chose two objective functions[1], reflecting different selections of target quantities. The first target quantity was the total ocean uptake, and the corresponding objective function is given by $[\Sigma\ C(S_{oc})]^{0.5}$ where the subscript indicates the submatrix containing only ocean sources. The sum is over all elements of the submatrix. The second objective function uses actually a set of target quantities, namely the regional ocean uptakes. It expresses the uncertainty of regional ocean uptake summed over all ocean regions and is given by *trace* $[C(S_{oc})]^{0.5}$. The difference between the two objective functions is the inclusion or exclusion of covariances.

The upper panel of Fig. 3.2 shows a plot of the first objective function along with the locations for adding up to three new stations to the network of Enting et al. (1995) as chosen by the optimisation algorithm. We see first that the simulated annealing algorithm performs well in finding the minimum in the function. Further, the positions of more than one station can also be predicted from the function map; for example the optimum locations for two stations are the two deepest minima on the map. The finding that the position of one new station does not distort the map sufficiently to influence the next choice greatly simplifies the problem if we are concerned with only small additions to the network. It allows us to optimise incrementally. The approach has been used by Patra and Maksyutov (2002) and Lane et al. (2004).

The second striking thing about the upper panel of Fig. 3.2 is the preference for land sites, despite the target quantity being ocean uptake. This arises from the propagation of information in the presence of global constraints. In the formulation of Enting et al. (1995), the global growth rate is specified and given a fairly small uncertainty. The land and ocean uptakes must sum to this figure. Thus an improvement in knowledge of one of these gives a corresponding improvement in knowledge of the other. Such an improvement is best achieved by constraining the flux with the highest posterior uncertainty in the current set-up. That is the Amazonian region and so this is chosen. The next favourite region is also over land. This coupling of regions was the major conclusion of the study.

The final thing to note about Fig. 3.2 is the radical difference between the two objective functions. Unsurprisingly, when asked to optimise the summed regional uncertainty (second objective function) rather than uncertainty of the total ocean uptake (first objective function), the algorithm chooses to improve estimates for the least constrained ocean regions. This is quite a different network to the first example. It is clear that quantitative network design requires precisely formulated questions. Requirements like 'the best knowledge of biosphere function in a region' will produce very different networks depending on how they are mathematically formulated. It is also obvious that the network depends on the prior uncertainties in use. If, for example, some external information greatly reduced the prior uncertainty on

---

[1] In network design, we are dealing with a nested optimisation problem. The inverse problem of solving the cost function of (3.5) is nested into the problem of optimising the network quality. This quality is expressed by a second function, which we will call objective function in order to avoid confusion with (3.5).

the Amazon region, the posterior uncertainty would (according to 3.2) also decrease and possibly eliminate the region from the favoured list. The influence of the prior cannot be circumvented by using an objective function such as the uncertainty reduction (ratio of posterior to prior uncertainty). Here one has the opposite problem so that a very weak prior will yield a more dramatic reduction with the addition of further stations.

After these caveats, one is left with robust general findings about the behaviour of optimal networks. For example, Rayner et al. (1996) did show that the existence of global constraints couples information from disparate regions. Thus improvements in knowledge of Amazonian fluxes improves knowledge of both total land and ocean fluxes. A corollary is that the 'hot-spot' strategy of choosing the least known region for improved observations is globally efficient.

Rayner et al. (1996) also touched on the impact of different descriptions of data uncertainty on network design. Rather than the uniform data uncertainty used to generate Fig. 3.2, they used an uncertainty proportional to the atmospheric signature of the terrestrial biospheric flux meant to approximate the model error contribution $C(d_{mod})$ to the overall uncertainty in (3.3). The different choice did not overturn the main findings, but the choice of scaling for the new error term was arbitrary. Gloor et al. (2000) greatly expanded this aspect of the study and produced networks that compromised between the strongest possible observations of a region (often gained by placing a station in the centre) and large data uncertainty.

Gloor et al. (2000) made an interesting contrast with the earlier study of Rayner et al. (1996). First they rejected the use of prior information at all. As such, they required larger numbers of stations (around 150) to saturate the information needed for their inversion. They also studied a wider range of sampling strategies than did Rayner et al. (1996), especially aerial profiles and upper tropospheric transects. Again it is the generalities of the study rather than specific information on placement that will endure. Particularly (and controversially) they noted that the 'fence post' strategy of surrounding a target region with measurement sites was less effective than sampling of the concentration gradients among neighbouring regions by placing stations near their centres. They also noted that aerial profiles were more useful than airborne transects since measurements in the upper troposphere were only weakly connected to surface sources by atmospheric transport.

These early studies were made in the context of inversion configurations common at the time. In particular, they solved for a relatively small number of regions and used data taken as monthly or annual averages. Both of these have since changed and require some revising of earlier results. The use of large regions in these early inversions gave measurements an unnaturally large influence since a measurement could constrain parts of the large region that might never be connected to it by atmospheric transport. Kaminski et al. (2001) noted the dangers of this for biasing inversions and current practice is to increase the number of regions, preferably to the full spatial resolution of the transport model. This must have implications for network design. A hint of this was given by Patra et al. (2003), who noted that the range of preferred sites in their network optimisation expanded as they increased the number of regions in their inversion. They performed an incremental inversion in which sites were added one at a time and the optimal site list for n sites was used for the placement of site $n + 1$. In general, the approach suggested by the optimisation is to place a site in each region. Obviously, this is impossible in the context of inversions at the resolution of the transport model. And even that resolution can only provide a discrete approximation of the actual two-dimensional flux field. Much of the underdetermined nature of the inverse problem is, thus, hidden by suppressing most of its degrees of freedom (Kaminski and Heimann 2001). It is unclear yet how this affects the optimal network.

The other major change in inversion formulation since the initial network design studies is the trend towards using data at higher time resolutions. In the context of formal inversions this was pioneered by Law et al. (2002, 2003). They showed that the differential sampling afforded by synoptic variations in advection could act as a more precise regional constraint than the highly diffusive monthly mean responses. Further, provided the biases due to the use of large regions could be avoided, the estimates were usefully accurate, that is errors were consistent with their uncertainties. This work still used the "large-region" approach although with considerably higher spatial resolution in the source space than was traditional. Peylin et al. (2005) have since demonstrated inversions of such high-frequency data in combination with the transport model resolution of sources.

In a network design context, Law et al. (2003) investigated a network of high-frequency monitoring stations intended to elucidate the regional patterns of carbon flux within Australia. They used 12 regions over Australia, still much coarser resolution than the transport model but unusually high for the large-region approach. They used incremental optimisation to design the network. They noted again the tendency to place a station in each region.

Another application field for network design is the evaluation of spaceborne sensors providing integrals of carbon dioxide over vertical columns. This type of application does not require an optimisation algorithm. It was pioneered by Rayner and O'Brien (2001), who estimated the sensor precision required by a transport inversion to achieve a specified minimim posterior uncertainty for surface flux field. A more recent example of this approach is provided by Houweling et al. (2004) who evaluated the performance of a set of existing and planned satellite instruments.

What has not yet been tried is the network design problem with inversions at the transport model resolution and high-frequency data. The problem appears computationally challenging as the transport model Jacobian M in (3.1) and (3.2) is large. Here the alternative formulation (3.4) helps. In the case where we wish to optimise the constraint on an integrated flux we can use (3.8), where N is a linear operator. Further we note that in the network optimisation calculations only M and $C(d)$ change as the network changes. Substituting (3.4) into (3.8) we note that most terms like $D(N)C(x_0)$ can be precomputed. If we are considering additions to a fixed network, we can also replace $C(x_0)$ by the posterior covariance from that network so most of the term $MC(x_0)M^T$ is also precomputed. We are then left with inverting matrices of the dimension corresponding to the additional observations, which may be feasible for diurnal or even hourly data.

An accurate specification of the uncertainty contribution from model error in (3.3) is notoriously difficult. Quantifying the impact of differences in modelled transport is, however, less difficult and provides a popular surrogate for model error. An ideal environment for this task is provided by the TRANSCOM project (see http://www.purdue.edu/transcom) of the International Geosphere-Biosphere Programme (IGBP), which compares forward simulations and inverse simulations of a set of transport models. Gurney et al. (2002) describe the TRANSCOM inversion experiment, in which 16 different transport models are used with all other inputs to the inversion held constant. They reported two uncertainty expressions as covariance matrices

$$C^B_{j,k} = \frac{1}{16} \sum_{i=1}^{16} (S^i_j - \overline{S}_j)(S^i_k - \overline{S}_k)$$

(3.9)

and

$$C^W_{j,k} = \overline{C_{j,k}}$$

(3.10)

where i counts the 16 models, j and k their 22 source regions. The overbar denotes the ensemble average across the models and the superscripts B and W stand for between uncertainty and within uncertainty, respectively. Equation 3.9 expresses the uncertainty in actual estimates from the inversion while (3.10) expresses the average posterior uncertainty.

Patra et al. (2003) used the above-sketched incremental optimisation approach to extend the Gurney et al. (2002) network interatively. A site was added to the current network when it performed best at reducing the average posterior uncertainty over all source regions for any of the models or the within uncertainty (3.10).

Rayner (2004) also based a network design study on the Gurney et al. (2002) set-up (with one additional model, as described by Gurney et al. 2003). He defined the sum of the between (3.9) and within (3.10) uncertainties as total uncertainty and used that as an additional objective function. The inclusion of the variance of actual estimates from (3.9) immediately circumscribes the study since one must have real data to calculate these. One is, hence, left choosing networks from a discrete list of stations. Rayner (2004) used the list of 110 compiled by Gurney et al. (2002) from which they chose the 76 used in that study. The use of discrete lists also demands a change in algorithm since simulated annealing requires continuous fields. Rayner (2004) used the techniques of genetic algorithms. Briefly, these maintain populations of potential solutions which are allowed to 'breed', 'mutate' and 'compete'. The iterations of the scheme can be compared to generations of a population and we hope that the overall fitness of the population will improve. See http://csep1. phy.ornl.gov/CSEP/MO/MO.html and references therein for more details.

Rayner (2004) optimised networks of 76 (like TRANSCOM) and 110 stations and with the within-uncertainty or total-uncertainty objective functions. Figure 3.3 shows the cases for 76 stations. One might expect that the networks for total uncertainty would be smaller than for within uncertainty as the optimisation rejected

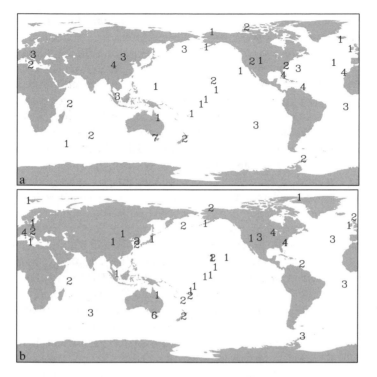

Fig. 3.3 Optimised networks for within-model (**a**) and total (**b**) uncertainty for a 76-site network. Numbers indicate the number of times the site is included. Figure taken from Rayner (2004).

stations with large model–model differences. In fact, the opposite is the case, the networks generally dispersed in the presence of model–model difference. It appears that we should consider model–model difference like a small-scale heterogeneity. For example two models may place the maximum response to a flux at each of two neighbouring stations. A network designed to reduce the impact of model–model difference will average across this heterogeneity. If we optimise only for within uncertainty, we risk large between uncertainty and ultimately large total uncertainty. The need for real data limits the use of this approach, however, Rayner (2004) noted that stations with highly variable (among models) responses to input fluxes would be penalised in an optimisation of total uncertainty. This generality seems likely to survive beyond the details of the study.

3.4 CCDAS

The previous section has demonstrated network design concepts for the special case of atmospheric transport inversions. Transport inversions deliver posterior information on the exchange fluxes at the Earth's surface over the atmospheric sampling period. Information on the processes behind the fluxes can enter the inversion only via the priors, for instance by using the output of a process-based model as a prior flux field. The transport inversion, however, has no means of conserving the process dynamics. Thus the posterior flux field will generally be inconsistent with the output of process-based models. The second significant restriction of transport inversions consists in the fact that possible target quantities (for use in 3.7 and 3.8) can only be diagnostic. Extending the modelling system by a process model that can prognose the process dynamics avoids these two restrictions. The corresponding inverse model is usually termed CCDAS. Another benefit of including a process model is that it enables access to additional observations (see examples below). The CCDAS control variables can be initial or boundary conditions as well as parameters in the formulation of the processes. The capability of estimating process parameters provides a means to directly improve our process understanding. The present section generalises the design concepts that the previous section introduced and illustrated for transport inversions to the design of networks that provide observations to a CCDAS.

Kaminski et al. (2002) present an early CCDAS version built around the Simple Diagnostic Biosphere Model SDBM, (Knorr and Heimann 1995) coupled to the atmospheric transport model TM2 (Heimann 1995). CCDAS was later upgraded (Kaminski et al. 2003; Scholze 2003; Rayner et al. 2005) by replacing SDBM with the Biosphere Energy Transfer Hydrology Scheme (Knorr 1997, 2000). In contrast to the diagnostic SDBM, which is driven by observed vegetation index data, BETHY is a prognostic model that uses a set of meteorological driving data to integrate the model state forward in time.

Adding the terrestrial component to the transport model renders the composite model non-linear. Hence, we must now use Eqs. 3.5 and 3.6 to infer posterior information on the control variables of a CCDAS. While the CCDAS built around SDBM has 24 control variables, the one around BETHY has 57 to about 1,000, depending on the set-up. For a control space of that size, (3.5) is most efficiently minimised via an interative gradient algorithm. At the minimum of the cost function, CCDAS evaluates the cost function's Hessian to approximate the posterior uncertainties via (3.6). Typical CCDAS target quantities are regional and global means of fluxes such as NEP, averaged over diagnostic or prognostic integration periods. The target quantities' posterior uncertainties are derived via (3.8). The Jacobian matrix $D(N)$ is the derivative of the target quantities with respect to the control variables. Efficient derivative code providing the gradient (adjoint model), the Hessian and the Jacobian are generated automatically from BETHY's source code via the automatic differentiation tool TAF (Giering and Kaminski 1998). This automation has proven useful, as it allows quick updates of CCDAS after modifications of BETHY.

The above-cited CCDAS applications run globally and assimilate atmospheric carbon dioxide flask samples provided by a global network and (in a less formal pre-step) remotely sensed vegetation index data. There are a number of assimilation systems that differ from CCDAS in the assimilation approach or in the type of data that are assimilated. For instance, the system of Pak (2004) uses the same approach but assimilates eddy flux measurements of heat and carbon into a one-point version of the CSIRO Atmosphere Biosphere Land Exchange (CABLE) model. Vukićerić et al. (2001) also minimised (3.5) by a gradient method, using the adjoint of their k-model to provide the gradient of J with respect to 16 parameters. They assimilated atmospheric temperature and carbon dioxide but did not calculate posterior uncertainties.

Lacking the adjoint, gradient information is usually approximated by finite differences. This means each component of the control vector is perturbed in turn, and the perturbation's impact on the cost function is evaluated in a separate model run. This impact may be dominated by higher order effects, if the perturbation is too large. If it is too small, the impact is dominated by numerical noise. The approximate gradient usually slows down the convergence of the gradient algorithm. The computational cost of the approximation typically limits the complexity of the model, its spatial resolution or the number of control variables that can be treated. Wang et al. (2001) used finite difference approximations for the gradient and Hessian (actually via Jacobians) with respect to up to seven model parameters for assimilation of eddy flux measurements into a one-point version of the CSIRO Biospheric Model (CBM). Santaren et al. (2003) applied the same strategy to estimate five parameters of a one-point version of their model ORCHIDEE from eddy flux measurements. Williams et al. (2005) used an ensemble Kalman filter embedded in a finite difference gradient algorithm to estimate 14 control variables of a box model from eddy flux and carbon stock measurements.

The CCDAS approach finds the most likely value for the control parameters by maximising the posterior PDF. This is possible for a large variety of distributions. However, the identification of the inverse Hessian with the covariance, key to the network design application, requires the assumption of a Gaussian posterior. Randerson et al. (2002) avoid such assumptions by directly sampling a three parameter posterior PDF which quantifies the fit of their coupled biosphere-transport model to flask samples of atmospheric carbon and its isotopic composition. The model domain are the northern high latitudes, and in complexity the model is similar to the above-mentioned SDBM. Knorr and Kattge (2005) applied a guided Monte Carlo sampling method (Metropolis et al. 1953) of the 14 (C4) and 23 (C3) posterior parameter PDFs, for one-point set-ups of BETHY. Their posterior PDFs are generally close to Gaussian.

Barrett (2002) used a genetic algorithm to estimate 22 parameters of the conceptual VAST model from measurements of NPP and carbon pool sizes over Australia but did not calculate posterior uncertainties.

The Optimisation InterComparison project (OptIC, see http://www.globalcarbonproject.org/ACTIVITIES/OptIC.htm) evaluates different optimisation techniques for estimating four parameters of a simple terrestrial biosphere model.

In a CCDAS context, the network design problem can be tackled very much in the same manner as for a pure atmospheric transport inversion. The basis is again the specification of a target quantity (see Fig. 3.4). The CCDAS can then quantify

Fig. 3.4 Evaluating a network control vector in terms of the posterior uncertainty of the target quantity. Oval boxes denote data, and rectangular boxes denote processing.

the quality of a given network via (3.6) and (3.8) by the target quantity's posterior uncertainty. In contrast to the case of transport inversion, this target quantity can also be a prognostic one, that is a network can be optimised in order to reduce the uncertainty in a particular aspect of the model prediction.

Next, a set of candidate networks needs to be defined. The methodological framework of Sect. 3.2 requires the network to provide its observations in the form of a mean value and a covariance matrix of uncertainties. Now, one of the major tasks of network design is to explore and evaluate not yet existing, fictive networks. This means for many candidate networks only a fraction of the observations is real and the remainder fictive. The most convenient way of accounting for fictive observations is to ignore them in (3.5) and then assume in (3.6) that the observed value equals the value that the model simulates from the posterior mean. The inclusion of a fictive observation then affects only the posterior uncertainty and not the posterior mean value. This means, for each candidate network, we need to extend the real observations' covariance of uncertainty (3.3) by the components corresponding to the fictive observations.

The strategy for solving the network design problem will usually be selected according to the set of candidate networks. If it comprises only a few elements, the posterior uncertainty of the target quantity can be evaluated for each of them and the networks be ranked accordingly. As an example imagine the extension of an existing network for budgeting the Amazon rain forest by a flux tower. It is likely that logistic and political constraints limit the set of candidate networks to just a few.

Larger sets of candidate networks have to be searched by an algorithm that is more efficient than testing candidate by candidate. Such search (or optimisation) algorithms typically operate on a subset of R^n. Hence, the set of candidate networks is to be parametrised; that is it is represented as the range of a function of a vector of network[2] control variables. For instance, when the network design problem consists in adding a fixed number of atmospheric sampling sites to a given network, these control variables can be the vector of the station's coordinates on the globe. The control vector could also have components with a discrete domain, for example when the set of candidate networks contains elements that differ by a data stream that can be switched on or off.

The task of the optimisation algorithm is to search the domain of the control vector for an element such that the corresponding network minimises the target quantity's posterior uncertainty. Section 3.3 provides examples for applications of optimisation algorithms to network design. Rayner et al. (1996) and Rayner (2004) demonstrated use of simulated annealing and a genetic algorithm. Another important class of optimisation algorithms are the powerful gradient algorithms (Gill et al. 1981). The principal functioning of these optimisation algorithms is sketched in Fig. 3.5. Starting by a first guess of the control vector, the algorithms work interatively through the following loop:

---

[2] To avoid confusion we prepend the term 'network' and thus distinguish the control variables determining the network from those determining the behaviour of the model inside CCDAS (denoted by x in 3.5).

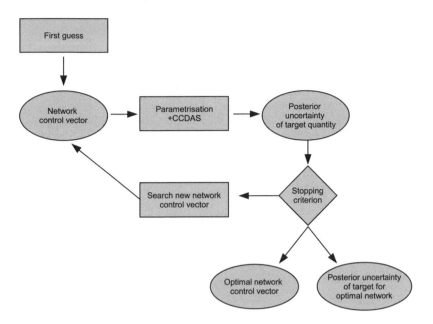

Fig. 3.5 Flow diagram for an iterative optimisation algorithm. Oval boxes denote data, and rectangular boxes denote processing. Figure 3.4 details box 'parametrisation + CCDAS'.

1. Perform at least one evaluation of the target quantity's posterior uncertainty as a function of the control vector (as depicted in Fig. 3.4). This evaluation involves the composition of the network from the network control vector and then a CCDAS run for that network. Gradient algorithms require, in addition, at least one evaluation of the gradient of the posterior uncertainty with respect to the control vector.
2. Check for convergence and exit in case of convergence. A suitable convergence criterion might be a threshold for the target quantity's posterior uncertainty. Gradient algorithms typically use a threshold for the gradient, which approaches zero as the posterior uncertainty reaches a minimum.
3. Propose new network control vector.

The outputs of the optimisation algorithm are the optimal network control vector, and thus the optimal network, plus the corresponding posterior uncertainty of the target quantity.

A first demonstration of network design in a CCDAS context is provided by the above-sketched study of Kaminski et al. (2002). They did, however, not enter into an optimisation loop but tested just two candidate networks. Both networks differ by a flux measurement in the model's broadleaf evergreen (BE) biome. The study did not select a particular target quantity but looked instead at the posterior uncertainties of the control variables. Figure 3.6 shows the posterior values and uncertainties of the soil model parameter Q10 for both networks.

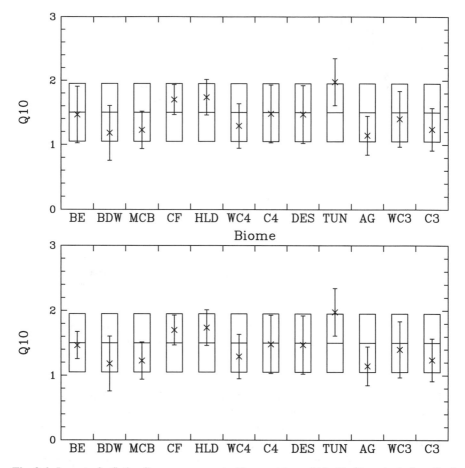

Fig. 3.6 Impact of a fictive flux measurement with uncertainty of $10\,gC/m^2/year$ in the broadleaf evergreen (BE) biome onto the posterior uncertainty in the soil respiration parameter Q10 as quantified by a CCDAS based on the Simple Diagnostic Biosphere Model Knorr and Heimann L995 and using atmospheric carbon dioxide samples. Upper panel: Without fictive measurement. Lower panel: With fictive measurement. Figure taken from Kaminski et al. (2002).

The above approach can be easily extended to network design problems where more than one target quantity are to be considered. The network can then be optimised for a function, for example a weighted sum, of the posterior uncertainties of the individual uncertainties. Recall the above-mentioned example of Rayner et al. (1996), who chose the regional ocean fluxes as target quantities and the sum of squares as weighting function. This means they optimised their atmospheric network to minimise the sum of squares of the regional ocean fluxes posterior uncertainties.

3.5   Perspective and Recommendations

The previous section has made it clear that a CCDAS constitutes an ideal tool for network design, as it is able to rigorously quantify the performance of an observational network in terms of the posterior uncertainty in specified target quantities. These target quantities may be regional or global net fluxes integrated over present or future periods, as required by policy makers as a basis for their decisions. Alternative target quantities more relevant for model developers may be fluxes resulting from particular processes such as photosynthesis or heterotrophic respiration integrated over the distribution of plant functional types.

The previous section has, however, also made it clear that there are hardly any CCDAS applications to network design. The primary reason is that as yet only few of these systems exist, and none of them can yet handle all available streams of observations relevant to the carbon cycle. Our first recommendation is, hence, to upgrade the existing CCDASs (or build new ones) so as to handle the available data streams. A CCDAS-based network design tool is being built (see http://imecc. ccdas.org) within the Integrated Infrastructure Initiative IMECC under the Sixth Framework Programme of the European Commission.

Restricting the focus to the atmospheric component of the carbon cycle, the situation is fortunately much better. There are a number of atmospheric transport inversion systems in operation, and some of them have been applied in design studies for atmospheric networks, including air- and space-borne instrumentation. Most of our insights are, hence, based on the experience from these studies, and the following remarks generalise these insights to CCDAS-based network design.

One of these generalised findings is that the optimal network usually depends strongly on how one chooses the target quantity that formalises the quality of the network: For instance, it is likely that reducing uncertainties in the twenty-first century's terrestrial net flux either over Europe or globally requires significantly different networks. When no decision can be made between competing target functions, it may be useful to optimise for a weighting function of their posterior uncertainties, or use a multi-criteria optimisation.

Also, the optimal network depends on the model underlying the CCDAS. This has an interesting implication. Assume a network is to be constructed to provide observations for assimilation into a particular CCDAS. We then recommend to use that very CCDAS for the network design, because optimality of the network is guaranteed only with respect to the CCDAS it has been designed with. Of course, continual improvements in modelling and delays in building network infrastructure mean that the optimal network will always be slightly out of date. This is another reason why general properties of the optimal network are more reliable than specific choices.

The reason for the dependency of the optimal network on the model is that each model can only approximate the truth, that is, there is model error. Another generalisation from transport inversion systems is that incorporating model error into CCDAS will impact the optimal network. A further recommendation is, hence, first to quantify this uncertainty most accurately and second to reduce it by improving the models underlying the CCDASs. Both are simplified by keeping the CCDAS flexible with respect to the process formulations in the model. Future studies must also pay more careful regard to the statistical coherence of the modelling systems we use.

Acknowledgement This work was supported in part by the European Commission under contract numbers FP6-511176-2 (CarboOcean), FP6-505572 (CarboEurope IP), and FP6-026188 (IMECC).

References

D. J. Barrett. Steady state turnover time of carbon in the Australian terrestrial biosphere. *Global Biogeochem. Cycles* 16(4), 2002. doi:10.1029/2002GB001860.

I. G. Enting. *Inverse Problems in Atmospheric Constituent Transport*, Cambridge University Press, 2002.

I. G. Enting, C. M. Trudinger, R. J. Francey, and H. Granek. Synthesis inversion of atmospheric CO_2 using the GISS tracer transport model. Tech. Paper No. 29, CSIRO Div. Atmos. Res., 1993.

I. G. Enting, C. M. Trudinger, and R. J. Francey. A synthesis inversion of the concentration and $\delta^{13}C$ of atmospheric CO_2. *Tellus*, 47B:35–52, 1995.

R. Giering and T. Kaminski Recipes for adjoint code construction. *ACM Trans. Math. Software*, 24(4):437–474. 1998.

P. E. Gill, W. Murray, and M. H. Wright. *Practical Optimization*. Academic Press, New York, 1981.

M. Gloor, S. M. Fan, S. Pacala, and J. Sarmiento. Optimal sampling of the atmosphere for purpose of inverse modeling: A model study. *Global Biogeochem. Cycles* 14:407–428, 2000.

K. R. Gurney, R. M. Law, A. S. Denning, P. J. Rayner, D. Baker, P. Bousquet, L. Bruhwiler, Y.-H. Chen, P. Ciais, S. Fan, I. Y. Fung, M. Gloor, M. Heimann, K. Higuchi, J. John, T. Maki, S. Maksyutov, K. Masarie, P. Peylin, M. Prather, B. C. Pak, J. Randerson, J. Sarmiento, S. Taguchi, T. Takahashi, and C.-W. Yuen. Towards robust regional estimates of CO_2 sources and sinks using atmospheric transport models. *Nature*, 415:626–630, 2002.

K. R. Gurney, R. M. Law, A. S. Denning, P. J. Rayner, D. Baker, P. Bousquet, L. Bruhwiler, Y.-H. Chen, P. Ciais, S. Fan, I. Y. Fung, M. Gloor, M. Heimann, K. Higuchi, J. John, E. Kowalczyk, T. Maki, S. Maksyutov, P. Peylin, M. Prather, B. C. Pak, J. Sarmiento, S. Taguchi, T. Takahashi, and C.-W. Yuen. TransCom 3 CO_2 inversion intercomparison: 1. Annual mean control results and sensitivity to transport and prior flux information. *Tellus*, 55B(2):555–579, 2003. doi:10.1034/j.1600-0560.2003.00049.x.

M. Hardt and F. Scherbaum. Optimizing the station distributions of seismic networks for aftershock recordings by simulated annealing. In *EOS Supplement, Transactions of the Fall Meeting, 1992*, page 351. A.G.U., 1992.

M. Hardt and F. Scherbaum. The design of optimum networks for after-shock recordings. *Geophys. J. Int.*, 117:716–726, 1994.

M. Heimann. The global atmospheric tracer model TM2. Technical Report No. 10, Deutsches Klimarechenzentrum. Hamburg, Germany (ISSN 0940-9327), 1995.

S. Houweling, F.-M. Breon, I. Aben, C. Rödenbeck, M. Gloor, M. Heimann, and P. Ciais. Inverse modeling of CO_2 sources and sinks using satellite data: A synthetic inter-comparison of measurement techniques and their performance as a function of space and time. *Atmos. Chem. Phys.*, 4:523–538, 2004.

T. Kaminski and M. Heimann. Inverse modeling of atmospheric carbon dioxide fluxes. *Science*, 294:5541, October 2001.

T. Kaminski, P. J. Rayner, M. Heimann, and I. G. Enting. On aggregation errors in atmospheric transport inversions. *J. Geophys. Res.*, 106:4703–4715, 2001.

T. Kaminski, W. Knorr, P. Rayner, and M. Heimann. Assimilating atmospheric data into a terrestrial biosphere model: A case study of the seasonal cycle. *Global Biogeochem. Cycles*, 16:1066, 2002. doi:10.1029/2001GB001463.

T. Kaminski, R. Giering, M. Scholze, P. Rayner, and W. Knorr. An example of an automatic differentiation-based modelling system. In V. Kumar, L. Gavrilova, C. J. K. Tan, and P. L'Ecuyer, editors, *Computational Science—ICCSA 2003, International Conference Montreal, Canada, May 2003, Proceedings, Part II*, volume 2668 of *Lecture Notes in Computer Science*, pages 95–104, Springer, Berlin, 2003.

W. Knorr. *Satellitengestützte Fernerkundung und Modellierung des globalen CO_2-Austauschs der Landvegetation: Eine Synthese*. PhD thesis, Max-Planck-Institut für Meteorologie, Hamburg, Germany, 1997.

W. Knorr. Annual and interannual CO_2 exchanges of the terrestrial biosphere: Process-based simulations and uncertainties. *Glob. Ecol. Biogeogr.*, 9:225–252, 2000.

W. Knorr and M. Heimann. Impact of drought stress and other factors on seasonal land biosphere CO_2 exchange studied through an atmospheric tracer transport model. *Tellus, Ser. B*, 47(4):471–489, 1995.

W. Knorr and J. Kattge. Inversion of terrestrial biosphere model parameter values against eddy covariance measurements using Monte Carlo sampling. *Global Change Biol.* 11:1333–1351, 2005.

R. M. Law, P. J. Rayner, L. P. Steele, and I. G. Enting. Using high temporal frequency data for CO_2 inversions. *Global Biogeochem. Cy.*, 16:1053, 2002. doi:10.1029/2001GB001593.

R. M. Law, P. J. Rayner, L. P. Steele, and I. G. Enting. Data and modelling requirements for CO_2 inversions using high frequency data. *Tellus*, 55B(2):512–521, 2003. doi:10.1034/j.1600-0560. 2003.0029.x.

R. M. Law, P. J. Rayner, and Y. P. Wang. Inversion of diurnally-varying synthetic CO_2: Network optimisation for an Australian test case. *Global Biogeochem. Cycles* 18(1):GB1044, 2004. doi:10.1029/2003GB002136.

N. Metropolis, A. W. Rosenbluth, M. N. Rosenbluth, A. H. Teller, and E. Teller. Equation of state calculations for fast computing machines. *J. of Chem. Phys.*, 21:1087–1092, 1953.

B. Pak. Parameter optimization using the adjoint of a biosphere model. Abstract A11F-02. *EOS Trans. AGU*, 85(47), December 2004.

P. K. Patra and S. Maksyutov. Incremental approach to the optimal network design for CO_2 surface source inversion. *Geophys. Res. Lett.*, 29(10):1459, 2002. doi:10.1029/2001GL013943.

P. K. Patra, S. Maksyutov, D. Baker, P. Bousquet, L. Bruhwiler, Y-H. Chen, P. Ciais, A. S. Denning, S. Fan, I. Y. Fung, M. Gloor, K. R. Gurney, M. Heimann, K. Higuchi, J. John, R. M. Law, T. Maki, P. Peylin, M. Prather, B. Pak, P. J. Rayner, J. L. Sarmiento, S. Taguchi, T. Takahashi, and C-W. Yuen. Sensitivity of optimal extension of CO_2 observation networks to model transport. *Tellus B*, 55(2):498–511, 2003.

P. Peylin, P. J. Rayner, P. Bousquet, C. Carouge, F. Hourdin, P. Ciais, P. Heinrich, and AeroCarb Contributors. Daily CO_2 flux estimate over Europe from continuous atmospheric measurements: Part 1 inverse methodology. *Atmos. Chem. Phys.*, 5:3173–3186, 2005.

J. T. Randerson, C. J. Still, J. J. Balle, I. Y. Fung, S. C. Doney, P. P. Tans, T. J. Conway, J. W. C. White, B. Vaughn, N. Suits, and A. S. Denning. Carbon isotope discrimination of arctic and boreal biomes inferred from remote atmospheric measurements and a biosphere-atmosphere model. *Global Biogeochem. Cycles* 16(3), 2002. doi:10.1029/2001GB001435.

P. J. Rayner. Optimizing CO_2 observing networks in the presence of model error: Results from transcom 3. *Atmos. Chem. Phys.*, 4:413–421, 2004.

P. J. Rayner, I. G. Enting, and C. M. Trudinger. Optimizing the CO_2 observing network for constraining sources and sinks. *Tellus*, 48B:433–444, 1996.

P. J. Rayner and D. M. O'Brien. The utility of remotely sensed CO_2 concentration data in surface source inversions. *Geophys. Res. Lett*, 28:175–178, 2001.

P. Rayner, M. Scholze, W. Knorr, T. Kaminski, R. Giering, and H. Widmann. Two decades of terrestrial carbon fluxes from a Carbon Cycle Data Assimilation System (CCDAS). *Global Biogeochem. Cyc.*, 19, 2005 doi:10.1029/2004GB002254.

D. Santaren, P. Peylin, N. Viovy, and P. Ciais. Parameter estimation in biogeochemical surface model using nonlinear inversion: Optimization with measurements of a pine forest. *Geophy. Res. Abs.* 5:12020, 2003.

M. Scholze. *Model Studies on the Response of the Terrestrial Carbon Cycle on Climate Change and Variability*. Examensarbeit, Max-Planck-Institut für Meteorologie, Hamburg, Germany, 2003.

A. Tarantola. *Inverse Problem Theory—Methods for Data Fitting and Model Parameter Estimation*. Elsevier Science, New York, 1987.

T. Vukićević, B. H. Braswell, and D. Schimel. A diagnostic study of temperature controls on global terrestrial carbon exchange. *Tellus*, 53B(2):150–170, 2001.

Y. P. Wang, R. Leuning, H. Cleugh, and P. A. Coppin. Parameter estimation in surface exchange models using non-linear inversion: How many parameters can we estimate and which measurements are most useful? *Glob. Change Biol.*, 7:495–510, 2001.

M. Williams, P. A. Schwarz, B. E. Law, J. Irvine, and M. R. Kurpius. An improved analysis of forest carbon dynamics using data assimilation. *Glob. Change Biol.*, 11:89–105, 2005.

Chapter 4
Quantifying Fossil Fuel CO$_2$ over Europe

Ingeborg Levin and Ute Karstens

4.1 Introduction

Europe is responsible for more than 25% of global fossil fuel CO$_2$ emissions (Marland et al. 2006), and these emissions account for about 30–50% of the observed CO$_2$ variability in this region (see Sect. 4.2.1). To balance greenhouse gases over Europe, therefore, also requires quantification of CO$_2$ emissions from fossil fuel (i.e. coal, oil and natural gas) burning. Reliable continuous observations of the fossil fuel CO$_2$ component are needed in order to validate emission-based model simulations and finally allow for robust estimates of the biogenic part in the observed atmospheric CO$_2$ variations. Fossil fuel emissions in Europe are very heterogeneously distributed in space with hot spots in highly industrialised and populated regions. The temporal variability comprises seasonal as well as diurnal cycles with a strong coupling to ambient temperature variations (i.e. domestic heating), the current economic situation (i.e. industry) and other factors such as the general meteorological conditions or holiday periods (i.e. traffic and industry). All these parameters need to be accurately modelled if fossil fuel emissions shall be estimated in a realistic and quantitative way from bottom-up information on the respective sources (see Reis et al. 2008).

'Monitoring' fossil fuel CO$_2$ in the atmosphere is in principle possible via radiocarbon (^{14}C) observations in atmospheric CO$_2$. ^{14}C is the radioactive carbon isotope which is naturally produced in the atmosphere via cosmic ray–induced reactions of neutrons with atmospheric nitrogen. The radioactive half life of ^{14}C is 5,730 years. The natural equilibrium level of atmospheric ^{14}CO$_2$ is established by production in the stratosphere and subsequent decay in all carbon-exchanging reservoirs, that is, the atmosphere, the biosphere and the oceans. This natural level has been disturbed in the last century by man's activities, not only via the ongoing input of fossil fuel CO$_2$ into the atmosphere known as Suess effect (Suess 1955), but also by artificially produced ^{14}C during nuclear detonations in the atmosphere, mainly in the 1950s and early 1960s (^{14}C bomb effect). CO$_2$ from burning of fossil fuels, due to its age of several hundred million years, is free of ^{14}C; adding fossil fuel CO$_2$ to the atmosphere, therefore, not only leads to an increase of the CO$_2$ mixing ratio but also to a decrease of the ^{14}C/^{12}C ratio in atmospheric CO$_2$. From a ^{14}CO$_2$ measurement at a polluted sampling site, for example, in the boundary layer on the European continent, we can therefore directly calculate the regional fossil fuel CO_2 surplus if the undisturbed background $^{14}CO_2$ level is known.

Particularly in Europe, long-term quasi-continuous $^{14}CO_2$ measurements exist at a number of marine, mountain as well as urban and rural sites (Levin et al. 1980, 1985, 1989, 2003; Kuc 1986; Meijer et al. 1995). These measurements allow determining the regional fossil fuel CO_2 offset at the stations when using the observations at, for example, Mace Head (Western Ireland, clean marine air) or Jungfraujoch (Swiss high altitude 'free troposphere' site) as background reference level. However, the $^{14}CO_2$ measurements need to be made quasi-continuously at very high precision to yield accurate estimates of fossil fuel CO_2. This is an expensive undertaking and, thus, not yet realised at many sites.

Because of the lack of such precise and frequent $^{14}CO_2$ observations, other tracers that could possibly serve as surrogates for fossil fuel CO_2 have been discussed in the literature. Among these are carbon monoxide (CO) and acetylene (C_2H_2), common by-products of (fossil fuel) combustion processes, sulphur hexafluoride (SF_6), a purely artificial and very inert gas, which has a similar spatial source distribution as fossil fuel CO_2 (associated with energy distribution), and others. In particular, the use of CO as a quantitative tracer of fossil fuel CO_2 has been investigated in recent years in a number of experimental and modelling studies (Bakwin et al. 1997; Potosnak et al. 1999; Gamnitzer et al. 2006; Rivier et al. 2006; Turnbull et al. 2006).

The aim of this chapter is to review the state of the art of quantifying fossil fuel CO_2 over Europe. We will first provide an estimate of the relative signal of fossil fuel emissions over Europe using emissions inventories and atmospheric transport modelling and compare them to the respective signals from biogenic (and oceanic) sources and sinks. Then we will give an overview on the existing $^{14}CO_2$ measurements and respective ^{14}C-based fossil fuel CO_2 estimates, including their seasonal variability and temporal trends. We will further discuss the quality of the surrogate tracer CO, and finally present a sensitivity test of a proposed method, which allows to estimate the uncertainty of using continuous CO measurements together with ^{14}C-calibrated $CO/CO_2(foss)$ ratios to estimate fossil fuel CO_2 at high temporal resolution (Levin and Karstens 2007). Finally, we will make a recommendation for a fossil fuel CO_2 monitoring network for the European continent.

4.2   State of the Art of Quantifying Fossil Fuel CO_2 over Europe

4.2.1   The Atmospheric CO_2 Signals as Derived from Bottom-Up Emissions Inventories and an Atmospheric Transport Model

In order to estimate biogenic sources and sinks of CO_2 over Europe from atmospheric observations and inverse modelling, the fossil fuel CO_2 component is normally calculated with the same model from bottom-up emissions and subtracted from the respective atmospheric observations to obtain the pure biogenic (and oceanic) signal. This requires precise emissions inventories. For Western Europe, two spatially disaggregated emissions inventories for fossil fuel CO_2 (and CO) are available today. (1) The Emission Database for Global Atmospheric Research (EDGAR), which provides annual mean emissions for the base years 1990, 1995 and 2000 on a global $1° \times 1°$ grid (Olivier et al. 2005). (2) At the Institute of Energy Economics and Rational Use of Energy (IER, University of Stuttgart, Germany), a new high-resolution emissions inventory on a $50 km \times 50 km$ grid for the year 2000 and for the greater part of Europe was constructed based on United Nations Framework Convention on Climatic Change (UNFCCC) statistics (Reis et al. 2008). Here, national totals were disaggregated in time and space, separately for each sector, according to typical diurnal, weekly and seasonal cycles and spatial distributions of point, line and area sources, respectively. Especially, the high temporal resolution of this IER data set is a substantial improvement compared to EDGAR because it allows a more realistic simulation of the temporal characteristics in the atmospheric tracer mixing ratios. In the modelling studies presented here, the IER emissions inventories were complemented by the EDGAR emissions inventories in countries where IER data were missing to cover the entire model domain.

In order to estimate atmospheric mixing ratios over Europe for 2002 at hourly resolution as caused by the different European CO_2 sources and sinks, we used here the regional atmospheric transport model REMO (Chevillard et al. 2002). The same model is later on applied to simulate the temporal and spatial distribution of CO as well as for the sensitivity tests in Sect. 4.2.6 of this chapter. In this set-up of REMO, the horizontal grid resolution is $55 km \times 55 km$ and the model domain covers the area north of 30° N. For the simulation of the total atmospheric CO_2 mixing ratio, the biosphere–atmosphere exchange, as described by the terrestrial biosphere model BIOME-BGC (Churkina et al. 2003) as well as ocean fluxes (Takahashi et al. 1999), was included. For CO surface emissions from sources other than fossil fuel burning, that is, biomass burning, soil and ocean emissions, as well as oxidation of methane and volatile organic compounds (VOCs), photochemical processes for CO destruction in the atmosphere and soil uptake were taken into account. For the meteorological part analysis, data from the European Centre for Medium-Range Weather Forecasts (ECMWF) were used for initialisation and as lateral boundary information with a time resolution of 6 h. For more details of the modelling parameters and source/sink distributions, see Gamnitzer et al. (2006).

Figure 4.1 (upper panels) shows the comparison of CO_2 observations made by the German Umweltbundesamt at the Schauinsland observatory located on a mountain ridge at about 1,205 m above sea level (asl) in the Black Forest in Germany (47°55′N, 7°54′E) (Schmidt et al. 2003) with REMO model estimates for a typical winter (February) and a typical summer (July) month in 2002. The second row of panels in Fig. 4.1 shows the REMO-simulated CO_2 offsets compared to Jungfraujoch values originating from biogenic sources and sinks, fossil fuel emissions (EDGAR inventory) and ocean fluxes. It is obvious from these model estimates that even at

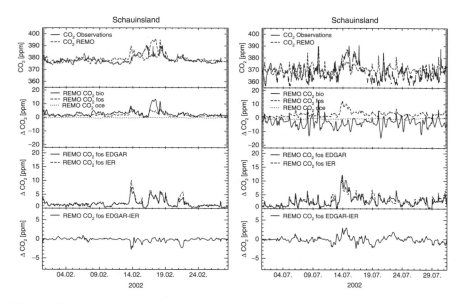

Fig. 4.1 Comparison of continuous hourly CO_2 observations at the mountain station Schauinsland with respective REMO model estimates for February and July 2002. The upper panels show the observations in comparison with the model results including all CO_2 flux components; the second panels show separately simulated components of the total CO_2 signal (ocean, terrestrial biosphere, and EDGAR fossil fuel emissions). The third rows compare the simulated fossil fuel CO_2 components if based on the two different inventories used in this study, IER and EDGAR, while the bottom panels show the differences between these two.

this quite remote mountain site, the regional fossil fuel CO_2 signal is of similar magnitude as that from biospheric sources. This is the case not only in winter but also in summer where the biospheric signal is dominated by the net continental sink. Note that in particular in winter, the observed CO_2 variability seems to be mainly driven by transport of CO_2 to the station from essentially co-located biogenic and fossil fuel sources. The third row of panels in Fig. 4.1 shows the comparison of the regional fossil fuel component when using EDGAR respectively IER emissions and the lowest panels the differences between EDGAR and IER regional fossil fuel CO_2 simulated in REMO. These results show that just simply using the different emissions inventories already causes ΔCO_2(foss) differences up to 4 ppm or about 30% of the signal itself.

Figure 4.2 shows respective comparisons for the regional polluted Heidelberg site located in the Western outskirts of Heidelberg (49°25′N, 8°41′E, 115 masl), a medium-sized city in the highly populated and polluted upper Rhine valley. Quasi-continuous CO_2 (and other trace gas) measurements in Heidelberg are made on air collected from the roof of the institute's building at about 30 m above local ground. Although the spatial resolution of REMO is not well suited to simulate the heterogeneous

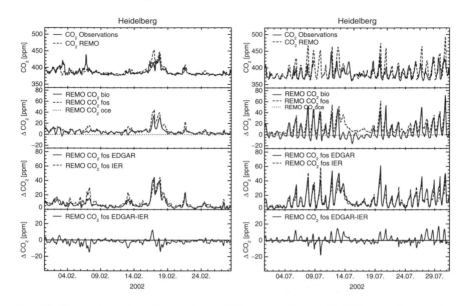

Fig. 4.2 Comparison of continuous hourly CO$_2$ observations at Heidelberg with respective REMO model estimates for February and July 2002. The upper panels show the observations in comparison with the model results including all CO$_2$ flux components; the second panels show separately simulated components of the total CO$_2$ signal (ocean, terrestrial biosphere and EDGAR fossil fuel emissions). The third rows compare the simulated fossil fuel CO$_2$ components if based on the two different inventories used in this study, IER and EDGAR, while the bottom panels show the differences between these two.

source distribution influencing the Heidelberg observations, the agreement between model results and observations (upper panels in Fig. 4.2) is surprisingly good. The Heidelberg sampling site, located 30 m above ground at some distance to the largest emissions of the city, seems to sample air which is often representative for the respective REMO grid box (see also discussion on this point by Gamnitzer et al. 2006). As expected, in Heidelberg, the fossil fuel CO$_2$ offset in winter is higher than the biospheric one, however, still of comparable magnitude. Differences in CO$_2$(foss) from REMO simulations, when based on different inventories (lowest panels in Fig. 4.2), can be as large as 20 ppm, particularly during strong night-time inversion situations in summer. Here, the influence of the diurnal cycle in the IER emissions is particularly strong.

Altogether the examples presented here for Schauinsland and Heidelberg illustrate the importance of the task to determine fossil fuel CO$_2$ over Europe correctly, as it contributes about 50% to the total observed continental CO$_2$ signal. Only using emissions from different emissions inventories (but the same transport model) can cause differences of more than 30% in the simulated CO$_2$(foss) component, and we are not yet able to decide which of the inventories is correct.

4.2.2   Long-Term Observations of $^{14}CO_2$ in Europe

4.2.2.1   The European $^{14}CO_2$ Network and Measurement Techniques Applied

Two-weekly integrated $^{14}CO_2$ measurements in continental Europe have started already in 1959 at the station Vermunt in the Austrian Alps (Levin et al. 1985). These measurements mainly focused on monitoring the transient of bomb $^{14}CO_2$ in the northern hemisphere troposphere as a basis to study the dynamics of the global carbon cycle (Naegler and Levin 2006). In the 1980s and 1990s, $^{14}CO_2$ measurements were extended towards more polluted or semi-polluted regions (e.g. Levin et al. 1980, 1989; Kuc 1986; Meijer et al. 1995) with an additional focus on determining fossil fuel CO_2 in these areas. Measurements are made by conventional counting technique (Kromer and Münnich 1992), which requires CO_2 from large samples of 10–20 m³ of air. In order to obtain these large CO_2 amounts, some enrichment process is necessary, that is, chemical absorption of CO_2 in basic solution (Levin et al. 1980; Meijer et al. 1995) or, alternatively, physical adsorption on molecular sieve (Kuc 1986).

Figure 4.3 shows the REMO-simulated monthly mean excess fossil fuel CO_2 level over Europe (using Jungfraujoch as the background reference) at about 130 m above ground in February 2002. The distribution of fossil fuel CO_2 was estimated using the fossil fuel CO_2 emissions from the EDGAR inventory for the year 2000. There is a general increase of fossil fuel CO_2 over Europe from West to East of

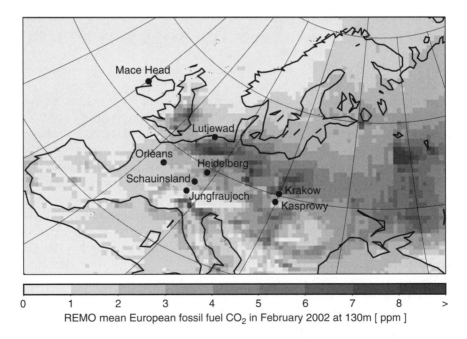

Fig. 4.3 Monthly mean fossil fuel CO_2 over Europe at 130 m above ground in February 2002 as simulated with REMO using the EDGAR emissions inventory for 2000. The stations where quasi-continuous $^{14}CO_2$ measurements are performed are marked.

about 4–6 ppm with hot spots of up to 15 ppm in highly populated and industrialised regions. The locations of the European $^{14}CO_2$ monitoring stations are also shown in Fig. 4.3. The current network consists of two marine Atlantic clean air sites [Mace Head, Ireland and Izaña, Canary Islands (not shown)], one moderately polluted coastal site in the Netherlands (Lutjewad), three urban sites (Orléans, Heidelberg and Krakow), two mountain stations (Schauinsland, Black Forest, Germany and Kasprowy Wierch, High Tatra, Poland), and one high-altitude Alpine site (Jungfraujoch, Swiss Alps).

4.2.2.2 Measurements of Fossil Fuel CO$_2$ over Europe

The two-weekly integrated long-term observations of $^{14}CO_2$, which are available at some of the European sites, provide an excellent measure of the seasonal cycle and the long-term trend of fossil fuel CO$_2$ over Europe. As an example, Fig. 4.4 (upper

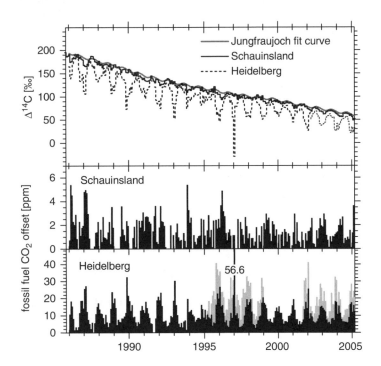

Fig. 4.4 Monthly mean $\Delta^{14}CO_2$ measurements in Heidelberg and at Schauinsland station in comparison to the continental reference level over Europe as derived from observations at Jungfraujoch (upper panel). $\Delta^{14}C$ concentrations are per mil deviations of the $^{14}C/^{12}C$ ratio from National Bureau of Standards (NBS) oxalic acid standard activity (Stuiver and Polach 1977). Fossil fuel CO$_2$ component at Schauinsland (middle panel) and Heidelberg (lower panel) as calculated from the respective $\Delta^{14}C$ difference to the continental reference level according to Eq. 4.3. In the case of Heidelberg, also the total CO$_2$ offset compared to marine background air (GLOBALVIEW-CO$_2$ 2005) is plotted (grey shaded). Note that the biospheric contribution in summer (minimum offsets) is larger than that from fossil fuels while it is of the same magnitude as the fossil fuel component during winter.

panel) shows the monthly mean $^{14}CO_2$ measurements at Heidelberg and Schauinsland stations (Levin et al. 2003; Gamnitzer et al. 2006) together with the continental $^{14}CO_2$ reference level. To estimate this reference level, we have used the quasi-continuous $^{14}CO_2$ observations at Jungfraujoch (Levin and Kromer 2004), assuming that this site at an elevation of 3,450 masl in the Swiss Alps most of the time monitors the undisturbed $^{14}CO_2$ level in the free troposphere over Europe. We observe a small seasonal variation of $^{14}CO_2$ at this background site which, to a large extent, is caused not only by the seasonal input of (natural) $^{14}CO_2$ from the stratosphere to the troposphere, by interhemispheric exchange, but also by $^{14}CO_2$ disequilibrium fluxes between the atmosphere and the biosphere (Hesshaimer 1997; Randerson et al. 2002; Naegler 2005). A small addition to the seasonal amplitude is caused by variations of the fossil fuel CO_2 component in background air at mid-northern latitudes (Naegler 2005). A harmonic fit curve (Nakazawa et al. 1997) was calculated through the Jungfraujoch observations and assumed to represent the continental reference level over Europe.

Compared to the continental reference level, $\Delta^{14}CO_2$ at Schauinsland and Heidelberg is almost always lower because of dilution of the $^{14}C/^{12}C$ ratio with ^{14}C-free fossil fuel CO_2 at these sites (Fig. 4.4, upper panel). In Heidelberg, mean negative departures of $\Delta^{14}C$ values from the background are about 30–60‰ in winter and between 10‰ and 30‰ in summer. At the mountain site Schauinsland, the corresponding departures in summer are only between 2‰ and 6‰ while in winter the fossil fuel effect is between 10‰ and 15‰. The measured CO_2 mixing ratio CO_{2meas} at the sites consists of three components, a background component CO_{2bg}, a biospheric component CO_{2bio} and a fossil fuel component CO_{2foss}. The $\Delta^{14}C$ of these components are $\Delta^{14}C_{meas}$, $\Delta^{14}C_{bg}$, $\Delta^{14}C_{bio}$ and $\Delta^{14}C_{foss}$, respectively. To calculate the regional fossil fuel CO_2 component, we can formulate two balance equations, one for the CO_2 mixing ratio and one for the $^{14}C/^{12}C$ ratios [proportional to $(\Delta^{14}C_i + 1,000)$] (Levin et al. 2003):

$$CO_{2meas} = CO_{2bg} + CO_{2bio} + CO_{2foss} \tag{4.1}$$

$$CO_{2meas}(\Delta^{14}C_{meas} + 1,000) = CO_{2bg}(\Delta^{14}C_{bg} + 1,000) + CO_{2bio}$$
$$(\Delta^{14}C_{bio} + 1,000) + CO_{2foss}(\Delta^{14}C_{foss} + 1,000). \tag{4.2}$$

The fossil fuel term in Eq. 4.2 is zero as $\Delta^{14}C_{foss} = -1,000‰$. Setting $\Delta^{14}C_{bio}$ equal to $\Delta^{14}C_{bg}$, and combining both equations leads to the fossil fuel CO_2 component to be calculated from Eq. 4.3:

$$CO_{2foss} = CO_{2meas} \frac{\Delta^{14}C_{bg} - \Delta^{14}C_{meas}}{\Delta^{14}C_{bg} + 1,000} \tag{4.3}$$

[Note that the $\Delta^{14}C$ difference of the total CO_2 respiration flux (autotrophic and heterotrophic) compared to recent atmospheric $\Delta^{14}C$ levels can be estimated to lie between +15‰ and +25‰ (Naegler, 2005), which, if neglected as was done here, causes a possible mean bias in CO_2(foss) of less than 0.2 ppm]. At Schauinsland, the CO$_2$ mixing ratio is measured continuously since 1972 by the German Umweltbundesamt (Schmidt et al. 2003). For Heidelberg, CO$_2$ mixing ratios are available only from 1995 to 1998 and from 2002 onwards. For the periods without CO$_2$ observations, we approximate mixing ratios in Heidelberg using the mean seasonal offset from background air measured for 1995–1998 and 2002–2004 and adding the actual background mixing ratio for 48° N from GLOBALVIEW-CO$_2$ (2005). Fossil fuel CO$_2$ is then calculated from Eq. 4.3 for Schauinsland and Heidelberg as displayed in Fig. 4.4. The long-term mean fossil fuel CO$_2$ component at Schauinsland is 1.4 ppm with slightly higher values in the winter half year (1.7 ppm) than in summer (1.0 ppm). In Heidelberg, the long-term mean fossil fuel CO$_2$ component is almost one order of magnitude larger than at Schauinsland (10.9 ppm). The very pronounced seasonality of fossil fuel CO$_2$ in Heidelberg (winter: 14 ppm, summer: 6.5 ppm) is caused by a significant seasonality of the emissions, but mainly by seasonally varying atmospheric dilution of the emissions. The Rhine valley, particularly during winter, often experiences strong inversion situations causing large pile-ups of ground level emissions in the atmospheric surface layer. It is interesting, however, to note here that even in winter about half of the total CO$_2$ pile-up is due to biogenic emissions (Fig. 4.4, lowest panel). This was also estimated in the REMO modelling study. Similar to the Heidelberg and Schauinsland stations, we are able to estimate (monthly mean) fossil fuel CO$_2$ levels for other ^{14}CO$_2$ stations in Central Europe (not shown).

In summary: The current network of ^{14}C-based fossil fuel CO$_2$ observations is very sparse and unevenly distributed with important sites missing in South-Western, North-Eastern and Southern Europe. Mean observed CO$_2$(foss) levels are on the order of a few, up to a few tens of ppm, comparable to the REMO model simulations (Fig. 4.3).

4.2.3   How Well Do Model Estimates of the Fossil Fuel CO$_2$ Mixing Ratio over Europe Compare to Observations?

The regional fossil fuel CO$_2$ mixing ratios derived from ^{14}CO$_2$ observations at Jungfraujoch, Schauinsland and Heidelberg have been compared to estimates from different models in an intercomparison study by Geels et al. (2006). Five transport models were involved in this study: the global models TM3 and LMDZ and the regional models HANK, DEHM and REMO. TM3, LMDZ and HANK are coarser resolution models with grids of 1°×1° or larger, while DEHM and REMO have an approximate resolution of 0.5°×0.5° over Europe. The results for July and December 1998 are displayed in Fig. 4.5. Also included are pure model comparisons for a number of sites where no ^{14}CO$_2$ observations exist. Plotted is the fossil fuel CO$_2$ offset relative to the results for Mace Head, a marine background station at the Western coast of Ireland. In all model simulations, the same emissions inventory (EDGAR3.0 for the year 1990) was used.

Except for the Heidelberg and Tver site, in July 1998 all model estimates agree within better than 1 ppm. At Schauinsland, the models generally underestimate

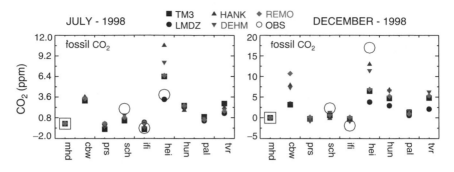

Fig. 4.5 [Adopted from Fig. 4.4 of Geels et al. (2006)] East–West transect of the fossil fuel CO_2 component over Europe in July and December 1998 as estimated with five different transport models with the same underlying emissions. Note that TM3, LMDZ and HANK are coarser resolution models with grids of $1° \times 1°$ or larger (TM3) while DEHM and REMO have an approximate resolution of $0.5° \times 0.5°$ over Europe (*obs* Observations; *mhd* Mace Head, 53°19′N, 9°53′W, 26 masl; *cbw* Cabauw tower, 51°58′N, 4°55′E, 213 m above ground; *prs* Plateau Rosa, 45°56′N, 7°42′E, 3,480 masl; *sch* Schauinsland, 47°55′N, 7°55′E, 1,205 masl; *jfj* Jungfraujoch, 46°33′N, 7°59′E, 3,580 masl; *hei* Heidelberg, 49°24′N, 8°42′E, 115 masl, 30 m above ground; *hun* Hegyhatsal tower 46°57′N, 16°39′E, 115 m above ground; *pal* Pallas, 67°58′N, 24°07′E, 560 masl; *tvr* Tver tower, 56°57′N, 32°57′E, 29 m above ground).

CO_2(foss) by about a factor of 2 while at Heidelberg all models except LMDZ overestimate fossil fuel CO_2 by up to a factor of 2.5. In December 1998, the differences in the fossil fuel CO_2 offset between models are very large (note the different scale compared to July), in particular at the ground level sites Cabauw, Heidelberg, Hungary and Tver. In Heidelberg, all models now underestimate fossil fuel CO_2, in some cases (REMO, LMDZ and TM3) by up to a factor of 3. One could argue that for all models, the spatial resolution is too coarse to reliably estimate mixing ratios at a site with a very heterogeneous source distribution. However, also at the remote Russian site Tver, with little direct influence from fossil fuel CO_2 emissions, large differences between models occur (up to a factor of 3). This is worrying because a station such as Tver may be one of the key sites in Europe for regional inversions of CO_2 observations and respective quantification of regional biogenic fluxes.

Figure 4.6 shows a more detailed comparison of fossil fuel CO_2 mixing ratios simulated with REMO and observed for individual two-weekly integrated samples collected in Heidelberg and at the Schauinsland mountain station in 2002. Again REMO was run with the two different emissions inventories from EDGAR and IER. While for Schauinsland in 2002 REMO results compare within 30% with the observations (for both emissions inventories), in Heidelberg in 2002 REMO generally overestimates regional CO_2(foss), sometimes by more than a factor of 2. This overestimation may be partly explained by transport deficiencies of the model (see Sect. 4.2.5), but Gamnitzer et al. (2006) could also show that the inventories must be wrong in the Heidelberg catchment area, particularly in summer.

These model–data comparisons of fossil fuel CO_2 clearly show that there are large discrepancies between different models using the same emissions inventory and between models and observations, which can be partly explained by model

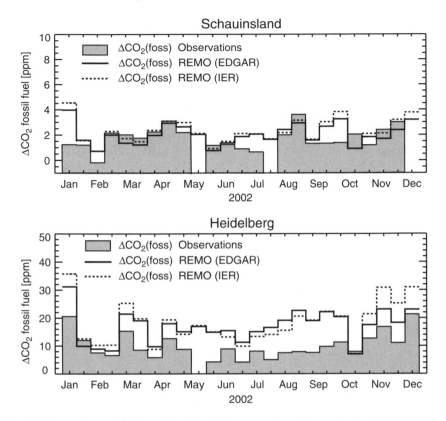

Fig. 4.6 Comparison of ^{14}C-based fossil fuel CO$_2$ observations (two weekly integrated samples from the year 2002) with respective REMO simulation results for the same time intervals at Schauinsland (upper panel) and Heidelberg (lower panel).

transport deficiencies but also point to deficiencies in the emissions inventories. Although the stations available to date are far from ideal for this model–data comparison, this underlines the importance of model validation and improvement of emissions inventories (as well as model transport) and the necessity of ongoing verification with [14]CO$_2$ observations. Such verification is necessary in particular at those sites that provide key CO$_2$ concentration records for European CO$_2$ inversions.

4.2.4 Which Precision and Temporal Resolution of [14]CO$_2$ Measurements Is Required to Reliably Monitor Fossil Fuel CO$_2$ over Europe ?

Our current network of two-weekly integrated [14]CO$_2$ observations in Europe has shown to be very sparse (compare Fig. 4.3) and provides mean fossil fuel CO$_2$ concentrations only on a monthly or at best on a weekly time resolution. Higher temporal resolution (i.e. hourly) is in principle possible if ^{14}C analyses are made by Accelerator Mass Spectrometry (AMS) technique, which allows ^{14}C analyses on air samples of less than 1 L volume. However, for large (conventional counting) but also small volume samples (AMS analysis), the costs of $^{14}CO_2$ measurements are high so that—at least for continuous long-term monitoring programs—higher than weekly resolution is normally not affordable.

The second important point to mention here is the sensitivity of $^{14}CO_2$ measurements to determine fossil fuel CO_2: with the existing technology of high-precision conventional counting, the precision of ^{14}C measurements is limited to $\Delta^{14}C = 1–2‰$ (Kromer and Münnich 1992). For AMS analyses (where much smaller sample sizes are measured), the standard precision of single analysis is normally on the order of ±4‰. However, some laboratories recently reached a maximum precision of ±1.6‰ by analysis of multiple aliquots of the same sample (Turnbull et al. 2006). This does, however, require a large analytical effort and further increases costs of $^{14}CO_2$ analyses. Still, a better precision than ±1.5‰ is probably not realistic on a routine basis. If we further assume that the background $^{14}CO_2$ level (i.e. at the Jungfraujoch or at Mace Head, if continental European observations are concerned) can be determined also to about ±1‰ on a monthly mean basis (by fitting/averaging two-weekly samples), the fossil fuel CO_2 component on a weekly integrated sample can be determined at best to ±0.7 ppm. This is about half of the mean fossil fuel CO_2 component at a remote continental mountain site, such as Schauinsland in summer (Fig. 4.4).

However, although the sensitivity and temporal resolution of $^{14}CO_2$ observations to determine fossil fuel CO_2 are limited, $^{14}CO_2$ measurements have the unique advantage to allow direct estimates of fossil fuel CO_2 which is not possible with any other tracer (see Sect. 4.2.5). A dense network of (integrated) $^{14}CO_2$ observations is, therefore, indispensable if fossil fuel CO_2 shall be reliably monitored over Europe.

4.2.5   Surrogate Tracers to Quantify Fossil Fuel CO_2 on a Continuous Basis

The limited temporal resolution, the costs, and the rather low sensitivity of $^{14}CO_2$ measurements were the main reasons why a number of proxies for fossil fuel CO_2 have been suggested and used in the past to indirectly derive fossil fuel CO_2 mixing ratios in the atmosphere. The demands to a good proxy for fossil fuel CO_2 are (1) the sources of the tracer should be related in a unique way to those of fossil fuel CO_2, (2) the tracer should be easily and precisely measured continuously and (3) it should behave conservatively in the atmosphere or its sink processes should be well understood so that it is possible to model them accurately. None of the tracers mentioned above fulfils all these demands: While CFCs and SF_6 are chemically stable in the troposphere, their emissions are not directly co-located with those of fossil fuel CO_2, in particular on the local and regional scale. On a larger continental or hemispheric scale, SF_6 may be better suited (Rivier et al. 2006). In the case of CO, its major sources are very closely linked to those of fossil fuel CO$_2$: any fossil fuel combustion process with CO$_2$ as an end product is, to some extent, associated with the production of CO. However, the chemical sink of CO in the atmosphere has a strong diurnal and seasonal cycle, and CO has also an atmospheric source from oxidation of VOCs. Similarly important, the CO/CO$_2$ emission ratio of different fossil fuel sources, such as domestic heating or emissions from traffic, can vary by orders of magnitude (UNFCCC 2005; Olivier et al. 2005) questioning the use of this tracer for quantitative estimates, in particular in polluted or semi-polluted areas where the source distribution is heterogeneous and the source mix can be quite variable.

At least two experimental (Gamnitzer et al. 2006; Turnbull et al. 2006) and one modelling study (Rivier et al. 2006) have been pursued recently to investigate the suitability of different surrogate tracers. However, in the study by Rivier et al. (2006), no ^{14}CO$_2$ measurements were available providing reliable validation of the simulated fossil fuel CO$_2$ component. Turnbull et al. (2006) compare ^{14}C-, CO- and SF$_6$-based fossil fuel CO$_2$ mixing ratios on atmospheric samples collected in New England and Colorado. They find implausibly large differences between SF$_6$-based and ^{14}C-based estimates which disqualify SF$_6$ as a quantitative fossil fuel surrogate tracer. Their CO-based estimates agree better with ^{14}C-based fossil fuel CO$_2$ estimates but, as mentioned above, seasonally coherent biases point to other source and sink processes influencing regional CO mixing ratios which need to be taken into account.

A very detailed study using 3 years of quasi-continuous (two-)weekly integrated ^{14}CO$_2$ measurements as well as measured diurnal cycles of ^{14}CO$_2$ in combination with continuous CO$_2$ and CO measurements and the REMO model to simulate CO and CO$_2$ mixing ratios from emissions inventories was performed by Gamnitzer et al. (2006). The observational data used by Gamnitzer et al. (2006) are presented in Fig. 4.7 as monthly means and extended up to the end of 2004.

The upper two panels of Fig. 4.7 again show Δ^{14}CO$_2$ at Heidelberg and Jungfraujoch (background) as well as the ^{14}C-based regional fossil fuel CO$_2$ offset in Heidelberg (compare Fig. 4.4). The third panel shows monthly mean regional CO offsets relative to the marine background estimated from Atlantic stations in mid northern latitudes (Masarie, pers. comm., extrapolated to 2004), while the lowest panel gives the ratios of ΔCO/ΔCO$_2$(foss) calculated from the ΔCO$_2$(foss) and ΔCO data in the two middle panels. On a monthly mean basis, the locally observed ΔCO/ΔCO$_2$(foss) ratio in Heidelberg varies by only ± 21% and shows no systematic seasonal cycle. This indeed suggests that CO, at least for the Heidelberg site, may be well suited as a surrogate tracer for fossil fuel CO$_2$.

A more detailed comparison by Gamnitzer et al. (2006) of the Heidelberg observations for 2002 with REMO-simulated ΔCO and ΔCO$_2$(foss) offsets does, however, give a somewhat different picture. While the ratios of ΔCO/ΔCO$_2$(foss) show good agreement between observed [13.5 ± 2.5 ppb CO/ppm CO$_2$(foss)] and modelled values [EDGAR inventory: 13.0 ± 0.9 ppb CO/ppm CO$_2$(foss), IER inventory: 11.5 ± 1.1 ppb CO/ppm CO$_2$(foss)] individual offsets of ΔCO$_2$(foss) (see Fig. 4.6) but also of ΔCO (not shown) differ by up to a factor of 2 between model and observations. This may be either due to errors in the model transport or due to errors in

Fig. 4.7 Monthly mean values of $\Delta^{14}CO_2$ at Heidelberg and Jungfraujoch (upper panel), ^{14}C-based estimates of $\Delta CO_2(foss)$ and ΔCO in Heidelberg (middle panels). The lowest panel shows the ratios of regional $\Delta CO/\Delta CO_2(foss)$.

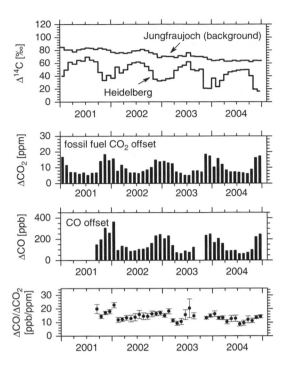

respective emissions inventories or both. Gamnitzer et al. (2006) have, therefore, corrected the model results for transport deficiencies by means of continuous ^{222}Radon observations and respective model results. But even after this ^{222}Radon correction, differences between observations and REMO simulations are as large as 100% for both components, $\Delta CO_2(foss)$ and ΔCO. For CO_2, these remaining differences are probably due to deficiencies in the emissions inventories, but for CO also, the non-fossil sources and sinks of CO have large uncertainties. Gamnitzer et al. (2006) have estimated the uncertainty of the CO production via VOC oxidation to more than 50%, and that of the soil sink of CO, which is not well quantified at all in the literature (Sanhueza et al. 1998), to up to 100%. They have performed a number of model scenarios quantifying possible biases in the CO model estimates associated with VOC oxidation and soil uptake which yield a total range of almost 100% in the modelled CO offset in Heidelberg and respective systematic changes of 50% in the modelled $\Delta CO/\Delta CO_2(foss)$ ratio. The good agreement between modelled and observed $\Delta CO/\Delta CO_2(foss)$ ratio in Heidelberg could, thus, be purely accidental.

Still the good agreement between modelled and observed $\Delta CO/\Delta CO_2(foss)$, whether accidental or real, allowed Gamnitzer et al. (2006) to correct modelled $\Delta CO_2(foss)$ with modelled and measured ΔCO in a similar way as the ^{222}Radon correction. With this correction, agreement of better than 20% root mean square (RMS) error for EDGAR emissions and better than 35% RMS error for IER emissions between observed (^{14}C-based) and simulated REMO plus emissions inventory derived $\Delta CO_2(foss)$ in Heidelberg could be achieved. The CO-correction or normalisation obviously takes care of some deficits of simulated transport in REMO but also of other CO modelling deficits. This example shows that in a situation where the CO/CO$_2$(foss) ratio of emissions is more or less correct continuous CO measurements (in combination with CO model results) can help to correct simulated ΔCO_2(foss) mixing ratios for model and partly also inventory deficiencies.

The implications of these CO studies are twofold: (1) when CO alone shall be used as a quantitative tracer for fossil fuel CO$_2$, the CO/CO$_2$(foss) ratio of emissions as well as its temporal changes need to be known accurately and (2) reliable estimates of the non-fossil sources and the sinks of CO are required as these may change atmospheric CO/CO$_2$(foss) ratios by up to 50%.

4.2.6 Can We Combine Integrated $^{14}CO_2$ and Continuous CO Observations to Improve Estimates of Continuous Regional ΔCO_2(foss)?

In the preceding section, we have seen that CO can be a valuable tracer for fossil fuel CO$_2$ if the CO/CO$_2$(foss) ratios of emissions from inventories in the catchment area of the site are more or less correct. Besides continuous CO observations, this also requires accurate modelling of atmospheric CO, and the reliability of the CO/CO$_2$(foss) ratio of the emissions must still be confirmed, for example, by comparison with (integrated) $^{14}CO_2$ measurements at the sites in question. But what is really needed to properly determine fossil fuel CO$_2$ over Europe is a simple observational approach not biased by large uncertainties of CO modelling and/or inventory data. Such an approach has been recently suggested by Levin and Karstens (2007): it is based on integrated $^{14}CO_2$ and continuous CO observations where ΔCO_2(foss) mixing ratios could then be estimated from the measured hourly ΔCO mixing ratios and a weekly mean ΔCO to ΔCO_2(foss)$^{14C\text{-based}}$ ratio according to

$$\Delta CO_2(foss)_{hourly} = \Delta CO_{hourly}^{meas} \frac{\Delta CO_2(foss)_{weekly}^{14C\text{-based}}}{\left\langle \Delta CO_{hourly}^{meas} \right\rangle_{weekly}}. \tag{4.4}$$

Levin and Karstens (2007) used REMO-simulated ΔCO_2(foss) and ΔCO records for the year 2002 to determine the uncertainty of this approach for different sites in Europe by assuming that these modelled mixing ratios represent the correct hourly mixing ratios of both gases. From the continuous hourly modelled ΔCO_2(foss) and ΔCO values, weekly means and subsequent weekly mean $\Delta CO/\Delta CO_2$(foss) ratios were calculated and these mean ratios were then applied to the original high-resolution modelled ΔCO record to re-calculate a high-resolution ΔCO_2(foss) record

$$\Delta CO_2(foss)_{hourly}^{re\text{-}calculated} = \Delta CO_{hourly}^{model} \frac{\left\langle \Delta CO_2(foss)_{hourly}^{model} \right\rangle_{weekly}}{\left\langle \Delta CO_{hourly}^{model} \right\rangle_{weekly}}. \tag{4.5}$$

The differences between the original high-resolution $\Delta CO_2(foss)$ record (i.e. the true record) and the ΔCO-based re-calculated $\Delta CO_2(foss)$ record (Eq. 4.5) provide a measure of the minimum uncertainty of any attempt to use hourly ΔCO *measurements* to derive an hourly $\Delta CO_2(foss)$ record according to Eq. 4.4. Figure 4.8 shows the results of this test for Heidelberg, separately with REMO simulations based on IER and those based on EDGAR emissions for February and July 2002. The $\Delta CO_2(foss)$ REMO records named 'original' are the originally calculated values assumed to represent the truth.

In addition to estimates from Eq. 4.5, we made here a second test where $\Delta CO_2(foss)$ was calculated in a similar way, but replacing the (REMO-simulated 'observed') mean ΔCO to mean $\Delta CO_2(foss)$ ratio by the mean ratio of the respective emissions in the catchment area (which we call '$\Delta CO_2(foss)$ inventory'):

$$\Delta CO_2(foss)_{hourly}^{inventory} = \Delta CO_{hourly}^{model} \frac{\langle CO_2(foss) \rangle_{inventory}}{\langle CO \rangle_{inventory}}. \tag{4.6}$$

For these inventory-based estimates, the mean ratios of emissions were calculated assuming a catchment area corresponding to eleven $1° \times 1°$ EDGAR grid boxes around Heidelberg [the same catchment area as in the Gamnitzer et al. (2006) study]. The CO emissions also included non-fossil sources and biogenic emissions.

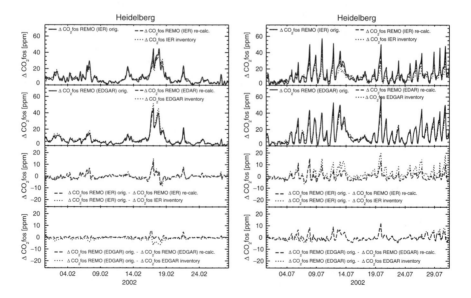

Fig. 4.8 Sensitivity tests of the uncertainty of ΔCO-based $\Delta CO_2(foss)$ estimates: Upper two panels: $\Delta CO_2(foss)$ estimated from 'atmospheric' ΔCO records and weekly mean $\Delta CO/\Delta CO_2(foss)$ ratios calculated from original 'atmospheric' results according to Eq. 4.5 and from $CO/\Delta CO_2(foss)$ emissions ratios directly, according to Eq. 4.6 in comparison to original 'atmospheric' $\Delta CO_2(foss)$ records (plotted for IER and EDGAR inventories separately). Lower two panels: respective differences.

The RMS deviations of ΔCO_2(foss) from the true values when using the weekly mean modelled ratios and Eq. 4.5 for Heidelberg are about 18% in the case of IER emissions and about 14% in the case of EDGAR emissions. Deviations for EDGAR are smaller because they are caused only by differences of the CO/CO_2(foss) emissions ratios in the different parts of the catchment area (which the original REMO calculations are taking care of) and the processes influencing the chemistry of CO (production by VOC oxidation and destruction by OH). The larger deviations in the IER results are caused by both spatial and temporal (i.e. diurnal) variations of the CO/CO_2(foss) ratio of emissions (these vary by about 10% in winter and up to 20% in summer), which again are taken care of in the original REMO results. If only the pure inventory ratios are used to estimate ΔCO-based ΔCO_2(foss) (Eq. 4.6), the RMS errors increase to about 30% in the case of IER emissions and to about 20% in the case of EDGAR emissions. This illustrates the importance of atmospheric chemistry processes as well as soil uptake influencing CO mixing ratios in the continental troposphere. At other European sites, the relative RMS error for the application of atmospheric ratios (according to Eq. 4.5) increased with decreasing fossil fuel CO$_2$ offset showing maximum uncertainties at Schauinsland station of about 40% for the IER and about 35% for EDGAR emissions inventories. This is due to the fact that the non-fossil sources of CO become increasingly important with decreasing fossil fuel component.

In summary, and in the real world, if continuous CO measurements are available at a CO$_2$ monitoring site as well as, for example, weekly integrated $^{14}CO_2$ observations to obtain a good estimate of regional ΔCO_2(foss) and of the mean $\Delta CO/\Delta CO_2$(foss) ratio, we will be able to estimate regional fossil fuel CO$_2$ at hourly resolution with an uncertainty of about 20–40% [the uncertainty of the ^{14}C-based determination of fossil fuel CO$_2$ of about ±1 ppm (see Sect. 4.2.4) must be added to this number]. Note, however, that this estimate would be purely based on observations and would not be biased by uncertainties of emissions inventories or in modelling of atmospheric CO!

4.3 Conclusions

How should an observational network for continuous monitoring of the fossil fuel CO$_2$ component over Europe look like? From the studies by Turnbull et al. (2006) and in particular by Gamnitzer et al. (2006), we have learned that fossil fuel CO$_2$ estimates based solely on emissions inventories can be largely biased, in the case of the Heidelberg measurement site by up to a factor of 2. Moreover, the model comparison study by Geels et al. (2006) showed that large differences of up to a factor of 2 or even more exist in monthly estimates of the fossil fuel component in atmospheric CO$_2$ between different models, even when using the same fossil fuel emissions inventories as input data. With a fossil fuel CO$_2$ signal over Europe being of the same magnitude as the respective biospheric signals, this is a non-satisfactory situation if we try to understand causes and trends in the biosphere–atmosphere CO$_2$ exchange in this region.

Continuous CO observations together with CO model estimates as applied by Gamnitzer et al. (2006) have the potential to correct model-derived ΔCO_2(foss) mixing ratios yielding final uncertainties of 20–35% if the CO/CO_2(foss) ratios of the emissions inventory used are reliable. Such a CO correction does, however, require not only modelling of the fossil fuel CO_2 component but also modelling of CO with all its uncertainties in non-fossil CO sources including oxidation of VOC, CO sinks in the atmosphere as well as soil deposition. For such an approach, errors associated with uncertainties in model transport also have to be taken into account, which will additionally complicate the interpretation and application of the CO correction method. Taking advantage of CO observations can then correct for some shortcomings related to transport and other unaccounted-for processes and may be one step forward towards improvement of estimating fossil fuel CO_2 based on modelling studies. The success of such an approach does, however, fully depend on the reliability of the temporally and spatially variable CO/CO_2(foss) ratios available from emissions inventories. In fact, the observed CO/CO_2(foss) ratio in the Heidelberg catchment area has changed already over the course of the last 5 years by more than 20%.

A more reliable approach towards a better quantification of high-resolution fossil fuel CO_2 over Europe, therefore, seems to be the observation-based approach suggested by Levin and Karstens (2007), namely, for example, weekly integrated high precision $^{14}CO_2$ measurements combined with precise continuous CO observations. This approach will be accurate to within 20–40%, depending on the site, with additional random weekly mean biases of about 1 ppm caused by the $^{14}CO_2$ measurement error. This approach seems to be a fair compromise of costs and affordable observational effort.

We, therefore, recommend high-precision weekly integrated $^{14}CO_2$ observations as well as precise continuous CO measurements at *all* continental CO_2 measurement sites which shall be used for model inversions of CO_2 sources and sinks. These observations will allow estimating a high-resolution record of fossil fuel CO_2 mixing ratios at these sites, and finally determining the solely biogenic CO_2 component required to invert for continental biogenic CO_2 sources and sinks and their long-term changes.

Acknowledgements We wish to thank the personnel at the Schauinsland and Jungfraujoch stations for collecting the atmospheric $^{14}CO_2$ samples, as well as Frank Meinhardt at UBA Schauinsland for making available the continuous atmospheric CO_2 measurements from this site. The members of the Heidelberg Radiocarbon Laboratory are gratefully acknowledged for their ongoing effort to analyse all $^{14}CO_2$ samples presented here. Camilla Geels made available to us Fig. 4.5. Dietmar Wagenbach (IUP) and Tobias Naegler (IUP) have helped with fruitful discussions. This work was funded by the European Union under contracts No. EVK2-CT-1999-00013 (AEROCARB) and GOCE-CT2003-505572 (CarboEurope-IP).

References

Bakwin, P. S., Hurst, D. F., Tans, P. P., and Elkins, J. W. 1997. Anthropogenic sources of halocarbons, sulfur hexafluoride, carbon monoxide, and methane in the south-eastern United States. J. Geophys. Res. 102:15,915–15, 925.

Chevillard, A., Karstens, U., Ciais, P., Lafont, S., and Heimann, M. 2002. Simulation of atmospheric CO$_2$ over Europe and western Siberia using the regional scale model REMO. Tellus 54B:872–894.

Churkina, G., Tenhunen, J., Thornton, P., Falge, E. M., Elbers, J. A., Erhard, M., Grünwald, T., Kowalski, A. S., Rannik, Ü., and Sprinz, D. 2003. Analyzing the ecosystem carbon dynamics of four European coniferous forests using a biogeochemistry model. Ecosystems 6:168–184.

Gamnitzer, U., Karstens, U., Kromer, B., Neubert, R. E. M., Meijer, H. A. J., Schroeder, H., and Levin, I. 2006. Carbon monoxide: A quantitative tracer for fossil fuel CO$_2$? J. Geophys. Res. 111:D22302, doi:10.1029/2005JD006966.

Geels, C., Gloor, M., Ciais, P., Bousquet, P., Peylin, P., Vermeulen, A. T., Dargaville, R., Aalto, T., Brandt, J., Christensen, J. H., Frohn, L. M., Haszpra, L., Karstens, U., Rödenbeck, C., Ramonet, M., Carboni, G., and Santaguida, R. 2006. Comparing atmospheric transport models for future regional inversions over Europe. Part 1: Mapping the CO$_2$ atmospheric signals over Europe. Atmos. Chem. Phys. Discuss. 6:3709–3756.

GLOBALVIEW-CO$_2$ 2005. Cooperative Atmospheric Data Integration Project—Carbon Dioxide. CD-ROM, NOAA CMDL, Boulder, Colorado (Also available on Internet via anonymous FTP to ftp.cmdl.noaa.gov, Path: ccg/co2/GLOBALVIEW).

Hesshaimer, V. 1997. Tracing the global carbon cycle with bomb radiocarbon, Ph.D. thesis, University of Heidelberg.

Kromer, B., and Münnich, K. O. 1992. CO$_2$ gas proportional counting in Radiocarbon dating—review and perspective. In Radiocarbon after four decades, eds. R. E. Taylor, A. Long and R. S. Kra, pp. 184–197. New York: Springer-Verlag.

Kuc, T. 1986. Carbon isotopes in atmospheric CO$_2$ of the Krakow region: A two-year record. Radiocarbon 28:649–654.

Levin, I., Münnich, K. O., and Weiss, W. 1980. The effect of anthropogenic CO$_2$ and ^{14}C sources on the distribution of ^{14}CO$_2$ in the atmosphere. Radiocarbon 22:379–391.

Levin, I., Kromer, B., Schoch-Fischer, H., Bruns, M., Münnich, M., Berdau, D., Vogel, J. C., and Münnich, K. O. 1985. 25 years of tropospheric ^{14}C observations in Central Europe. Radiocarbon 27:1–19.

Levin, I., Schuchard, J., Kromer, B., and Münnich, K. O. 1989. The continental European Suess effect. Radiocarbon 31:431–440.

Levin, I., Kromer, B., Schmidt, M., and Sartorius, H. 2003. A novel approach for independent budgeting of fossil fuels CO$_2$ over Europe by ^{14}CO$_2$ observations. Geophys. Res. Lett., 30(23):2194, doi:10.1029/2003GL018477.

Levin, I., and Kromer, B. 2004. The tropospheric ^{14}CO$_2$ level in mid-latitudes of the Northern Hemisphere (1959–2003). Radiocarbon 46:1261–1272.

Levin, I., and Karstens, U. 2007. Inferring high-resolution fossil fuel CO$_2$ records at continental sites from combined ^{14}CO$_2$ and CO observations. Tellus B, doi:10.1111.j1600.0889.2006.00224.x 59B,245–250.

Marland, G., Boden, T.A., and Andres, R. J. 2006. Global, Regional, and National CO$_2$ Emissions. In Trends: A Compendium of Data on Global Change. Carbon Dioxide Information Analysis Center, Oak Ridge National Laboratory, U.S. Department of Energy, Oak Ridge, Tenn., U.S.A.

Meijer, H. A. J., van der Plicht, J., and Gislefoss, J. S. 1995. Comparing long-term atmospheric ^{14}C and ^{3}H records near Groningen, the Netherlands with Fruholmen, Norway and Izaña, Canary Islands ^{14}C stations. Radiocarbon 37(1):39–50.

Naegler, T. 2005. Simulating bomb Radiocarbon: Implications for the Global Carbon Cycle. Ph.D. thesis, University of Heidelberg.

Naegler, T., and Levin, I. 2006. Closing the global bomb radiocarbon budget. J. Geophys. Res. 111, D12311, doi: 10.10292005JD006758.

Nakazawa, T., Ishizawa, M., Higuchi, K., and Trivett, N. B. A. 1997. Two curve fitting methods applied to CO$_2$ flask data. EnvironMetrics 8:197–218.

Olivier, J. G. J., van Aardenne, J. A., Dentener, F., Ganzeveld, L., and Peters, J. A. H. W. 2005. Recent trends in global greenhouse gas emissions: Regional trends and spatial distribution of key sources. In Non-CO_2 Greenhouse Gases (NCGG-4), ed. A. van Amstel (coord.), pp. 325–330. Millpress, Rotterdam, ISBN 9059660439. Information available online at www.rivm.nl.

Potosnak, M. J., Wofsy, S. C., Denning, A. S., Conway, T. J., Munger, J. W., and Barnes, D. H. 1999. Influence of biotic exchange and combustion sources on atmospheric CO_2 concentrations in New England from observations at a forest flux tower. J. Geophys. Res. 104:9561–9569.

Randerson, J. T., Enting, I. G., Schuur, E. A. G., Caldeira, K., and Fung, I. Y. 2002. Seasonal and latitudinal variability of troposphere $\Delta^{14}CO_2$: Post bomb contributions from fossil fuels, oceans, the stratosphere, and the terrestrial biosphere. Global Biogeochem. Cycles 16, doi:10.1029/2002GB001876.

Reis, S., Pfeiffer, H., Theloke, J., and Scholz, Y. 2008. Temporal and spatial distribution of Carbon emissions. This issue, Chapter 5.

Rivier, L., Ciais, P., Hauglustaine, D. A., Bakwin, P., Bousquet, P., Peylin, P., and Klonecki, A. 2006. Evaluation of SF_6, C_2Cl_4 and CO as surrogate tracers for fossil fuel CO_2 in the United States using models and continuous atmospheric observations on high towers. J. Geophys. Res. 111, D16311, doi:10.1029/2005JD006725.

Sanhueza, E., Dong, Y., Scharffe, D., Lobert, J. M., and Crutzen, P. J. 1998. Carbon monoxide uptake by temperate forest soils: The effects of leaves and humus layers. Tellus 50B:51–58.

Schmidt, M., Graul, R., Sartorius, H., and Levin, I. 2003. The Schauinsland CO_2 record: 30 years of continental observations and their implications for the variability of the European CO_2 budget. J. Geophys. Res. 108(D19), 4619, doi:10.1029/2002JD003085.

Stuiver, M., and Polach, H. A. 1977. Discussion: Reporting of ^{14}C data. Radiocarbon 19:355–363.

Suess, H. E. 1955. Radiocarbon concentration in modern wood. Science 122:415.

Takahashi, T., et al. 1999. Net sea-air CO_2 flux over the global oceans: An improved estimate based on the sea-air pCO_2 difference. Proceedings of the 2nd CO_2 in Oceans Symposium, Tsukuba, JAPAN, January 18–23.

Turnbull, J. C., Miller, J. B., Lehman, S. J., Tans, P. P., Sparks, R. J., and Southon, J. 2006. Comparison of $^{14}CO_2$, CO, and SF_6 as tracers for recently added fossil fuel CO_2 in the atmosphere and implications for biological CO_2 exchange. Geophys. Res. Lett. 33, L01817, doi: 10.1029/2005GL024213.

UNFCCC (United Nations Framework Convention on Climate Change) 2005. *Greenhouse gases database*. Bonn, Germany. Available online at ghg.unfccc.int, Download on 10. Feb. 2005.

Chapter 5
Temporal and Spatial Distribution of Carbon Emissions

Stefan Reis, Heiko Pfeiffer, Jochen Theloke, and Yvonne Scholz

5.1 Introduction

In the early stages of research in climate change, the focus was on the quantification of global and regional carbon cycles. At this stage, the accurate determination of location and time of emissions played a less prominent role. But with the growing need for verification experiments and prior information for atmospheric transport models (Chaps. 3, 12) to underpin and support policy development, and thus the application of climate and atmospheric dispersion models, it became evident that both anthropogenic and biogenic carbon emissions had to be provided in a spatial and temporal resolution matching the requirements of said models.

The methodology for the temporal and spatial resolution of anthropogenic emissions, in general, has been around for some time. Some of the techniques described in the following sections have, for instance, been developed for the application of atmospheric dispersion models to quantify acid deposition and ambient air concentrations of tropospheric ozone. One major activity during the 1990s in this field has been the work within the EUROTRAC subproject Generation and Evaluation of Emission Data (GENEMIS), which has been documented in Friedrich and Reis (2004).

5.2 Carbon Emissions from Fossil Fuel Use

5.2.1 Quantification of Anthropogenic Carbon Emissions

In comparison to, for instance, emissions of non-methane volatile organic compounds from solvent use, emissions of carbon dioxide and carbon monoxide from the combustion of fossil fuels are less difficult to quantify. On the other hand, methane emissions originate from a portfolio of sources, most of them non-combustion activities, and hence need different approaches for a temporal and spatial distribution. This section will focus on the three main substances, CO_2, CH_4 and CO, and address anthropogenic sources of emissions only. Other contributions, for instance, from non-methane volatile organic compounds or halogenated hydrocarbons (e.g. HFC) are not included in the analysis.

5.2.1.1 Carbon Dioxide

Complete combustion of fossil fuels releases carbon stored in coal, oil and natural gas into the atmosphere. In combination with oxygen from the air needed for the combustion process, CO_2 is formed and emitted. The largest emissions stem from power generation and transport, followed by industrial combustion and production, for instance, cement and iron and steel production.

5.2.1.2 Carbon Monoxide

CO is a product of incomplete combustion, when fossil fuels are converted in an atmosphere lacking oxygen necessary for complete combustion. The ratio of CO production varies with the availability of oxygen in the combustion chamber and technologies, such as lean engine concepts and combustion chamber optimisations can influence the amount of CO formed significantly. In comparison to CO_2, small combustion sources and transport contribute relatively higher shares of CO emissions because of the optimised combustion processes usually found in large combustion plants.

Table 5.1 shows typical CO_2/CO ratios for selected emission sources, and indicates how different combustion processes can be in terms of energy conversion and relative emissions. The figures given are average values based on current emission factors and are only indicative, that is, they may vary significantly between processes, fuels and environmental conditions. Other sources, for instance, small combustion

Table 5.1 Typical CO_2 to CO ratios for selected emission sources. Values for vehicles are averaged over different driving modes (highway, rural and urban) as well as across vehicle size and engine categories and are only indicative

Source	EF CO_2	EF CO	CO_2/CO ratio
Public power plants	[kg/kWh$_{el}$]		
Hard coal	972.03	0.81	*1198*
Lignite	1039.80	1.19	*868*
Natural gas	538.87	0.49	*1087*
Oil	904.06	0.33	*2734*
Road Transport	[g/km]		
Passenger car gasoline, EURO IV standard	186.89	1.32	*171*
Passenger car diesel, EURO IV standard	151.91	0.23	*809*
Passenger car gasoline, pre-EURO I standard	188.07	13.05	*16*
Passenger car diesel, pre-EURO I standard	169.31	0.48	*387*
Heavy-duty vehicle, EURO V standard *(from 2008)*	718.42	0.43	*1723*

processes in household space heating, show ratios between 17 (hard coal coke) and 2722 (natural gas) and even for an individual passenger car, the ration can vary by a factor of 2–3 depending on driving patterns and load factor.

5.2.1.3 Methane

Methane emission sources vary with regions, global estimates assuming about one quarter stemming from enteric fermentation in animals, another quarter originating from gas, oil and coal production. Other major sources include municipal waste disposal in landfills and wastewater treatment as well as rice cultivation. These processes are not combustion related and some of them are biogenic/geogenic sources, with significant uncertainties.

5.2.1.4 Emission Factors and Activity Rates

Common practise for the determination of emission values is the use of emission factors (EF) and activity rates (A), calculating Emission (E) as

$E = EF \times A$
EF = the specific amount of a substance emitted per specific activity unit
A = for example, the unit of thermal energy converted, annual mileage
E = resulting emission

While this approach is straightforward, its implications for temporal and spatial distribution of emissions are profound. In most cases, the temporal and spatial patterns, for example, for the creation of a country emission inventory, are of less importance; hence emission factors and activity rates are used without expressing

Table 5.2 CO_2 emissions from non-EU countries in Europe for the year 2000 in Tg, comparing EDGAR inventory data and emissions reported to UNFCCC, where available

	EDGAR	UNFCCC	Difference
Croatia	61.7	19.4	−68.5%
Macedonia (FYROM)	10.3	N/A	N/A
Turkey	237.5	223.8	−5.8%
Subtotal EU candidate countries	*309.5*	*243.2*	
Russian Federation	1821.9	N/A	N/A
Belarus	54.4	N/A	N/A
Ukraine	418.7	295.5	−29.2%
Albania	3.6	N/A	N/A
Bosnia & Herzegovina	43.1	N/A	N/A
FR of Yugoslavia	70.1	N/A	N/A
Norway	95.5	41.5	−56.5%
Switzerland	45.5	43.9	−3.5%
Subtotal non-EU countries	*2552.9*	*382.0*	

them as a function of space and time. However, in most cases, activity rates will be dependent on factors, such as the time of the year, meteorological parameters, and show distinctive temporal patterns (diurnal, weekly, monthly or annual). At the same time, meteorological and behavioural patterns of activity rates may change with locations. The methodologies for the temporal and spatial resolution are described in the Sect. 5.2.2.

Uncertainties are subject to significant variability because of the large number of influencing factors in the determination of emissions, either as national annual totals, or with high temporal and spatial resolution. Typically, only rough assessments of uncertainty levels are available (see e.g. Table 5.2 below).

5.2.2   Magnitude of Annual Emissions in Europe

Emission inventories are compiled for different purposes and in the case of carbon emissions from fossil fuel use, there are three main activities to be mentioned:

• The Emission Database for Global Atmospheric Research[1] (EDGAR)
• Data submissions of signatories to the UNFCCC[2]
• Data submissions to the EMEP[3] programme

EDGAR covers the widest range of substances on a global scale, while UNFCCC data are limited to greenhouse gases and EMEP solely collects data relevant under the United Nations Economic Commission for Europe (UNECE) Convention on Long-Range Transboundary Air Pollution (CO, but excluding greenhouse gases).

Figure 5.1 displays a comparison of CO_2 emission date for the EU27 (27 European Union member states; note that Cyprus and Malta have not been listed for space reasons). In total, EDGAR has a slightly higher emission figure for the EU27 (9.2%) than the UNFCCC inventory. However, for some individual countries, the difference between both inventories is larger, for example, Belgium (17.7%), The Netherlands (30.1%), Bulgaria (44.2%) and Latvia (23.8%). For other European countries, the comparison is less straightforward, as many of these are not reporting to UNFCCC. Table 5.3 indicates several gaps in reporting to UNFCCC (due to non-ratification of the Kyoto protocol), and hence missing significant sources in Europe, for instance, 1822 Tg from the Russian Federation alone. These non-EU countries contributed about 35% of the total CO_2 emissions of 7364 Tg in the year 2000.

In a similar way, Fig. 5.2 compares methane emissions as reported under the UNFCCC with EDGAR inventory data, with some slight differences in overall emission levels, which most likely originate from different emission sources being

---

[1] http://www.mnp.nl/edgar/

[2] http://unfccc.int, emission data excluding land-use and land-cover change (LULUCF) have been used for this comparison.

[3] http://webdab.emep.int/

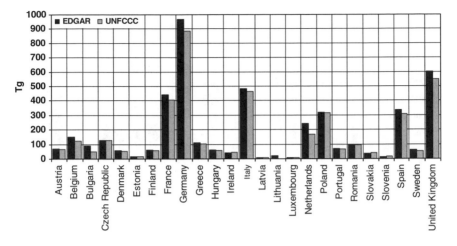

Fig. 5.1 CO_2 emissions for the year 2000 in the EDGAR database (Source: EDGAR 32FT2000) and as reported to the UNFCCC for the EU27.

Table 5.3 CH_4 emissions from non-EU countries in Europe for the year 2000 in Tg, comparing EDGAR inventory data and emissions reported to UNFCCC, where available

	EDGAR	UNFCCC	Difference
Croatia	164.2	121.0	−26.2%
Macedonia (FYROM)	59.3	N/A	N/A
Turkey	1199.5	2349.9	95.9%
Subtotal EU candidate countries	*1243.0*	*2471.0*	
Russian Federation	23753.2	N/A	N/A
Belarus	771.0	546.6	−29.1%
Ukraine	3910.9	3661.1	−6.4%
Albania	110.7	N/A	N/A
Bosnia & Herzegovina	135.1	N/A	N/A
FR of Yugoslavia	368.9	N/A	N/A
Norway	415.3	235.9	−42.1%
Switzerland	208.8	179.5	−14.1%
Subtotal non-EU countries	*29673.9*	*4623.1*	

excluded from the framework convention, which are comprised in the EDGAR inventory figures. For CH_4, the differences for individual countries are more pronounced, with an EDGAR arriving at an overall lower total emission figure (5.5%) than UNFCCC. For some of the larger countries in particular, figures are significantly higher in the UNFCCC dataset, for example, the France (36.2%), UK (53%), Portugal (73.9%) and Ireland (282.8%). Other countries show higher emissions in the EDGAR inventory, for instance, Poland (31.6%), the Czech Republic (25.5%) and Germany (14.7%). While it is beyond the scope of this chapter to analyse the

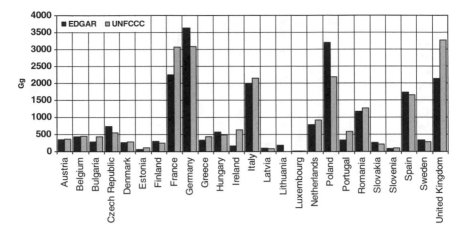

Fig. 5.2 CO_2 emissions for the year 2000 in the EDGAR database (Source: EDGAR 32FT2000) and as reported to the UNFCCC for the EU27.

Table 5.4 CO emissions from non-EU countries in Europe for the year 2000 in Tg, comparing EDGAR inventory data and emissions reported to EMEP

	EDGAR	UNFCCC	Difference
Croatia	597.9	395.2	−33.9%
Macedonia (FYROM)	216.5	83.5	−61.4%
Turkey	3826.8	3919.5	−5.2%
Subtotal EU candidate countries	*4641.2*	*4398.2*	*−5.2%*
Russian Federation	27893.6	10242.5	−63.3%
Belarus	718.2	724.8	0.9%
Ukraine	13276.0	2276.3	−82.9%
Albania	283.0	124.5	−56.0%
Bosnia & Herzegovina	1166.9	95.7	91.8%
FR of Yugoslavia	827.1	383.3	−52.9%
Norway	1624.9	564.0	−65.7%
Switzerland	649.1	401.3	−38.2%
Subtotal non-EU countries	*46456.9*	*14818.3*	*−68.1%*

reasons for these differences in detail, it is fair to assume that different methodologies for emission calculations in different countries, in particular, for non-combustion sources, are a likely cause (Table 5.4).

In the case of non-EU countries, EDGAR figures are between 6.4% and 42.1% higher than UNFCCC data, with the exception of Turkey, where UNFCCC emissions are almost twice as high as in the EDGAR database.

For carbon monoxide, EMEP data submitted by countries (complemented and gap filled by independent expert estimates) have been compared with the EDGAR emissions (CO is not reported to the UNFCCC) in Fig. 5.3. The overall agreement between both inventories is quite good (3.7% difference in EU27 totals), however,

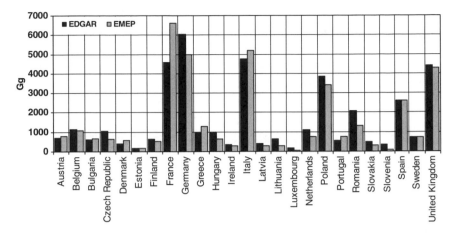

Fig. 5.3 CO emissions for the year 2000 in the EDGAR database (Source: EDGAR 32FT2000) and as reported to EMEP[4] for the EU27.

a significant spread between individual countries can be observed, ranging from −72% to 44.7% with EDGAR showing a tendency to have lower emissions, with few exceptions.

Similar to the case of methane, CO emissions are marked more by process-related factors, than CO_2 emissions and the differences in accounting for these, for instance, in the case of calculating cold-start emissions from vehicles or emissions from small combustion sources is likely to be the cause for significant differences between inventories.

A rough calculation of the carbon content in these emissions, based on the EDGAR dataset, as it is the most comprehensive for the European situation, gives an estimate of total C emissions for the year 2000 of 2.1 Pg. Carbon dioxide marks the dominating contribution (96%), even though its carbon content is only 27%, while CH_4 and CO each contribute ~2%.

For Europe, two spatially disaggregated emissions inventories for CO_2 and CO from fossil fuel are available today. (1) EDGAR provides annual mean emissions for the base years 1990, 1995 and 2000 on a $1° \times 1°$ grid. (2) The Institute of Energy Economics and Rational Use of Energy (IER, University of Stuttgart, Germany) constructed a new hourly emissions inventory on a $50\,km \times 50\,km$ grid for the year 2000 based on national submissions to the UNFCCC. The two products significantly differ at the local to regional scale but the reasons are not yet fully understood. To further explore these, the methodologies on which the generation of the inventories are based need careful assessment. For instance, differences may occur

---

[4] For consistency, the *Expert Emissions* dataset of EMEP was used, which can be accessed online at http://webdab.emep.int/

because of the different coverage of emissions that have to be reported under the UNFCCC, excluding emissions from national bunkers, international shipping and suchlike. At this stage, choosing one or the other product as prior information in atmospheric models results in uncertainties as large as the uncertainty in the atmospheric transport itself (Chap. 3).

5.3   Concepts for Spatial and Temporal Disaggregation

5.3.1   Spatial Emission Patterns

National emission data, as, for instance, reported to the IPCC, can be further resolved by intersection with spatially explicit datasets, such as land-use data or digital road and railway maps. Whenever available, however, emission datasets with a better resolution (i.e. district [NUTS[5] 2] or municipality [NUTS 3] level instead of national or regional) should be used for intersection with land-use data to further improve the spatial resolution. Emission data on NUTS 2 or NUTS 3 level can be generated using statistical information with the respective level of resolution as allocation parameters. Such allocation parameters can be the number of inhabitants, number of employees in different branches, number of farms and animals and suchlike. Appropriate parameters have to be determined for every emission activity and correlated to emission amounts to be usable as allocation parameters.

According to their geographic structure, *point*, *line* and *area* emission sources can be distinguished. Emissions from each source type require different information for spatial resolution.

5.3.1.1   Large Point Sources

Typically, the following emission sources are considered large point sources. Their geographical location is clearly defined by their coordinates and the following selected sources have been identified and characterised for air pollutant emissions mainly

- Power plants with a thermal capacity $\geq 300\,MW$
- Refineries
- Nitric acid and sulphuric acid production plants
- Iron & steel plants with a production capacity of $\geq 3\,Mt/a$
- Paper & pulp industry with a production capacity of $\geq 100\,kt/a$
- Automotive paint shops with a capacity of $\geq 100,000$ cars per annum
- Airports with at least 100,000 landing-take-off-cycles per annum
- All other sources emitting more than $1000\,Gg\;SO_2/NO_x$ or more than $300\,Tg/a\;CO_2$.

---

[5] NUTS—*Nomenclature des unités territoriales statistiques*, official statistical nomenclature for spatial data in the European Union.

The last category includes, among others, district heating plants, waste incineration plants and various industrial production and combustion plants. Even though these sources have been categorised for emissions of air pollutants originally, many are relevant with regard to fossil fuel combustion and hence this information is equally relevant for the spatial resolution of carbon emissions.

The geographical coordinates of point sources can be taken, for instance, from centrally available information (e.g. the *European Pollutant Emission Register*, EPER[6], respectively the upcoming *European Pollutant Release and Transfer Register*, EPRTR) or can be determined by surveys with the help of country experts. For some specific sectors or emission sources, information is as well collected in the frame of legally binding regulations, such as the EC *Large Combustion Plants Directive* (LCPD) and *Directive on Integrated Pollution Prevention and Control* (IPCC).

Line Sources

Routes of transportation of passengers and cargo are line sources: roads, railways, ships and pipelines. For these line sources, geographic information system (GIS)-based vector data are available, mostly for trans-national railways, inland waterways, highways and federal roads. In addition to that, the INSPIRE[7] directive of the European Commission aims at the harmonisation of spatial datasets across Europe.

Road Traffic and Railway Traffic

Because of the high density of roads in urban areas, urban traffic is most often treated as an area source. Rural and highway traffic is attributed to line sources with the aid of a digital European road network[8]. The average length of a street section between two points of the digital road network is 1.5 km. Emissions from railway traffic can as well be allocated using the Environmental Systems Research Institute (ESRI) digital European network.

Shipping

Shipping includes activities at sea, in port and on inland waterways. The coastline separates national from international shipping. Emissions from national shipping are mostly much lower than that from international shipping. They usually only account for 0–2% of total CH_4 and N_2O emissions in European countries, but their share in CO_2 emissions may add up to 40% in some countries. It has to be noted

---

[6] http://www.eper.ec.europa.eu/eper/

[7] http://www.ec-gis.org/inspire/

[8] http://www.esri.com/library/brochures/pdfs/transporteuro.pdf

that international shipping activities are projected to increase steadily and significantly. For some pollutants (e.g. SO_2), international shipping is expected to be the main single contributor by 2030.

Emissions from inland waterway transportation can be located with the aid of land-use data. Emissions of SO_2, NO_x, non-methane volatile organic compounds (NMVOCs) and CO have previously been calculated by EMEP (Jonson et al. 2000) with a spatial resolution of 50×50 km from which the spatial distribution of emissions of greenhouse gases can be derived as emissions are proportional to fuel consumption.

Air Traffic

Emissions from air traffic occur during LTO's (landing-and-take-off-cycles, limited by a height of 1000 m and including activities on ground like taxi and idle) and during cruise (any air traffic activity above 1000 m). LTO's are allocated to airports and attributed to the category of large point sources.

Spatial disaggregation of cruising activities might in the simplest case be based on repetitive flight schedules assuming that flight routes are always in linear distance between take-off and landing site. According to Kalivoda et al. (1997), much more detailed data about air traffic movements are available from the European Organisation for safety of air navigation (EUROCONTROL) which will have to be examined in order to verify their usability.

Area Sources

All emission sources that can neither be attributed to point nor line sources are classified as area sources. Area sources comprise all activities which are diffuse, are attributed to a large number of small individual sources or have indistinct spatial patterns, for example, urban traffic, household emissions and small industrial and commercial plant emissions. National emission data (and where available, data on NUTS 2 or 3 level) can be spatially resolved by intersection with CORINE land-use data that are available for the European Union area with a resolution of 250 m $\times 250$ m. This dataset was generated in the scope of the CORINE[9]—Programme. The most recent dataset for the year 2000 can be obtained from the European Topic Centre on Land Use and Spatial Information[10].

5.3.2   Temporal Emission Patterns

Most often, no direct information about the temporal distribution of emissions is available from emission sources. The temporal behaviour of emission sources can be

---

[9] Coordination, Information, Environment

[10] http://terrestrial.eionet.europa.eu/CLC2000

represented by surrogate statistical data like fuel use, working hours, traffic counts and suchlike. Distinct time curves can be derived for individual installations, types of sources and whole source categories (see Table 5.5 for an overview of indicators).

For stationary sources, indicator data reflecting temporal patterns of operation are most often easier to obtain, than, for instance, for road transport or individual industrial processes. Figure 5.4 displays exemplary graphs for the hourly resolution

Table 5.5 Overview of different indicators that can be applied to distribute emission data over time

Sector	Indicator data for monthly resolution	Indicator data for daily resolution	Indicator data for hourly resolution
Power plants	Fuel use	Load curves	Load curves
Industrial combustion	Fuel use, temperature, degree days, production	Working times, holidays	Working times
Commercial, institutional and residential combustion	Fuel use, degree days	User behaviour	User behaviour
Refineries	Oil throughput, fuel use	Working times, holidays	Working times, shift times
Industrial processes	Production	Working times, holidays	Working times, shift times
Road transport	Traffic counts	Traffic counts	Hourly traffic counts
Air transport	LTO cycles, passenger and freight numbers	LTO cycles, passenger and freight numbers	LTO cycles, passenger and freight numbers

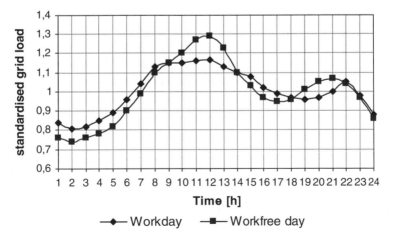

Fig. 5.4 Standardised grid load time-variation curves for a workday and a work-free day in Germany (based on load data from the Union for the Co-ordination of Transmission of Electricity, UCTE).

of power plant load for different types of days, however, the liberalisation of the
energy market and the variable production of power across the European grid at
any time has made it more difficult to obtain information on power plant operation
recently.

5.4 State-Of-The-Art and Current Capabilities of Fossil Fuel Source Mapping

5.4.1 Top–Down

One of the most used global emission inventory systems, the EDGAR[11] dataset,
uses a simple top–down mapping approach to distribute country total emission
estimates. EDGAR uses a population map (based on the GEIA-Li total population
map) modified to allocate all population within National Aeronautics and Space
Administration (NASA)–Goddard Institute of Space studies (GISS) country cells
and additional small entities. National emissions from fossil fuel combustion are
distributed to the 1×1 degree grid, distinguishing urban and rural population in
order to account for the allocation of industry and power generation emissions versus agricultural emissions.

The advantage of this approach is obvious, as population distribution data are
readily available for current and past years, and projections for future development
are likely reasonably accurate to allow for scenario calculations. However, the
drawbacks are that in particular patterns of emissions, which are not directly related
to the location of human population are not captured and point, line and area
sources are virtually indistinguishable (for instance, to reflect the height of the
emission source for modelling).

Clearly, at the ambition level of building a global inventory and at a resolution
of 1×1 degree, the top–down approach is robust and the availability of ex-post and
ex-ante data for emission distribution is a positive side effect.

In most cases, a relatively coarse spatial resolution will fit the purpose of modellers, for instance, for global climate change models and suchlike. But, where a
more accurate representation of emissions is required, for example, as input data
for atmospheric dispersion and chemistry transport models to assess the influence
of emissions on ambient air quality, more advanced methods are needed to distribute emissions. Hence, additional information to distribute emissions—again
starting from national total emission inventories—is used to map emission from
individual sources and source sectors. Based on methodology as described above
in Sect. 5.3, information on point, line and area sources is typically applied to
allocate specific sector emissions to grid cells. Figure 5.5 gives an example of this

---

[11] http://www.mnp.nl/edgar

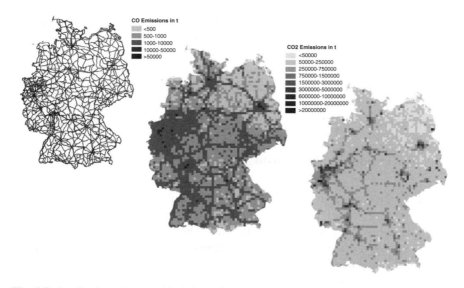

Fig. 5.5 Application of geographic information system (GIS) datasets to establish a top–down distribution of emissions, exemplary for a detailed road network of motorways and other major roads (left) for Germany in 2000. The centre figure displays the spatial distribution of CO and CO_2 is given on the right. With road transport being one of the dominating source groups for CO and CO_2, the road network is clearly visible, while small combustion sources for CO, in particular, lead to a higher influence of area sources compared to the case of CO_2.

approach, illustrating the influence of, for example, information on a detailed road transport network on the high-resolution mapping of emissions.

While the strength of this more detailed approach is evident, it does require considerably more data and information, such as traffic counts for major roads, location and operation of point sources and suchlike. This kind of information is most often available on national, district or municipality level, yet only ex-post and with considerable time lag; hence it is not suitable for projections.

5.4.2 Bottom–Up

Top–down approaches will provide a sufficient level of detail for most applications and data needs. For very high resolutions, both temporal and spatial, however, using bottom–up methodology for the generation of an inventory with built-in high resolution is preferable. This also applies to projections, when emission source strength is varied according to anticipated changes in behavioural patterns and economic development scenarios.

This implies combining information on emission factors and activity rates already at the highest temporal resolution available and requires statistical datasets reflecting local aspects, for example, the use of solid fuels in residential combustion or the utilisation of stationary or mobile equipment. Because of the extremely high data requirements, bottom–up approaches have rarely been applied in a consistent manner, with the exception of a few case studies, for instance, the Augsburg experiment (Slemr et al. 2002), which served as a test bed to validate emission models and chemistry-transport models against detailed measurements for a mid-sized town in southern Germany.

5.5   Uncertainties, Gaps and Robustness

In general, the uncertainties associated with the temporal and spatial resolution of emission data are closely related to those of generating emission inventories in the first place. Emission factors and activity rates, which comprise the main input factors, are of varying quality, depending on the statistical systems in place for reporting and collecting these datasets.

S = small (10%); M = medium (50%); L = large (100%); V = very large (>100%) '-' not applicable/negligible.

Various methods exist to assess the uncertainties along the chain of calculations for the generation of emission inventories, for example, using error propagation or Monte-Carlo simulations. These have been extensively discussed in literature (see, for instance, Kühlwein and Friedrich 2000, or Vogel et al. 2000) and can be found as well in the IPCC report[12] Good Practice Guidance and Uncertainty Management in National Greenhouse Gas Inventories. It is difficult to assess, to what extent the application of methods for temporal and spatial distribution of emissions increases the uncertainty range in comparison to the uncertainty associated with the national total figures. Provided, the same datasets are used and detailed methods are applied, the overall annual emission total of, for instance, a bottom–up emission inventory in an hourly resolution should result in the same emission figure, with the same uncertainty bandwidth. The accuracy spatial and temporal variation, though, is entirely dependent on the quality and completeness of information used for the distribution. This quality varies by source and region, with large point source activities (e.g. power plants) in Western Europe most likely being well documented in high temporal resolution, while private road transport emission variations or sources in Eastern Europe being more uncertain (Table 5.6).

Finally, recent developments, in particular, in the way public power generation is operated in Europe, with the deregulation of the energy markets, has introduced an additional source of variation. Before the liberalisation, national power grids and electricity generation capacities were operating in a significantly more predictable environment, hence making assessments of hourly, daily, weekly or seasonal

---

[12] http://www.ipcc-nggip.iges.or.jp/public/gp/english/

Table 5.6 Indication of uncertainty estimate for greenhouse gases

Emission source category	Activity data	Emission factor			Total emissions		
		CO_2	CH_4	N_2O	CO_2	CH_4	N_2O
Fossil fuel use							
Fossil fuel combustion	S	S	M	M	S	M	M
Fossil fuel production	S	M	M	–	M	M	–
Industry/solvent use							
Iron & steel production	S	–	S	–	–	S	–
Non-ferro production	S	–	S	–	–	S	–
Chemicals production	S	–	S	L	–	S	M
Cement production	S	S	–	–	S	–	–
Solvent use	M	–	–	–	–	–	–
Miscellaneous	V	–	–	–	–	–	–
Landuse/waste treatment							
Agriculture	S	–	L	L	–	L	L
Animals (excreta/ruminants)	S	–	M	L	–	M	L
Biomass burning	L	S	M	L	L	L	L
Landfills	L	–	M	–	–	L	–
Agricultural waste burning	L	–	L	L	–	L	L
Uncontrolled waste burning	L	–	–	–	–	–	–
All sources	–	–	–	–	S	M	L

Source: Olivier et al. (1999).

production more robust. In the new situation, the combination of time and location of power generated is difficult—if possible at all—to predict, and even ex-post evaluations are often restricted by production data being seen as commercially sensitive and thus not readily disclosed for statistical purposes. At the same time, technologies introduced, for instance, in telematics for road toll schemes and suchlike, which may help to gather datasets to improve the spatial and temporal resolution of traffic-related emissions in the future.

5.6 Strategy Towards Pan-European GHG Regularly Updated Emission Mapping

5.6.1 Constraints on the Development and Setup

Overall, the methodologies for the generation of emission datasets in high temporal, spatial and substance resolution has been established based on experience gathered from work on air pollutant emissions. For the time being, improvements of the resolution are mainly limited by computing time, but even more so by the availability and quality of input data. However, some constraints have to be noted.

For instance, the data collection on large point sources needs to be improved, for example, by reporting stack heights (usually information readily available) to better distinguish high and low sources and other parameters which are vital for modellers, but are typically not included as reporting schemes are designed based on the requirements from a regulatory viewpoint. Without this information, the plume rise and thus the effective induction of substances into the atmospheric layer within the dispersion model cannot be captured properly.

Secondly, there is a clear lack of reporting requirements for activity data on a recurrent, detailed sectoral basis. These would be required to facilitate the temporal resolution of individual industrial or power generation activities and even more so, for highly variable sources, such as private and public transport.

5.6.2   Future Directions for Enhanced Mapping

Different trends can be observed with regard to the availability of indicators for temporal and spatial distribution. While activities on mapping and earth observation increase the resolution of, for example, land-use data and the harmonisation efforts in the mapping realm, such as the European INSPIRE[13] initiative provide increasingly detailed datasets which can serve to distribute emissions spatially, the development is different for the information required to generate temporally resolved emission datasets. There, for instance, the deregulation of energy markets creates additional challenges for accurate spatial/temporal allocation of energy-related emissions. In addition to that, reporting requirements have increased significantly because of new directives and conventions entering into force, which has led to a gradual harmonisation of reporting guidelines. In this process, the high level of sectoral detail which had marked the reporting structure for air pollutants was watered down to match the coarser and mainly energy conversion-driven structure of the IPCC with significantly less detail.

Hence, two possible ways forward can be envisaged:

On the one hand, a relatively coarse distribution based on land-use data and using generic time curves to split annual total emissions into daily, weekly, monthly and seasonal patterns. This approach would be suitable for large-scale modelling applications, where a higher level of detail in the input datasets would not be required. On the other hand, for local or regional applications, for instance, to conduct verification experiments, it is vital to achieve greater accuracy not only with regard to the total amount of emissions over a year. In this case, a bottom–up construction of a spatially and temporally explicit emission inventory ought to be

---

[13] http://inspire.jrc.it/

considered. This would guarantee that no detailed information on emission factors and activity rates and their spatial and temporal variation would be lost by taking the detour via an inventory compiled and these data then distributed again.

5.6.3 Requirements for Ancillary Information for Scaling

As outlined above, the main requirements to improve the situation for—in particular—the temporal disaggregation of emission data would be a more comprehensive and consistent collection of activity data. In addition to that, the current trends in simplifying reporting structures to alleviate the administrative burdens of national experts, who have to report different datasets under several conventions and directives would need to be reversed. The reason for this is straightforward: with less-detailed information on emission source sectors retained in the emission inventories as they are compiled, a meaningful splitting of emissions is lacking vital indicators. If, for instance, emissions from power generation are not reported by the fuel types used in power plants, it is not possible to even distinguish base, medium and peak load plants with very different utilisation patterns.

Hence, including ancillary data for modelling purposes into the inventories, which are as of now driven by regulatory needs rather than the efficiency of data gathering for modelling would be beneficial. This could contribute to a significant improvement of model output quality, as well as reduce uncertainties to a large extent.

5.7 Conclusions

The temporal and spatial resolution of emissions from fossil fuels use are based on methodologies developed in the course of the last two decades, mainly aimed at the application to air pollutant emissions. However, the requirements are the same, as the determination of location and time of emission is a vital requirement when trying to verify model results with measurements. This is even more relevant, where measurements are marked by a high level of sophistication and detail, and where both anthropogenic sources and natural and biogenic sources may contribute to the overall ambient concentration of trace gases (for a more detailed description of the methodology, see Friedrich et al. 2003).

While methods are well developed and can be readily applied, the main problems lie in the considerable amount and detail level of information required, in particular, for high-resolution data generation. With reporting requirements often placing considerable burdens on the national experts compiling inventories, the level of detail in reporting that would facilitate the temporal and spatial distribution of emission data are not currently included in compulsory reporting. Only within the EMEP programme, some activity data reporting requirements have been introduced in recent years to satisfy the needs of Integrated Assessment Modelling.

Regarding the quality of emission inventories, in general, and temporally and spatially distributed emission data, in particular, uncertainties of indicator/surrogate data used for the distribution are often unknown. Typically, measurement errors are assumed to be at 20–30%, for example, determining emission factors. In addition to that, distributed emissions are subject to errors because of averaging of results for emission factors, incomplete information and suchlike (cf. Winiwarter and Rypdal 2001 and Rypdal and Winiwarter 2001 as well).

Uncertainty analysis needs to address uncertainty in the magnitude of emissions, as well as the uncertainty in trends of emissions (for future projections), and account for potential correlations between influencing factors. In this context, it is vital to be aware, that increased resolution does not always equal increased accuracy or quality of calculations.

References

Friedrich R, Freibauer A, Gallmann E, Giannouli M, Koch D, Peylin P, Pye S, Riviere E, San Jose R, Winiwarter W, Blank B, Kühlwein J, Pregger T, Reis S, Scholz Y, Theloke J, Vabitsch A (2003) Temporal and Spatial Resolution of Greenhouse Gas Emissions in Europe. Report from a Workshop held in Stuttgart. (http://gaia.agraria.unitus.it/ceuroghg/reportws3.pdf).

Friedrich R, Reis S (eds.) (2004) Emissions of Air Pollutants—Measurements, Calculation, Uncertainties—Results from the EUROTRAC Subproject GENEMIS. Springer Publishers.

Jonson, JE, Tarrasón L, Bartnicki J (2000) Effects of International Shipping on European Pollution Levels (EMEP/MSC-W, Note 5/00), Research Note no. 41 DNMI, Oslo.

Kalivoda MT, Kudrna M, Feller R (1997) Methodologies for Estimating Emissions from Air Traffic. Deliverable N° 18 for the CEC Funded Project 'Methodologies for Estimating Air Pollutant Emissions from Transport (MEET)'. http://www.inrets.fr/infos/cost319/MEETD eliverable 18.PDF

Kühlwein J, Friedrich R (2000) Uncertainties of modelling emissions from road transport. Atmospheric Environment 34, 4603–4610.

Olivier JGJ, Bouwman AF, Berdowski JJM, Veldt C, Bloos JPJ, Visschedijk AJH, Van der Maas CWM, Zandveld PYJ (1999) Sectoral emission inventories of greenhouse gases for 1990 on a per country basis as well as on 1° × 1°. Environmental Science and Policy 2, 241–264.

Rypdal K, Winiwarter W (2001) Uncertainties in greenhouse gas inventories—Evaluation, comparability and implications. Environmental Science and Policy 4, 107–116.

Slemr F, Baumbach G, Blank P, Corsmeier U, Fiedler F, Friedrich R, Habram M, Kalthoff N, Klemp D, Kühlwein J, Mannschreck K, Möllmann-Coers M, Nester K, Panitz HJ, Rabl P, Slemr J, Vogt U, Wickert B (2002) Evaluation of modelled spatially and temporally high resolved emission inventories of photo smog precursors for the city of Augsburg: The experiment EVA and its major results. Journal of Atmospheric Chemistry 42, 207–233.

Vogel B, Corsmeier U, Vogel H, Fiedler F, Kühlwein J, Friedrich R, Obermeier A, Weppner J, Kalthoff N, Bäumer D, Bitzer A, Jay K (2000) Comparison of measured and calculated motorway emission data. Atmospheric Environment 34, 2437–245.

Winiwarter W, Rypdal K (2001) Assessing the uncertainty associated with National greenhouse gas emission inventories: A case study for Austria. Atmospheric Environment 35(25), 5425–5440.

Chapter 6
Issues in Establishing In Situ Atmospheric Greenhouse Gas Monitoring Networks in Europe and in Regions of Interest to Europe

Euan Nisbet, Phillip O'Brien, C. Mary R. Fowler, and Aodhagan Roddy

6.1 Introduction: General Problems of In Situ Greenhouse Gas Monitoring

The atmospheric concentration of greenhouse gases can be measured in situ to great precision. However, can the emission and uptake fluxes of these greenhouse gases be inferred from these measurements? Just as a wolf sniffs the wind, so sources of emissions can be measured at every scale from local to global and then quantified by modelling. However, as monitoring equipment is usually static, measurements only apply to air masses which have passed through the station. Several problems emerge:

- Emission sources can be numerous and hard to distinguish. Seasonality may be muted. Thus, very high standards of precision are needed at the in situ stations accompanied by very careful inter-comparison between stations.
- At the risk of pushing the wolf analogy too far, pack hunters have advantages over solitary predators. A network of strategically placed in situ sampling stations can monitor and assess emissions within a region far more effectively than a whole series of independent stations acting alone.

In building an effective network of complementary in situ monitoring stations, the first problem is to ensure that the data from all the stations are *inter-compared*. This is not easy even for a single integrated network, but has been achieved by a multi-national network in Europe with careful round-robin programmes (e.g. Meth-MonitEUr 2005 for methane).

Local and specific emissions within access ('sniffing range') of stations can be determined by various techniques. Under certain conditions, the local natural ground source emission of ^{222}Rn can be used as a proxy for assessing footprint and inversion height—the atmospheric concentration of ^{222}Rn can, in effect, provide a measure of the volume into which the emission is diluted. Similarly on a local scale, 'ratios of mixing ratios' can be used very effectively (e.g. Levin et al. 1999; Lowry et al. 2001). If CO_2 emission inventories are relatively well-known in a city (e.g. from petrol sales and heating emissions), then the $CH_4:CO_2$ ratio gives the CH_4 emission (which may be harder to inventory) and $N_2O:CO_2$ can provide an independent check. The underlying assumption in both ^{222}Rn and CO_2 studies is that there is adequate knowledge of the emission rate of the proxy parameter.

Regional studies must be able to observe changes as measured against a changing background. Although the basic background information must come from global networks, such as the US NOAA (National Oceanic and Atmospheric Administration) programme, these have limitations particularly since their coverage of the tropical landmass is poor. Strong meteorological control on natural and anthropogenic sources, with meteorologically linked feedbacks, requires continued campaigns of process studies on both regional and local scales.

6.1.1   Key Issues

The difficulties of establishing greenhouse gas monitoring networks (Nisbet 2007) are frequently discussed at specialist workshops and meetings (e.g. Worthy and Huang 2005) but rarely mentioned in the scientific literature. Logistical problems such as shipping gas cylinders or maintaining high-quality data sets in remote locations are central problems of monitoring, yet tend to be taken as trivial by those using results.

Some key issues relevant to the establishment of a European network are

- Purpose of monitoring
- Choice of concentrations and isotopes to be measured
- How to measure
- Site location.

Then there are the data quality issues:

- Calibration of stations
- Inter-comparison around the network.

Next come the deeper issues:

- Usefulness: types of result
- Robustness of science
- Robustness of financial and political support.

Arguably, the robustness issues are the most important. The technology and methodology must be strong so that the network delivers good dependable measurements. Financial support for a network must be steady and sustainable—better not to attempt too much than to imperil long-term time series by a programme that is initially too ambitious.

Long-term monitoring of environmental systems is not glamorous and may not immediately deliver first-author research publications. But, it is vital if we are to understand and assist the health of our planet. It is impossible to go back and take measurements in hindsight. Change can only be observed if long-term monitoring time series are protected.

6.2 Purpose of Monitoring

6.2.1 Why Monitor?

6.2.1.1 To Understand the Global Biosphere

The global biosphere cannot be understood scientifically unless the carbon transfers can be measured. Methane has a lifetime of roughly a decade and CO_2 emissions are in the air for centuries. Thus, carbon gases have a 'topography' across the planet that is comparable to the distribution of temperature across the planet (although with the difference that there is a strong N-S gradient). Although carbon gases are well mixed on annual or decadal scales, there is strong seasonal and latitudinal variation from which much information can be derived.

At present, this can only be achieved with in situ monitoring that provides high-quality regional and global data sets from a number of carefully inter-compared sites. Satellite data (e.g. the Orbiting Carbon Observatory, or OCO, Sciamachy) are becoming a vital adjunct to in situ measurement but for many years, satellite measurements will remain both less precise and less accurate than ground-based measurements. Moreover, satellite measurements need to be 'ground-truthed': in situ and satellite work are mutually supportive.

In the late 1950s, the Mauna Loa record first began to show the 'breathing' of the planet (e.g. Keeling and Whorf 2005). Now 50 years later, total global carbon-gas budgets are reasonably well-known, though surprises still emerge and major sources and sinks remain poorly quantified. Studies are beginning to assess regional-scale budgets. On a local scale, carbon can be tracked isotopically from anthropogenic sources (e.g. urban power stations) and distinguished from other sources. For example, European landfill emissions of methane can be constrained by isotopic monitoring. This is a powerful method—grass burning in Southern Africa can be detected by sampling methane in New Zealand (Lowe et al. 1994).

With an inter-compared in situ monitoring record from a modest European and global network, coupled with the wide geographic scope of satellite monitoring, and supported by detailed local, regional, and global modelling, it should be possible to track and to quantify by sources and seasons the fluxes of carbon gases from major natural sources and sinks, not only in the wealthy northern lands but also in the tropics and oceans. Although single sites are very useful (e.g. Biraud et al. 2000, 2002; Ryall et al. 2001), an integrated network is needed to support regional inversion.

6.2.1.2 To Verify Emission Declarations

CO_2 and CH_4 emissions are declared under the UN Framework Convention for Climate Change (UNFCCC) and the Kyoto Treaty. Presently, carbon emissions are self-declared. This is done by the emitters at almost all levels, from farmer to European Union, by collecting 'bottom-up' statistics (e.g. numbers of eructating cows, tons of coal burnt). Such a self-declaration system is analogous to income tax being declared and paid with no checking mechanism. If the Kyoto process is to succeed, there is an urgent need for an auditing process that is independent of information provided by the emitters. That auditing process is atmospheric monitoring (Nisbet 2005).

'Trust but verify', said Ronald Reagan, citing an old Russian saying. Verification ensured the success of the Nuclear Test Ban Treaty, which set up the World-Wide Standardised Seismograph Network to detect nuclear tests (and was unexpectedly a significant factor in the understanding of Plate Tectonics). UNFCCC and Kyoto lack verification: there is no independent top-down (atmospheric) validation of emission declarations. The aim of verification should be to quantify national emissions by source type, season and location, and within nations to provide independent checks on the veracity of national inventory declarations of emissions.

6.2.2   What to Monitor? What Proportion of the Effort Should go to Each Species? What Should be the Balance Between In Situ and Satellite Monitoring?

The purpose of monitoring dictates the species selected. Chlorofluorocarbons (CFCs) are already well monitored under Montreal protocol rules. CFC sources are now largely controlled. What should be monitored in situ under a Kyoto-focussed programme? CO_2 and CH_4 mixing ratios are obvious targets, but beyond them, the choices are complex given the limited finance available. How much support should be given to isotopic monitoring? Does monitoring tropospheric ozone need support, or is it sufficient to study methane. What about hydrogen, which is an indirect greenhouse gas? What about N_2O, which is a 'low-hanging fruit' for emission controls but is badly neglected by monitoring?

Perhaps the best immediate answer is that the effort (defined as financial budget) should be apportioned between gases according to total radiative impact. On a short-term or present-day scale, methane has roughly half or more of the effective direct and indirect impact of carbon dioxide (Hansen and Sato 2001) and is much more easily reduced. If so, then methane monitoring should have half the budget allocated to CO_2. The next question concerns isotopic measurements—how much effort should go on these? That question is less easily answered. Isotopes offer insight into sources, so that not only can fluxes of gas be measured but also the specific types of source (e.g. methane gas leak or landfill) can be quantified. But isotopic work is expensive.

The best division of resources between in situ programmes and satellite programmes is also hard to decide. Many programmes have a rough rule-of-thumb of 50:50 allocation, which seems sensible. Europe has excellent satellite launching abilities that has led to strong support for satellite work and also for techniques, such as upward-looking FTIR (Fourier-Transform Infrared), that are used to ground-truth satellite data. However, this has had the effect of terminating or suppressing in situ work, even though that in situ work may be more accurate by a factor of 100. For example, sunlight-using satellite infrared observations can assess neither the Arctic in winter nor Europe by night. Also, they do not have the precision to observe absolutely the seasonal and diurnal methane cycles—only in situ chemical analysis is able to do this.

It is therefore essential that satellite and in situ observation work in tandem— both satellite and in situ studies need the support of the other.

6.2.3   Where to Monitor? Global or Regional or Local?

There is presently one major global monitoring network, the US NOAA network. NOAA has a good basic global network, better in the Pacific hemisphere than the Old World. Australia and France help on a semi-global scale while other national programmes (e.g. Germany, Japan, New Zealand) have long reach outside their own nations. There are important national programmes (e.g. Canada, China, Finland) but as a whole Europe is struggling to create integrated regional networks (but see Chap. 18). The Northwest Europe marine/coastal boundary layer is relatively well studied, as is the French sector of the Indian Ocean, but the Mediterranean and former Soviet Europe are not.

Monitoring networks need both rigorous calibration and inter-comparison between sites since without these, the data are useless. Badly managed sites and badly measured data are worse than useless since using poorly inter-compared or simply imprecise measurements in models make the models worse, not better, at capturing reality.

The NOAA monitoring programme very sensibly chose to set up a few continuous monitoring sites, mostly of marine boundary layer air and then supplemented this with flask collection. Mauna Loa (Scripps Inst.) is a site that samples free troposphere, but elsewhere the free troposphere can be monitored from aircraft, as in the CARIBIC programme.

Much of the globe's land area is not monitored. The Middle East and North Africa, major sources and consumers of oil and gas, have virtually no monitoring. In particular, the giant gas fields of the Gulf and Iran are little sampled for methane emission. The Indian subcontinent, potentially one of the largest coal-burning economies, has little monitoring apart from French efforts, and the huge source regions of Southeast Asia are essentially unmonitored.

Continuous monitoring in Africa is at the excellent station on Cape Point (South Africa). This is superbly located for observing the Southern Ocean, but less well located for sampling African air. It rarely, if ever, experiences air from the Congo (which is as far from Cape Town as Uganda is from Cyprus). There are only a few flask-sampling programmes in sub-Saharan Africa, apart from measurement on Mt. Kenya, now suspended. Much of Latin America is unmonitored. Generally, regional and local studies are badly needed in the tropics. In general, monitoring observations are not yet good enough to support regional modelling in many areas of emission.

Good local information and local meteorological knowledge are vital if in situ data are to be used in modelling. Global wind fields are now well-known. However, in large parts of Africa and other tropical regions, local meteorological data are not especially accurate and, in many places, long records have now been discontinued. Moreover, unmapped local sources may exist (e.g. new local landfills). Thus, interpreting both integrated and spot flask samples from these locations can be problematic.

6.2.4   Hotspots

In addition to collecting data from background sites for global and regional studies, there is also a need to monitor local high-emission 'hot-spots' (Levin et al. 1999; Lowry et al. 2001; Zinchenko et al. 2002). These have been much neglected in past work. For example, the commonly cited global average values for CO_2 and CH_4 mixing ratios are for marine background settings. If the landmasses are included, global values are higher.

Heavily populated areas are major source areas, where both point sources (e.g. power stations) and disseminated sources (e.g. cars) are significant. In such areas, either tall towers can be used to collect from various heights in the boundary layer or monitoring can be carried out at carefully located well-characterised sites. Location also needs care to ensure a well-distributed network that samples all the major source regions nearby: this means siting according to the wind spectrum.

In major cities, simple inexpensive monitoring (e.g. a single station running continuous real-time carbon gas and isotope measurements and ^{222}Rn) can, with good modelling, potentially assess emissions in real-time from an area with a foot-print of say 50–100 km, containing millions of people. The data are especially useful if assessed in conjunction with emission inventories. Alternately, they can be used to test inventories (Baggott et al. 2005). Emission inventories may be quoted to extremely high precision (e.g. London's inventory is quoted to nine significant figures or the nearest 100 kg (a single truck's diesel tank) (Buckingham et al. 1998)). However, they can at the same time be very inaccurate.

In poor-nation megacities, emission inventories are rare or may be absent. Measurement may be the simplest way of assessing emissions, but a potential major problem arises in many areas of major emissions, as knowledge of the height of the mixed layer is frequently not good enough to support reliable emissions studies. Here, ^{222}Rn or sodar can be very useful.

6.2.5   When to Monitor?

The obvious answer is frequently as possible, but this is not realistic since not every site can have a continuous instrument, and there will always be severe limitations on funding. Society may well prefer to support a medical clinic than a greenhouse gas station. Thus, any monitoring programme needs to be tailored very cautiously to sustainable long-term budgets. This is a key point and will be returned to later below—it is better to run a sustainable and reliable long-term Smart Car time series than to use erratic funding for stop-go Rolls-Royce programmes.

6.3   Criteria for Locating Monitoring Sites

Scientifically, the key problem is to locate monitoring sites in the best places to support modelling. How can the errors in representing actual greenhouse gases in the air be reduced in models? However, financial constraints often overwhelm scientific priorities. Given that financial support will only permit a certain number of sites, where are the places from which data sets will maximise the quality of the model and minimise errors compared to actual greenhouse gas distributions and sources? There are both intuitive and quantitative answers to this.

Intuitive answers would include classifying the air masses into broad categories—marine background air, continental interior background air, polluted hotspots, etc. However, the problem is more complex: specific sources will need to be identified if reduction programmes are to be assessed and this means that major source regions need to be studied with more care, either from within the region itself or downwind from it.

Quantitative answers come from the use of models to choose locations. Simple back-trajectory studies are particularly useful in assessing locations best suited to sample air from different source regions and background regions. Trajectory modelling can also be used to search for likely trajectories that cross two or more sampling points, and hence can directly sample more than one key emission area. Ideally, from new continuous sites, both major source locations and background air streams need to be accessible.

In order to determine which are the key locations for measuring particular sources, mesoscale forward models can be used, to search for the ability, from knowledge of carbon gases at surrounding points where measurement is already occurring, to reproduce greenhouse gas mixing ratios and isotopes at potential study sites. If the models are successful, there may be little benefit in sampling at a potential study site: the new data will add little to understanding.

Realistic sites are, however, not perfectly distributed and individual sites may appear poor in both intuitive and quantitative tests. The key need is for long-term sustained high-quality measurement. In practical terms, this cannot easily be delivered unless there is a strong long-term commitment to measurement. This will depend on managerial and political factors, not science. Thus, the USA, with a very long-term commitment to NOAA, can maintain a number of very high-quality baseline sites, while other nations may maintain one or two such sites and a few flask sites.

A long-term European network needs to be built on a regional/continental scale. The European dimension extends to the gas fields providing Europe's gas from

Northwest Siberia (a major methane source). Europe also has responsibility for sharing in African monitoring. The sampling strategy must reflect this extent: in other words, the network design must support modelling but yet not be so expensive that it cannot be financially supported for the long-term.

Future societal needs may need additional local-scale studies. This will require new stations. New methods and new instruments will be developed, each demanding new monitoring strategies. It is therefore vital for a sustainable network design to contain an integrated research component that will deliver continuous network improvement.

6.4   Present Site Location: European and Global Networks

While network sites should ideally be chosen on purely meteorological grounds, in practice this is unrealistic. The main site selection for US and northern European programmes was made long ago and the best future policy is to build on the existing network of stations. The equipment and skill investment should exploit the valuable heritage of time-series already accrued (Figs. 6.1 and 6.2).

Although these stations may not seem at first glance optimally sited in terms of atmospheric need, in pragmatic terms they are indeed optimally sited. This is because they have grown up naturally and have support over the long-term. Each station has sustaining factors. Some are close to key universities or institutes, or in

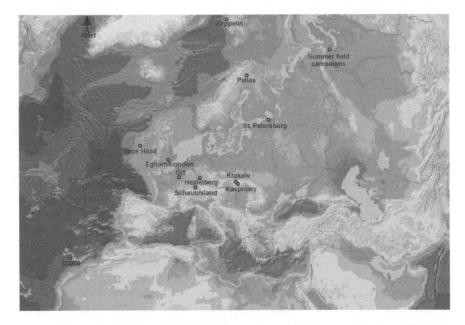

Fig. 6.1 Some existing CH_4 and $\delta^{13}C$-CH_4 measurement sites in Europe, Siberia and the Arctic. Reproduced from Meth-MonitEUr (2005), with kind permission from Meth-MonitEUr (2005) (*See Color Plate* 1).

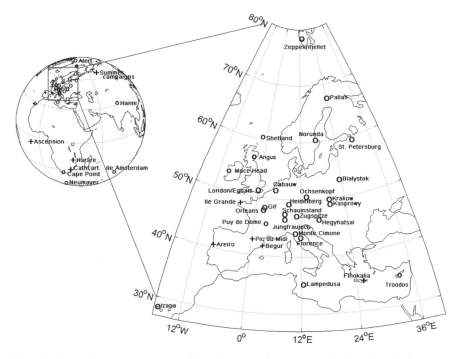

Fig. 6.2 Sites of some present and likely future carbon gas monitoring sites by European groups: circles, already in operation; crosses, proposed/under establishment. For global coverage, see map in Nisbet (2007).

locations of national interest. Many of these sites have a multi-year record of observation, representing a long-term financial investment. The records from the stations may not be well inter-compared or even well calibrated, but data exists. Hence, there is a need, as far as possible, to back-integrate these data sets by inter-comparison of archived air.

Local design of stations clearly depends on purpose. While tall towers on 'typical' flat land are ideal for characterising reasonably large footprints, major emissions often occur in port or river cities and locations with complex topographies (e.g. oilfields).

Microwave aerial design for cellphone telephony depends heavily on detailed local topographic modelling. This is a useful parallel. In assessing local emissions, there is a need for more thought about linking similarly sophisticated local topographic data into meteorological models of monitoring, so that footprint areas around continuous sites can be identified and fluxes assessed on a real-time basis.

The present array of nationally and internationally supported stations (Figs. 6.1 and 6.2) is strong in northwest continental Europe, and has Atlantic stations at Mace Head, Ireland, Ny Alesund, Svalbard, and Izaña (Tenerife), but is weak in southern and eastern Europe, the UK, and Iberia.

The far eastern Mediterranean, proximal to the Middle East oil and gas industry, is a major gap, as is the southern Mediterranean, near the large North African gas industry that supplies in many parts of Europe. In particular, the network lacks continuous stations in many key regions. Much is currently being done to improve this position. The Chiotto Tall Towers (part of CarboEurope) are adding important new geographic input (e.g. Bialystok, Angus). For methane, now there is also better integration between AGAGE-linked and NOAA-linked stations. Flask sampling is being used to extend geographic spread (e.g. French work in Pondicherry, South India, and a new Cyprus station in Geomon).

6.5   Developing A European Network—New Site Selection

Gaps in the network need to be filled to permit a full European understanding of regional methane budgets at a level that allows quantitative assessment of European sources. New sites should be co-sited with other gas and meteorological measurements. As mentioned above, there is a particular need for new stations in the Mediterranean and in Russia (Fig. 6.3).

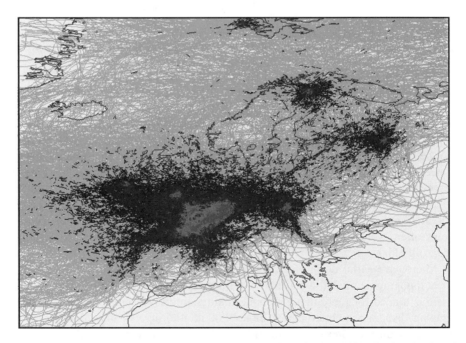

Fig. 6.3 A composite of daily 12.00 GMT 5-day back trajectories for 2002 indicating density of European methane coverage. Sites in Network are Heidelberg, Kasprowy, Mace Head, Ny-Ålesund, Pallas, Paris, Puy de Dome, RHUL, Schauinsland, and St. Petersburg. Red represents the highest density and yellow the lowest. Reproduced from NUI Galway, report to Meth-MonitEUr, with kind permission from Meth-MonitEUr 2005 (*See Color Plate* 2).

6.5.1 Trajectory Studies with the HYSPLIT 4 Model

Relatively simple conclusions can be reached about site locations using the HYSPLIT 4 (Hybrid Single Particle Lagrangian Integrated Transport) model, which was developed at the Atmospheric Research Laboratories, Maryland and designed to support a wide range of simulations related to the long-range transport, dispersion, and deposition of pollutants (Draxler 1992; Draxler and Rolph 2003; Rolph 2003). The calculation method is a hybrid between Eulerian and Lagrangian approaches: advection and diffusion calculations are made in a Lagrangian framework while concentrations are calculated on a fixed grid.

The network of existing methane sites has been assessed in terms of the mutual or shared-back trajectories, which allows assessment of the degree to which a particular station might contribute to a modelling programme for estimating regional emissions. When two (or more) sites share a back trajectory, then measurements from the first station can initialise the model and represent the air mass conditions prior to sampling at the second station (Figs. 6.3 and 6.4). Any differences between the sites are then ascribed to emissions, sinks and other processes that have occurred as the air travelled between sites.

Sites on the western fringe of Europe have relatively few shared-back trajectories arriving from other sites due to the prevailing westerly airflows in the North Atlantic region. However, they commonly share back trajectories with other stations further east,

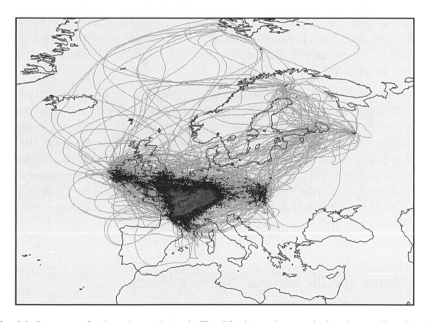

Fig. 6.4 Segments of trajectories as shown in Fig. 6.3 where prior to arrival at the sampling site, the air mass has passed within two degrees of another site in the network. Reproduced from NUI Galway, report to Meth-MonitEUr, with kind permission from Meth-MonitEUr 2005 (*See Color Plate* 2).

thereby the western stations can often act as the background or initialisation, from which to investigate the emissions in regions between stations. This is especially important as measurements made at western marine or coastal sampling stations tend to be representative of a larger region than measurements taken inland. This shared trajectory perspective shows the gaps in the coverage of the present network: the Mediterranean region and northern and southern Europe are poorly covered (Fig. 6.4).

Certain criteria can be identified to assist in selecting sites for new stations within the network.

(a) Prospective sites should be chosen on the basis that they should already have some limited air-sampling programme, or manned synoptic meteorological observations. A good example might be that the site is already part of the NOAA flask-sampling programme. The sites in Areiro (Portugal) and on the island of Lampedusa are good examples (Fig. 6.2).

(b) New sites should provide additional coverage to the present overall network. The sites of Areiro and Lampedusa would effectively add coverage of Iberia and southern France. Other potential sites in the eastern Mediterranean (e.g. Crete, Cyprus) appear largely independent of the larger network; typically, with air masses in this region neither originating nor finishing in Western Europe. Thus, they would extend the network significantly eastwards but not add to the knowledge of the present regions.

(c) As already mentioned, the site should not be already redundant—that is, the site must add new information, not already supplied within the network. New sites should target regions where at present coverage is sparse. However, this does not automatically exclude new sites which are geographically close to others, as topography has a profound influence on the movement of air masses: proximity of Alpine monitoring stations does not automatically make them good sites for monitoring northern Italy.

One purpose of the European monitoring network may be to provide data useful in predicting chemical weather in Europe. Figures 6.3 and 6.4 show that the present network does not allow for any robust statistics on how regional fluxes in eastern and southern Europe change with time. Moreover, with the exception of the heavily populated area in Northwest Europe, many locations may only be overpassed by one trajectory during an entire year (Fig. 6.3). For station-to-station trajectories, during the entire year, the region touched by more than one trajectory is yet smaller (Fig. 6.4). However, although rare, such individual trajectory studies are useful for case studies, especially if isotopic data are available.

6.6 Some Specific Issues in Network Design

It is assumed that long-term in situ stations should measure continuously for CO_2, CH_4, CO and (ideally) N_2O and H_2 mixing ratios. In addition, stations should gather local weather data (especially local winds in anticyclonic periods of nearly still air).

To be useful, all measurements should attain UN WMO Global Atmosphere Watch (GAW) quality standards, with regular inter-comparison with other stations. If stations do not supply these measurements, there is little purpose in monitoring except for specific campaign studies.

6.6.1 Source Identification—Trajectory Studies

Trajectory or transport models allow source identification in individual packets of air, in specific episodes. This is useful both in Kyoto-related studies (testing emission inventories and their seasonality) and also in biosphere studies. In testing inventories, such work is best at testing ratios of declared species rather than quantitatively determining fluxes, because of the uncertainties in local meteorological modelling and flux distribution. Individual studies collate to give semi-quantitative information about source characteristics and fluxes if multiple long-term trajectory records are gathered. The observational network thus needs enough stations to cover major sources and transport directions, as well as a few 'background' stations, especially those monitoring marine background inputs, and stations peripheral to the region of interest.

6.6.2 Isotopic Measurements

Isotopic measurements need precision that is adequate to differentiate between specific remote sources, and also to measure seasonality in the long-term record. Remote sources are studied using high-volume, high-precision spot sample studies (e.g. for methane, circa 30–100 L of air, better than 0.03‰ precision in $\delta^{13}C_{CH4}$). This high-precision work necessarily has infrequent sampling, if collected in flasks. Integrated sampling over 2-week periods remains valuable for characterising isotopic ratios in ambient air over large remote areas (e.g. Arctic).

Quasi-continuous isotopic measurement capability has recently been developed (Fisher et al. 2006). It is likely that continuous isotopic monitoring for CO_2 and CH_4 will soon be invaluable in quantifying emissions from major source regions. Low-volume intermediate-precision work (e.g. <100 mL, 0.05‰ precision in $\delta^{13}C_{CH4}$) can be run automatically every 30 min. However, these intermediate precision isotopic studies do not yet have adequate precision to monitor tropical seasonality for CH_4, nor to spot remote sources by back-trajectory study.

The issues of cost and reliability will be problematic if automatic isotopic monitoring instruments are placed at remote monitoring stations. Short-run continuous data sets have the same interpretive problems as the single station concentration data. However, in principle, we can create a relatively inexpensive isotopic network acquiring robust average gradients over time from a relatively sparse (e.g. several 100 km between stations) array of stations appropriately sited around the European region.

Although continuous measurements will be useful in areas of major sources such as urban centres, collection of integrated isotope samples (over say 2 weeks) remains necessary to determine regional gradients in a monitoring network. Integrated (large bag) sampling techniques have become 'unfashionable' but have clear assets in cost effectiveness when taking into account the limits of interpretive capacity the meteorology imposes.

For flask samples (event sampling) to be useful, they must be frequent. If collected too infrequently (e.g. the NOAA network one per week), data are not sufficient for rapid determination of gradient shifts in European weather systems. In a region with large variability on time-scales shorter than the sampling frequency, there will be a large element of randomness built into the data set. The flask sampling programmes are, nevertheless, extremely important both for giving global average pictures and most significantly to provide a completely independent verification of the performance of long-term monitoring instruments.

6.6.3 Vertical Measurement

Greenhouse gases are emitted on the surface. Vertical measurement (e.g. FTIR) is essential but the core of the Kyoto problem is on the ground. The circulation of the mid-troposphere and upper-troposphere around the planet in mid-latitudes takes a few weeks. This means that relatively sparse sampling at altitude is enough to define the fast-moving troposphere: more sampling will not necessarily provide much more information about sources. However, upward mixing from sources is rather slower—thus in order to ground-truth satellite data, regular from-ground upward-looking measurements will be needed. Such upward-looking vertical measurement, made from a few selected meteorologically simple sites, is vital for characterising the free troposphere.

The key problem with satellite studies is that they measure total column while the 'action' in greenhouse gases is in the mixed layer and low troposphere. At present, the usefulness of satellite measurements remains limited and there are significant problems in converting vertical column densities to mixing ratios. In particular, cloud parameterisation remains a problem, especially in the tropics. However (e.g. Frankenberg et al. 2006), the methodology is fast improving, the technique will be much more powerful when OCO is operating, and will be useful in sustaining global inversions. Thus, there is also a need for upward looking FTIR ground-truthing in difficult sites in the cloudy tropics, co-located with good in situ conventional measurement of carbon gases.

Satellite-mounted instruments, looking downwards, must have a high precision and also accuracy to be capable of detecting and discriminating between emission sources. The retrieval of information is improving rapidly, but the method remains unable to validate emission inventories. It is very difficult to determine diurnal and seasonal emission changes by total column observation. For methane, satellite measurements have a precision roughly two orders of magnitude poorer than ground measurements and can barely detect diurnal variation.

There are thus severe inherent problems with satellite studies. Infrared-based satellite work depends on sunlight and is thus not possible at night or during the polar winter. The Arctic is a key region of winter anthropogenic methane emission from the major gas fields, and is arguably the location of the greatest 'abrupt change' global climate risk (clathrate degassing). But it is dark in winter. In heavily populated areas, major emissions (e.g. heating) occur at night, compromising the usefulness of light-based sensing in the nocturnal hemisphere.

Summing up, in the near future, some satellite work (e.g. OCO) will indeed become very useful, dependent on better cloud parameterisation. Rigorous ground-truthing is essential, with a need for upward-looking FTIR instruments. Thus, in situ monitoring will be essential for the foreseeable future. Clearly, satellite data and vertical profiles are useful where available, but the backbone of a greenhouse gas sampling network must be built on surface monitoring stations unless new break-throughs in vertical sampling technology or remote sensing techniques occur.

6.7   Case Study—The South Atlantic

The South Atlantic is a region where better carbon gas monitoring is urgently needed. Le Quere et al. (2007) showed that the Southern Ocean sink of CO_2 has decreased between 1981 and 2004 by 0.03 PgC/year per decade (5–30% per decade), in spite of the large increase in atmospheric CO_2. They comment that data are 'particularly sparse in the Southern Ocean, where the magnitude of the CO_2 sink is heavily disputed, its inter-annual variability is unknown, and its control on atmospheric CO_2 during glaciations is firmly established but still not understood nor quantified'. There are very few stations in this region. The main data stream is from a single high-quality continuous station at Cape Point, supplemented by NOAA flask sampling from Halley Bay, Antarctica and Ascension Island.

This region is clearly a target for much better monitoring and is a European political responsibility. The main islands of the South Atlantic (Ascension, St. Helena, Falklands, S, Georgia, Tristan, Gough, and Bouvet) are all territories of European nations (UK and Norway), although UK-owned Gough Island is under South African administration.

Three sites were chosen for investigation—Ascension, Falklands, and South Georgia. Other sites considered were St. Helena, Tristan, and Gough Island. Each of these has merits, but in the context of the excellent Cape Point observatory, the study focussed on Ascension and the islands on the west of the ocean.

Ascension Island is located at 8°S, close to the submarine Mid-Atlantic Ridge. The island is a relict volcano, with a modest peak. The prevailing winds are the SE Trades, which are almost invariant. Only in *El Nino* conditions do occasional thunderstorms occur when the Inter-Tropical Convergence Zone drifts over the island. Figure 6.5 shows representative trajectories. It is clear that the low-level air sampled on Ascension is representative of the wider South Atlantic. Ascension Island is, therefore, an excellent site for monitoring the

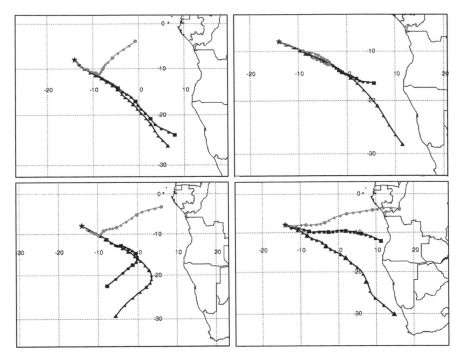

Fig. 6.5 Representative 5-day HYSPLIT back-trajectories for air arriving at Ascension Island on 16 Jan. 2006 (top left), 16 March 2006 (top right), 16 June 2006 (lower left), 16 Oct. 2006 (lower right). The three trajectories show air arriving at heights of 100 m (triangles), 500 m (squares), 1000 m (circles).

ambient South Atlantic, as far south as say 25–30°. There is however a steady flow of mid-troposphere (above 3 km) air from the east, including plumes from the biomass burning areas of Angola and Zambia/Congo. Upward-looking FTIR based on Ascension would monitor this, and may be useful in ground-truthing OCO. This air is also accessible via sampling on regular scheduled aircraft (Royal Air Force charter flights).

The Falklands Islands (Stanley 51°S) are a continental plateau. Figure 6.6 shows typical trajectories for air arriving at the Falklands. While much of this air arrives in the southwest to northwest sector, air also comes from the southern Pacific straight through the Drake Passage, directly from the south (Antarctica) and from the north (Southwest Atlantic). This is an excellent site for sampling air from southern South America, from the Drake Passage to the Antarctic penninsula region, from those parts of the South Atlantic to the north and east of the Falkland Islands, and from the southern Pacific across towards New Zealand.

South Georgia (54°S) is a remarkable ridge (up to 2 km high) straddling the westerly flow round Antarctica. Figure 6.7 shows typical trajectories. Much of the air comes through the Drake Strait between South America and the Antarctic Peninsula. However, the site also has access to air from eastern Argentina and air that has looped round the local Southern Ocean.

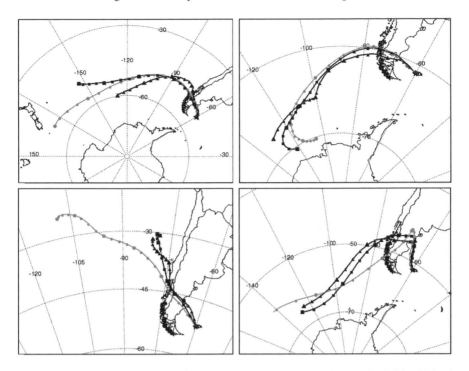

Fig. 6.6 Representative 5-day HYSPLIT back-trajectories for air arriving at the Falkland Islands on 16 March 2006 (top left), 18 March 2006 (top right), 16 May 2006 (lower left), 16 July 2006 (lower right). The three trajectories show air arriving at heights of 100 m (triangles), 500 m (squares), 1000 m (circles).

Each of these potential sites has specific advantages for improving South Atlantic monitoring. Logistics are particularly important in this region: there is a regular flight from Oxford–Ascension–Falklands, but in contrast, communication to South Georgia is less regular. Given the presence of the current Cape Point station and the flask sampling at Halley Bay, the prioritisation might be Ascension and Falklands first followed by South Georgia later.

6.8   The Need for Continuity

Long-term time series are needed to understand the great variability in the emissions and uptake budgets. Continuity is essential. Carbon-gas emissions come from many unexpected sources, and CO_2 sinks vary greatly year-on-year, while methane sinks (mainly OH) are more easily calculated. All these, sources and sinks, are closely related to climate and local meteorology. A cold spring can greatly delay carbon uptake over the whole of northern Eurasia, noticeably biassing the global pattern for a year. Anthropogenic sources and sinks are also meteorologically

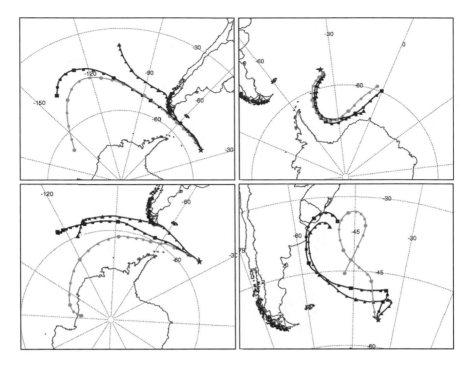

Fig. 6.7 Representative 5-day HYSPLIT back-trajectories for air arriving at South Georgia on 16 Feb. 2006 (top left), 16 June 2006 (top right), 16 July 2006 (lower left), 16 Nov. 2006 (lower right). The three trajectories show air arriving at heights of 100 m (triangles), 500 m (squares), 1000 m (circles).

biassed, both for heating and for human activities. Reservoirs of carbon can also respond to change—for example, peat carbon sequestration over the last 10,000 years has been immense but can be given back to the air if peat decays under global warming and drying. Similarly, clathrate storage of methane in the Arctic is very vulnerable to warming. Soil carbon stores in the tropics are large: these too are not only meteorologically dependent but also very directly controlled by human activity (e.g. deliberate fires). Many such stores can be released catastophically (eg. peat fires; large unstoppable underground coal fires in China). All these can only be tracked if long-term records are held.

Long-term change in major regions with large known carbon inventories (e. g. clathrate hydrate regions, forests, peatlands, and soil carbon in the tropics) can be studied by sustained measurement of regional fluxes and by assessing inventory changes. For example, back-trajectory studies from Ireland can assess the gas and isotopic mixes of Canadian forest fire outputs in Quebec. Very sudden change is probably easier to measure than slower but cumulatively larger change—thus emissions from a forest fire can be quantified more easily than a

10 year forest die-back. Fossil fuel burning is in principle probably one of the easier sources to quantify—isotopic ratios are known as are geographic locations.

Assessing the response of all these sources to climate change may be difficult. For example, quantifying future Arctic methane clathrate release may depend on comparative history control studies in areas where such release is not now occurring: but how to identify such areas? Present terrestrial inventory studies for emissions of the main greenhouse gases (CO_2, CH_4, N_2O) as well as CO and H_2 may still be very poorly defined.

6.9 Conclusion: Requirements for a Monitoring System

To be sustainable and cost-effective, a scientifically and socially useful European monitoring system needs

a. Mixing ratios measured by continuous high-precision ground-based monitoring from a network of stations with adequate spacing to cover the region
b. High-precision isotopic measurements from the same sites, either continuously or as integrated or very frequent spot samples
c. An associated long-term modelling and interpretation programme which would also routinely calculate total column estimates
d. High analytical quality, to WMO/IAEA GAW Expert Panel standards, the quality being audited and controlled via rigorous and regular inter-comparsion programmes
e. In addition to the continuous sites, flask sample programmes from distant sites (e.g. Africa), subject to the same QA/QC
f. Vertical profiling by upward-looking high-resolution FTIR, routinely assessed against the modelling results
g. Routine satellite ground-truthing against total column models and FTIR results
h. Regular aircraft monitoring to assess the free troposphere, plus regular but infrequent stratospheric sampling (e.g. Arctic, where the stratosphere is more easily accessible)
i. Monitoring programmes in key areas (e.g. major cities, major gas fields, or coal fields) for continuous assessment of major local sources

Most important of all is long-term financial support, so that once started, monitoring time-series are maintained and not abandoned.

Acknowledgements The authors gratefully acknowledge the NOAA Air Resources Laboratory (ARL) for the provision of the HYSPLIT transport and dispersion model and READY website (http://www.arl.noaa.gov/ready.html) used in this publication. We are grateful to Tuula Aalto for assistance with Fig. 6.2.

References

Baggott, S.L., Brown, L., Milne, R., Murrells, T.P., Passant, N., Thistlethwaite, G. and Watterson, J.D., 2005. UK Greenhouse Gas Inventory, 1990–2003. AEA Technology/ Department for Environment, Food and Rural Affairs. ISBN 0-9547136-5-6.

Biraud, S., Ciais, P., Ramonet, M., Simmonds, P., Kazan, V., Monfray, P., O'Doherty, S., Spain, T.G. and Jennings, S.G., 2000. European greenhouse gas emissions estimated from continuous atmospheric measurements and Radon 222 at Mace Head. *J. Geophys. Res.*, 105, 1351–1366.

Biraud, S., Ciais, P., Ramonet, M., Simmonds, P., Kazan, V., Monfray, P., O'Doherty, S., Spain, T.G. and Jennings, S.G., 2002. Sources of Carbon Dioxide, Methane, Nitrous Oxide, and Chloroform over Ireland inferred from Radon 222 and Radon 220 at Mace Head. *Tellus B*, 54, 41–60.

Buckingham, C., Clewly, L., Hutchinson, D., Sadler, L. and Shah, S., 1998. *London Atmospheric Emissions Inventory*. London Research Centre.

Draxler, R.R., 1992. Hybrid single-particle Lagrangian integrated trajectories (HY-SPLIT): Version 3.0—User's guide and model description. NOAA Tech. Memo. ERL ARL-195, 26 pp. and Appendices. [Available from National Technical Information Service, 5285 Port Royal Road, Springfield, VA 22161].

Draxler, R.R. and Rolph, G.D., 2003. HYSPLIT (Hybrid Single-Particle Lagrangian Integrated Trajectory) Model access via NOAA ARL READY Website (http://www.arl.noaa.gov/ready/hysplit4.html). NOAA Air Resources Laboratory, Silver Spring, MD.

Fisher, R., Lowry, D., Wilkin, O., Sriskantharajah, S. and Nisbet, E.G., 2006. High precision, automated stable isotopic analysis of atmospheric methane and carbon dioxide using continuous-flow isotope ratio mass spectrometry, *Rapid Communications in Mass Spectrometry*, 20, 200–208.

Frankenberg, C., Meirink, J.F., Bergamaschi, P., Goede, A.P.H., Heimann, M., Koerner, S., Platt, U., van Weele, M. and Wagner, T., 2006. Satellite cartography of atmospheric methane from SCIAMACHY on board ENVISAT: Analysis of the years 2003 and 2004. *J. Geophys. Res.*, 111, D07303, doi:10.1029/2005JD006235.

Hansen, J.E. and Sato, M., 2001. Trends of measured climate forcing agents. *Proc. Nat. Acad. Sci. USA*, 98, 14778–14783, www.pnas.org/cgi/doi/10.1073/pnas.261553698

Keeling, C.D. and Whorf, T.P., 2005. Atmospheric CO2 records from sites in the SIO air sampling network. In *Trends: A Compendium of Data on Global Change*. Carbon Dioxide Information Analysis Center, Oak Ridge National Laboratory, U.S. Department of Energy, Oak Ridge, Tenn., U.S.A.

Le Quere, C., Rödenbeck, C., Buitenhuis, E.T., Conway, T.J., Ray Langenfelds, R., Gomez A., Labuschagne C., Ramonet, M., Nakazawa T., Metzl, M., Gillett, N. and Heimann, M., 2007. Saturation of the Southern Ocean CO_2 sink due to recent climate change. *Science*, 316, 1735–1738. doi:10.1126/science 1136188.

Levin, I., Glatzel-Mattheier, H., Marik, T., Cuntz, M., Schmidt, M. and Worthy, D.E., 1999. Verification of German methane emission inventories and their recent changes based on atmospheric observations. *J. Geophys. Res.*, 104, D3, 3447–3456.

Lowe, D.C., Brenninkmeijer, C.A.M., Brailsford, G.W., Lassey, K.R., Gomez, A.J. and Nisbet, E.G., 1994. Concentration and ¹³C records of atmospheric methane in New Zealand and Antarctica: Evidence for changes in methane sources. *J. Geophys. Res.*, 99, 16913–16925.

Lowry, D., Holmes, C.W., Rata, N.D., Nisbet, E.G. and O'Brien, P., 2001. London Methane Emissions: Use of Diurnal Changes in Concentration and $\delta^{13}C$ to Identify Sources and Verify Inventories. *J. Geophys. Res.*, 106 (D), 7427–7448.

Meth-MonitEUr 2005. Final Report Section 6. Methane Monitoring in the European Union and Russia. European Union. http://www.gl.rhul.ac.uk/METH/MonitEUr/

Nisbet, E.G., 2005. Emissions control needs atmospheric verification. *Nature*, 433, 683.

Nisbet, E.G., 2007. Cinderella science. *Nature*, 450, 789–790.

Rolph, G.D., 2003. Real-time Environmental Applications and Display system (READY) Website (http://www.arl.noaa.gov/ready/hysplit4.html). NOAA Air Resources Laboratory, Silver Spring, MD.

Ryall, D.B., Derwent, R.G., Manning, A.J., Simmonds, P.G. and O'Doherty, S., 2001. Estimating source strengths of European emissions of trace gases from observations at Mace Head. *Atmos. Environ.*, 35, 2507–2523.

Worthy, D. and Huang, L., 2005. 12th WMO/IAEA meeting of experts on carbon dioxide concentration and related tracers measurement techniques. WMO Global Atmosphere Watch. Geneva: *World Meteorological Organisation* WMO TD 1275 (GAW 161).

Zinchenko, A.V., Paramonova,N.N., Privalov,V.I. and Reshetnikov, A.I., 2002. Estimation of methane emissions in the St. Petersburg region: An atmospheric nocturnal boundary layer budget approach. *J. Geophys. Res.*, 107, D20, 4416, doi:10.1029/2001JD001369.

Chapter 7
Estimating Sources and Sinks of Methane: An Atmospheric View

Peter Bergamaschi and Philippe Bousquet

7.1 Introduction

Methane (CH_4) is an important trace gas of the atmosphere. Its mixing ratio has increased by a factor of 2.5 compared to preindustrial levels (year 1800) (Etheridge et al. 1992) and reached almost 1,800 parts per billion (ppb) today (Dlugokencky et al. 2003; IPCC 2007). From ice core measurements, it is known that present atmospheric levels of CH_4 are unprecedented during at least the last 600,000 years (Petit et al. 1999; Spahni et al. 2005). Atmospheric CH_4 is the second most important anthropogenic greenhouse gas (GHG) after CO_2 (IPCC 2007). The direct radiative forcing of anthropogenic CH_4 is $0.48 \, W/m^2$, that is almost one third that of anthropogenic CO_2 ($1.66 \, W/m^2$) (IPCC 2007). Furthermore, CH_4 plays an important role in atmospheric chemistry, affecting the oxidizing capacity of the atmosphere and tropospheric ozone (O_3). The mean atmospheric lifetime of CH_4 is estimated to be 8.4 years on average (IPCC 2001).

CH_4 is emitted at the earth surface by a variety of natural and anthropogenic sources (Matthews and Fung 1987, 1991; Olivier and Berdowski 2001). The principal CH_4 production processes are: (1) biogenic CH_4 formation (by methanogenic bacteria under anaerobic conditions), (2) thermogenic formation, and (3) incomplete combustion of biomass or fossil fuels. The biogenic CH_4 formation occurs, for example, in wetlands, water-flooded rice paddies, landfills, and stomachs of ruminant animals, while thermogenic CH_4 formation is the most important process for generation of natural gas deposits (over geological timescales). Recently, it has been suggested that plants emit CH_4 also under aerobic conditions (Keppler et al. 2006). The experiments by Keppler et al. (2006) indicated that these emissions are related to a hitherto unknown nonmicrobial process in which the structural plant component pectin plays a central role. Keppler et al. (2006) estimated that these plant emissions may contribute significantly to the global CH_4 budget (62–236 Tg/year). However, consideration of the preindustrial CH_4 budget (Bergamaschi et al. 2006; Houweling et al. 2000), and a more detailed upscaling study (Kirschbaum et al. 2006) indicates that the upper estimate by Keppler et al. appears very unlikely.

The potential role of plants has now become further controversial, after a very recent study by Dueck et al. (2007) questioned the measurements given by Keppler et al.

Using an independent method based on ^{13}C-labeling and laser-based measurements, Dueck et al. did not find any evidence for substantial CH_4 emissions of plants under aerobic conditions. Clearly, further independent studies are required to resolve this issue.

The above-mentioned different principal CH_4 production processes are also reflected in clearly different compositions of the stable isotopes ($\delta^{13}C$, δD), providing isotopic fingerprints of the different source categories which can be utilized for further constraining the CH_4 budget (Bergamaschi et al. 2000b; Miller et al. 2002; Quay et al. 1999; Wahlen et al. 1987). Overall, total emissions of CH_4 are estimated to be 500–600 Tg CH_4/year, with large uncertainties on individual processes and on their location and variability (IPCC 2001; Wang et al. 2004).

The main sink of atmospheric CH_4 is the reaction with hydroxyl radical:

$$CH_4 + OH \rightarrow CH_3 \bullet + H_2O$$

This reaction is the first of an oxidation chain leading to CO and finally to CO_2.

Furthermore, depending on the atmospheric NO_x mixing ratios, this oxidation chain leads in large parts of the Northern Hemisphere to net photochemical production of O_3, while in the Southern Hemisphere, ozone destruction prevails (Crutzen 1974).

The destruction of CH_4 by OH in the troposphere represents about 90% of CH_4 loss in the atmosphere (IPCC 2001). The rest of the sink is due to an uptake of CH_4 by soils, reaction with Cl in the marine boundary layer (Platt et al. 2004), and to destruction in the stratosphere by reactions with OH, Cl, and $O(^1D)$ (IPCC 2001).

The production of OH radical in the troposphere is due to the reaction of excited atomic oxygen $[O(^1D)]$ produced by the photolysis of ozone ($\lambda < 320$ nm) with water vapor. The global mean OH concentration is estimated to be around 10×10^5 molecules/cm³ (8×10^5–12×10^5 cm⁻³) (IPCC 2001), with large spatial, diurnal, and seasonal variations depending mainly on the available radiation, ozone, and water vapor concentrations (Spivakovsky et al. 2000). As OH plays a major role in the removal of CH_4, but also of other atmospheric trace gases, such as CO and nonmethane hydrocarbons, quantification of annual global mean OH concentrations and their interannual to decadal changes are important. Such estimates are largely based on proxy methods, using trace gases (that react with OH) with relatively well-known emissions. In particular, methyl chloroform has been employed by several authors to infer OH fields on the basis of different methodologies (Bousquet et al. 2005; Dentener et al. 2003; Hein et al. 1997; Houweling et al. 1999; Krol and Lelieveld 2003; Krol et al. 2003; Prinn et al. 1995, 2001, 2005). Furthermore, [14]CO also has been used by several researchers (Manning et al. 2005).

The global CH_4 budget is given by Eq. 7.1,

$$\frac{dC}{dt} = E - S \tag{7.1}$$

where C is the global burden, E the total emissions, and S the total sinks.

The term which is best known is the change of atmospheric burden (from the very accurate atmospheric measurements). Furthermore, the CH_4 sinks also are believed to be relatively well known [OH sink estimates based on methyl chloroform are believed to be accurate within $\sim\pm10\%$, that is, $\sim\pm50\,Tg\;CH_4$/year; uncertainties of other minor sinks (soil sink, Cl in the marine boundary layer, and stratosphere) may add an additional uncertainty of $\sim\pm30\,Tg\;CH_4$/year]. Hence the total CH_4 sources can be relatively precisely constrained by Eq. 7.1. However, large uncertainties exist about the relative contributions to the total emissions from different source categories and their spatial emission distribution.

A variety of approaches have been used to estimate surface emissions of CH_4. They can be divided in "bottom-up" and "top-down" approaches. Bottom-up estimates generally use local measurements (e.g., wetlands, plants, and landfills) to assess emissions factors (g CH_4/m^2/day) that are extrapolated or integrated in biogeochemical models in order to get global emissions for a given process (Fung et al. 1991; Walter et al. 2001). They can also be based on statistical and economical models (e.g., gas, coal). More recently, coupling satellite measurements and biogeochemical models have permitted to improve the estimates of the location and the strength of biomass burning emissions (Van der Werf et al. 2003, 2004).

Top-down estimates are based on measurements of atmospheric CH_4 mixing ratios (including sometimes also measurements of its isotopic composition) and inverse atmospheric models. Mixing ratios of atmospheric CH_4 have been directly monitored since 1983 by the National Oceanic and Atmospheric Administration (NOAA) and other research organizations all over the world (Dlugokencky et al. 2001; GLOBALVIEW-CH4 2005). More than 80 sites measuring CH_4 on a weekly basis are available today. Some additional sites perform continuous measurements (Dlugokencky et al. 1995; Prinn et al. 2000; WMO 2003). Furthermore, CH_4 measurements from satellites became available recently, namely from the scanning imaging absorption spectrometer for atmospheric chartography (SCIAMACHY) instrument on environmental satellite (ENVISAT) which is in orbit since March 2002 (Buchwitz et al. 2005a, b; Frankenberg et al. 2005, 2006).

Spatial CH_4 gradients (and their changes with time) reflect not only the regional sources and sinks of CH_4, but also the action of atmospheric transport and chemistry. Optimizing surface emissions with atmospheric measurements using a chemical transport model (CTM) and prior estimates of sources and sinks is called a Bayesian inverse model. This top-down approach has been widely used to infer sources and sinks of CO_2 (Bousquet et al. 2000; Enting et al. 1995; Gurney et al. 2002; Rayner et al. 1999; Rodenbeck et al. 2003) and is increasingly used also for other trace gases. A particular challenge to apply inverse modeling to CH_4 (or other chemically active species) is the requirement to simulate the chemical sinks (mainly OH radicals) realistically. The spatial scale resolved by inversions is largely determined (1) by the density and the quality of atmospheric observations, (2) by the quality of the CTM, and (3) by the quality of prior information. In the case of CH_4, several source categories can overlay within the same geographical region. In an inverse procedure, separating different source categories mainly relies on their (1) different spatial emission patterns, (2) different temporal emission patterns, or (3) different isotopic signatures. The analysis of error correlations provides hints on the ability of inversions to properly separate regions or processes.

Several inverse modeling investigated global CH_4 sources. These studies optimized emissions from larger global regions or from different source categories, and focused on shorter timescales (1–2 years) or climatological averages (Bergamaschi et al. 2000a; Dlugokencky et al. 1995; Hein et al. 1997; Houweling et al. 1999; Mikaloff Fletcher et al. 2004a, b; Prinn et al. 2000; Wang et al. 2004). Recent developments focused on the refinement of spatial resolution of the inversions with the use of continuous in situ observations or satellite data, and on the extension of the inversion period to the full period of measurements (interannual analysis). In the following, we summarize three recent inverse modeling studies, which further extend our view on the global and the regional CH_4 budget:

1. On the basis of the atmospheric zoom model TM5, Bergamaschi et al. (2005) presented a coupled global–European inversion to derive CH_4 emission estimates for individual European countries. This study will be summarized in Sect. 7.2.
2. Recent satellite measurements from SCIAMACHY now provide comprehensive views on the global CH_4 distribution (Frankenberg et al. 2005, 2006). We will discuss some key results from a recent analysis on the basis of TM5 inverse simulations (Bergamaschi et al. 2007) in Sect. 7.3.
3. The interannual variability of CH_4 sources and sink during 1984–2003 has been investigated in a recent study by Bousquet et al. (2006). This study will be presented in Sect. 7.4.

7.2   Inverse Modeling of European CH_4 Sources

Only during the last few years, attempts have been made to derive top-down estimates on national scales, mainly on the basis of Lagrangian back trajectory or Lagrangian particle dispersion models (Manning et al. 2003; Vermeulen et al. 1999). These studies focused, however, on a limited spatial domain (e.g., Europe) and are not, or only weakly, coupled to the global tracer fields. Using the two-way nested atmospheric zoom model TM5 (Krol et al. 2005), Bergamaschi et al. (2005) presented a high resolution regional inversion which is fully consistent with the global inverse simulations.

This study used the "synthesis-inversion"/Green's function approach (Enting 2000; Heimann and Kaminski 1999), describing the total atmospheric CH_4 mixing ratio in space and time $\vec{d}_{model}(\vec{x}, t)$ as linear combination of n_{para} model runs for emissions from different European and global regions and emissions from different months $\vec{d}_{model,i}(\vec{x}, t)$, including one background run:

$$\vec{d}_{model}(\vec{p}, \vec{x}, t) = \sum_{i=1}^{n_{para}} p_i \, \vec{d}_{model,i}(\vec{x}, t) \qquad (7.2)$$

with scaling factors p_i (summarized as vector $\vec{p}$).

The definition of regions (European countries and larger global regions) is illustrated in Fig. 7.1.

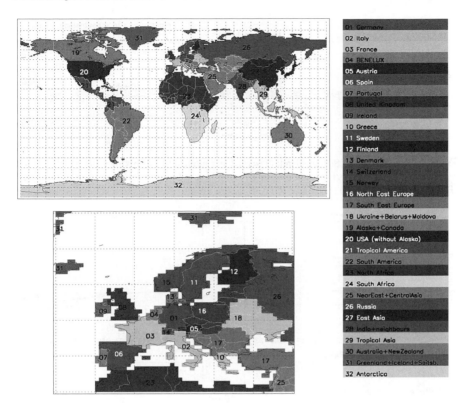

Fig. 7.1 Global and European regions used for the inversion (Bergamaschi et al. 2005) (*See Color Plate* 3)

Observational constraints are provided by high-frequency (quasi-continuous) measurements of atmospheric CH_4 mixing ratios at several western European monitoring sites (and some global sites), complemented by a comprehensive set of global flask measurements.

In general, all European sites are characterized by considerable synoptic variability, (Fig. 7.2) that is, variations due to varying origin of air masses, with typical timescales of a few days. Basing the inversion on daily mean values (instead of monthly means as used for most global inverse studies) directly exploits the information content of these synoptic events in order to derive the emissions from various regions or countries.

The available observational data put significant constraints on emissions from different regions. Within Europe, in particular, several Western European countries are well constrained (Bergamaschi et al. 2005). The inversion results suggest up to 50–90% higher anthropogenic CH_4 emissions in 2001 for Germany, France, and UK compared to the reported United Nations Framework Convention on Climate Change (UNFCCC) values (EEA 2003) (Fig. 7.3). A recent revision of the German inventory, however, resulted in an increase of reported CH_4 emissions by 68.5%

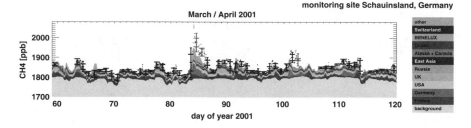

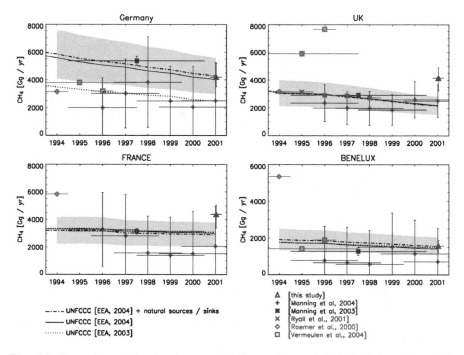

Fig. 7.2 Monitoring site Schauinsland: Observed and modeled CH_4 mixing ratios for the period March–April 2001. Colors highlight influence from different European countries or global regions (Bergamaschi et al. 2005) (*See Color Plate* 3).

Fig. 7.3 Comparison with other inverse modeling studies and United Nations Framework Convention on Climate Change (UNFCCC) values. UNFCCC estimates (black solid curve) are augmented by our bottom-up estimate of net natural sources (including soil sink), displayed by the black dash-dotted line. The gray-shaded area indicates an assumed 30% uncertainty of the total of UNFCCC and natural emissions. Colored symbols represent different top-down estimates, according to the legend below. For further details, see Bergamaschi et al. (2005) (*See Color Plate* 4).

(EEA 2004), being now in very good agreement with the top-down estimate. The top-down estimate for Finland is distinctly smaller than the a priori estimate, suggesting much smaller CH_4 emissions from Finnish wetlands (<1 Tg CH_4/year) than derived from the bottom-up inventory (~3 Tg CH_4/year).

The EU-15 totals are relatively close to the UNFCCC values (within 4–30%) and appear very robust for different inversion scenarios.

7.3 Use of Satellite Data for Inverse Modeling

Despite the extension of the ground-based measurements during the past two decades, many very important source regions are still poorly sampled by existing monitoring sites. Only recently, CH_4 measurements from satellites became available, namely from the SCIAMACHY instrument on ENVISAT which is in orbit since March 2002 (Buchwitz et al. 2005a, b; Frankenberg et al. 2005, 2006). In contrast to mid-infrared measurements by Intercerometric Monitor for Greenhouse gases (IMG) (which was in operation only from August 1996 to June 1997) (Clerbaux et al. 2003) and by follow-on instruments (AIRS: Atmospheric Infrared Sounder, IASI: Infrared Atmospheric Sounding Interferometer) (Turquety et al. 2004), SCIAMACHY CH_4 measurements are based on near-infrared solar absorption spectra and hence also sensitive toward the boundary layer and lower troposphere. Therefore, SCIAMACHY is very well suited for detection of signals directly related to emissions at the surface. Nevertheless, very high precision and accuracy of the column averaged mixing ratios (1–2%) is required to apply these measurements to inverse modeling (Meirink et al. 2006).

Frankenberg et al. (2006, 2005) presented CH_4 retrievals for 2003–2004, demonstrating that the north-south gradient as well as regions with enhanced CH_4 levels can be clearly identified from space. Large CH_4 enhancements are detected by SCIAMACHY, in particular, over the tropical regions of South America and Africa and over India and South East Asia.

A detailed evaluation of these retrievals using inverse model simulations have been presented by Bergamaschi et al. (2007). Furthermore, the study by Bergamaschi et al. (2007) presented an initial coupled inversion that simultaneously uses the NOAA surface observations and the new SCIAMACHY CH_4 retrievals and that includes a correction to compensate for the potential systematic bias of the SCIAMACHY data.

The results suggest significantly greater tropical emissions compared to either the a priori estimates or the inversion based on the surface measurements only. The large tropical CH_4 emissions derived from the SCIAMACHY observations are attributed in the inversion to greater emissions mainly from tropical wetlands, but at the same time also an increase of CH_4 emissions from termites, and a decrease of the soil sink are calculated, further enhancing net emissions from tropical regions. The derived large tropical CH_4 source may also support the hypothesis of CH_4 emissions from plants under aerobic conditions (Keppler et al. 2006), which are assumed to be located mainly in the tropics. However, as discussed in Sect. 7.1, further experimental studies are required to clarify the potential role of plants.

Emissions from rice paddies in India and Southeast Asia are relatively well constrained by the SCIAMACHY data and are slightly reduced by the inversion (Fig. 7.4).

This first coupled inversion demonstrates the usefulness of the combined use of satellite data and high-accuracy surface measurements. The SCIAMACHY data now provide information about CH_4 emissions from important source regions, which, so far, were poorly monitored by surface observations (in particular the tropical land masses of South America and Africa and Southeast Asia). On the other hand, the surface observations are very important as high-accuracy reference (accuracy of in situ CH_4 measurements in the order of 0.1%), which is used to

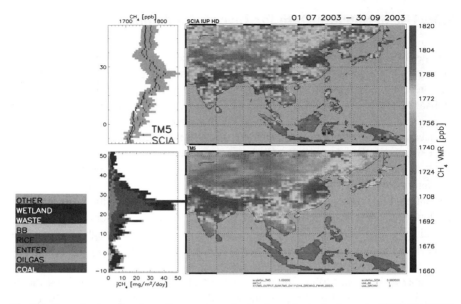

Fig. 7.4 Scanning imaging absorption spectrometer for atmospheric chartography (SCIAMACHY) measurements (right top) and TM5 simulations (right bottom) over Asia (*See Color Plate* 4).

correct some systematic latitudinal and seasonal bias of the satellite data. Because of their much higher accuracy, the surface data put more constraints on the inverse modeling systems in remote areas (far from significant CH_4 sources) than the SCIAMACHY data, while the latter provide important information on continental areas for which no surface measurements exist.

7.4 Interannual Variability of CH_4 Sources and Sinks: From Global to European Scale

7.4.1 Introduction: Observed CH_4 Growth Rate

The atmospheric growth rate of CH_4 presents large year-to-year variations superimposed on a slowing after 1991, which persisted until recent years (Fig. 7.5). Starting from nearly more than 12 ± 2 ppb/year during the 1980s, the growth rate of CH_4 abruptly dropped in 1992 after a peak in 1991, and has stagnated since then at 3 ± 4 ppb/year, on average. The volcanic eruption of Mount Pinatubo occurred in June 1991. The 1991 peak was attributed to reduced OH production (stratospheric injection of volcanic aerosols), and the drop to reduced wetland emissions (cooling effects afterward) (Walter et al. 2001). The economic collapse of Former Soviet Union (FSU) was a possible explanation for the persistence of low growth rate in the 1990s (Dlugokencky et al. 1994). Recent years were, nevertheless, marked by

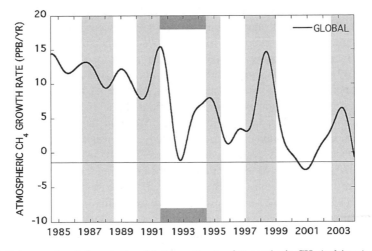

Fig. 7.5 Interannual variations in the global growth rate of atmospheric CH_4 (ppb/year) over the 1984–2003 period. The growth rates were calculated using data from the National Oceanic and Atmospheric Administration (NOAA) air sampling sites used in the inversion model (a maximum of 50 sites). Light-gray bands indicate El Niño conditions. Dark-gray squares indicate post-Pinatubo period.

two positive anomalies in 1997–1998 and in 2002–2003. Increased biomass burning emissions, predominantly in Southeast Asia and in boreal regions, were suspected by several studies of being a driving factor (Van der Werf et al. 2004). Overall, the hypotheses proposed to explain the variability in atmospheric CH_4 growth rates focused either on wetland emissions (Walter et al. 2001), on anthropogenic emissions (Dlugokencky et al. 1994), on wildfires (Langenfelds et al. 2002; Manning et al. 2005; Simmonds et al. 2005; Van der Werf et al. 2004), on OH photochemistry (Dentener et al. 2003; Dlugokencky et al. 1996; Wang et al. 2004), or even on interannual wind changes (Warwick et al. 2002). Different models were used but the contribution of each process has not yet been disentangled within a coherent framework.

We analyzed 20 years of atmospheric measurements of CH_4 and its ^{13}C isotope composition in an inversion model of atmospheric transport and chemistry to deduce variations of CH_4 emissions over the period 1984–2003. In the following, we briefly present the methodology used in this work and the results obtained from global to European scale for year-to-year variations of CH_4 surface emissions.

7.4.2 Method

Over the period 1984–2003, the CH_4 concentration responses to the action of OH sinks and regional surface sources were simulated each month with the three-dimensional chemistry-transport model INCA (INteractions Chemistry Aerosols)

(Hauglustaine et al. 2004). The model was forced with interannual analyzed winds (Uppala et al. 2005) and interannually varying OH concentrations (Bousquet et al. 2005). The inverse methodology is fully described in the study of Bousquet et al. (2005). CH$_4$ surface emissions are optimized each month for 11 land regions (Gurney et al. 2002), 1 global ocean region, and up to 10 processes over each land region (bogs, swamps, tundra, termites, fossil fuel and industry, gas, biofuel, ruminant animals, land-fills and waste, and soil uptake). This spatial partition (Fig. 7.6) allows performing both geographical-based and process-based analyses. We used the optimized interannual four-dimensional distribution of OH from the study of Bousquet et al. (2005).

Atmospheric CH$_4$ observations, from approximately weekly air samples collected in flasks, were inverted as monthly means. Uncertainties in the monthly means were taken from the GLOBALVIEW-CH$_4$ data product (GLOBALVIEW-CH4 2005) when available, or from submonthly variability in the measurements. In total, data from 68 sites from different networks were collected and used, with 75% contributed by the NOAA network. Atmospheric δ^{13}C-CH$_4$ flask data were used for the 1998–2004 period (Miller et al. 2002; Quay et al. 1999; Tyler et al. 1999).

No a priori interannual variability is assigned to regional fluxes, except for animal emissions. The a priori error on each monthly source is set to ±100%. Other soft constraints were put on anthropogenic emissions (Bousquet et al. 2006; Peylin et al. 2002). A control inversion was performed, supplemented by an ensemble of 17

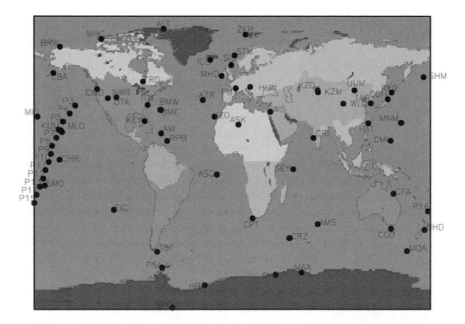

Fig. 7.6 Map of the regions and the air sampling sites used in the inversions (73 at maximum). The 11 continental regions used to partition the CH$_4$ correspond to those of the TRANSCOM3 model intercomparison experiment (Gurney et al. 2002). One global ocean region was used, the ocean being only a small net source of CH$_4$. Zero emissions are prescribed over Antarctica, Greenland, and innerseas and lakes (red color).

sensitivity inversions where different settings were varied individually (see Sect. 7.4.2). The ensemble of inversions is used in the following to assess the robustness of the results to each setting. For further details please refer to Bousquet et al. (2006).

7.4.3 Global and Hemispheric Results

The regional patterns of surface CH_4 emissions indicate that most of the global year-to-year variability lies in the tropics (Fig. 7.7). In contrast, the northern regions (NR = boreal and temperate North America, Asia, and Europe) show

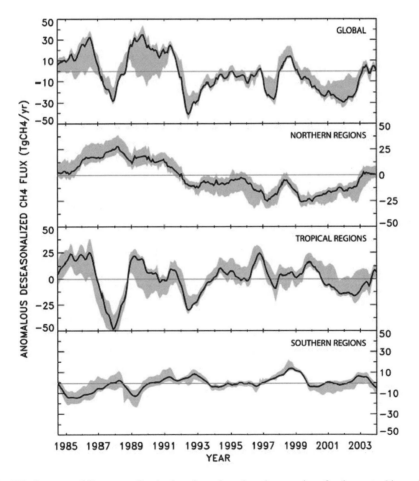

Fig. 7.7 Interannual flux anomalies broken down into three large regions for the control inversion (in Tg CH_4/year): northern regions (NR), tropical regions, and southern regions (SR). Surface emissions are plotted as deseasonnalized anomalies (12-month running mean subtracted), with a gray area representing the spread of an ensemble of 18 inversions with various settings.

smoother variations, yet with systematically less emissions in the 1990s than in the 1980s, except for 1997–1998 and 2002–2003. We found that the years 1987–1988, 1991–1992, 1997–1998, and 2001–2002 correspond to abnormally weaker CH_4 destruction by OH in the tropics. In the inversion, a compensation effect exists between the magnitude of methyl chloroform-derived OH changes and inverted CH_4 surface emissions, as the sum of the two must equal the observed atmospheric accumulation. Thus, biases in OH changes could account for some of the variability we attributed to wetlands. In the extreme case where OH interannual variability is set to be zero, the fluctuations of tropical wetland emissions are dampened by 50%, especially during the 1980s, when methyl chloroform data suggest large OH variability (Bousquet et al. 2005). Therefore, the inferred variability of CH_4 emissions critically depends on the variability of OH radicals for their amplitude but not for their phasing which appears more robust.

7.4.4 Process-Based Results

We present the inferred variability of three categories of emissions: natural and anthropogenic wetlands, biomass burning, and other emissions. We found that fluctuations in wetland emissions are the dominant contribution to interannual variability in surface emissions. Wetland emissions vary by ±12 Tg CH_4/year, explaining 70% of the global emission anomalies over the past two decades, against only 15% for biomass burning (Fig. 7.8). This result disagrees with previous studies suggesting a dominant role of fires (Langenfelds et al. 2002; Manning et al. 2005; Page et al. 2002; Simmonds et al. 2005; Van der Werf et al. 2004). The comparison with a simple wetland flux model (not shown in the figure) driven by interannually varying climate data (Uppala et al. 2005) and by first estimates of remotely sensed changes in flooded areas over the 1993–2001 period (Prigent et al. 2001) showed encouraging results on the amplitude and phasing of inverted wetland emissions. During the period 1993–2001, wetland emissions in the bottom-up model show a persistent negative trend of −2.5 Tg CH_4/year, in response to a dramatic decrease in flooded areas worldwide (at a rate of −1.1% per year for a mean area of 4.2×10^6 km²), mostly in temperate and tropical Asia and in tropical South America (Prigent et al. 2001). The inversion infers decreasing wetland emissions after 1993, but with a smaller trend (−0.6 Tg CH_4/year). Shrinking wetland areas may reflect recurrent dryness observed in the tropics after 1990 (Uppala et al. 2005), and northward after 1999 (Hoerling and Kumar 2003).

The inversion infers global variations in the biomass burning emissions in the order of ±3.5 Tg CH_4/year (Fig. 7.8). Over the 1996–2002 period, these anomalies are in very good agreement with independent estimates (Van der Werf et al. 2004) derived from remote sensing data (not shown in the figure). The members of the inversion ensemble that include $\delta^{13}C$-CH_4 observations agree best with the magnitude of the bottom-up anomalies (Van der Werf et al. 2004), but not with the phasing: these "isotopic" inversions place a strong

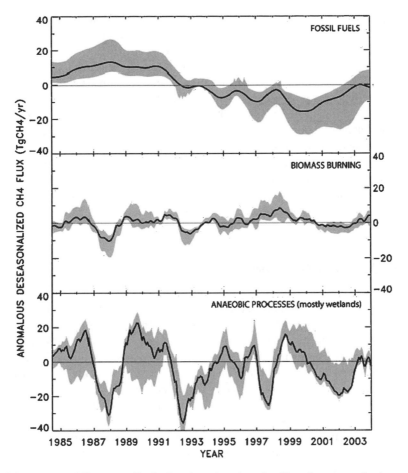

Fig. 7.8 Interannual flux anomalies broken down into three families of processes for the control inversion (in Tg CH$_4$/year): fossil fuels (coal, oil, gas, biofuel, and industry), biomass burning, and all other anaerobic processes (natural and anthropogenic wetlands, landfills, ruminant animals, termites). Surface emissions are plotted as deseasonnalized anomalies (12-month running mean subtracted), with a gray area representing the spread of an ensemble of 18 inversions with various settings.

(tropical) release of CH$_4$ by fires in 1997, six months earlier than inferred from the remote sensing data.

Other emissions, mainly anthropogenic, present smoother variations in time. However, we discovered that inferred anthropogenic emissions present significant long-term trends in the 1990s. After 1992, decreasing global emissions at a rate of -1.0 ± 0.2 Tg CH$_4$/year are required to match a small average growth rate of 3 ± 4 ppb/year in presence of (slightly) decreasing OH (Fig. 7.7). The inversion attributes this signal to decreasing anthropogenic emissions, in particular, to the northern fossil source (Fig. 7.8). This finding is consistent with reported declining

CH_4 emissions in the FSU (Dlugokencky et al. 1994) and with the EDGAR 3.2 (Emissions Database for Global Atmospheric Research) recent fossil CH_4 emissions inventory published for the 1990–1995 period (Olivier and Berdowski 2001). It is also consistent with the implementation of emission reduction measures to reduce landfill sources in developed countries during the 1990s (see, e.g., http://www.epa.gov/methane/sources.htm). After 1999, however, anthropogenic emissions increase again, especially in North Asia. This may reflect the booming Chinese economy. By 2003, we found that anthropogenic emissions recovered to their early 1990s levels. Without a coincident and important drop in northern wetland emissions after 1999 (Fig. 7.8) associated with dryer conditions (Hoerling and Kumar 2003), the growth rate of atmospheric CH_4 would therefore have increased much more rapidly. This suggests that the slow down in CH_4 growth rate observed during the early 1990s may represent only a temporary pause in the human-induced secular increase of atmospheric CH_4.

7.4.5 Northern Hemisphere Results

The discussion of inversion results at subcontinental to regional scales is limited by the capacity of the atmospheric network to infer independently the emissions of two regions. One possible estimate of this "separation power" is the posterior error correlation returned by the inversion model. Negative statistically significant error correlations indicate that the atmospheric network resolves only the sum of the process/regions. The annual error correlations between North America, Europe, and North Asia lay between −0.35 and −0.45 indicating a tendency to negative correlations but with only a small statistical significance. Therefore, it makes some sense to discuss the inverted emissions for these three subcontinental regions. This diagnostic could be refined by isolating the interannual part of the error correlation (Baker et al. 2006).

The year-to-year variability of the northern regions is found to be slightly dominated by North Asia, although North America and Europe also presents significant year-to-year variations (Fig. 7.9). Over the 1984–2003 period, Europe and North Asia present negative long-term trends in emissions of -1.5 ± 0.3 and -0.7 ± 0.4 Tg CH_4/year, respectively, whereas North America presents a small positive trend of 0.4 ± 0.2 Tg CH_4/year. The trend is continuous for Europe after 1989, representing a cumulative emission reduction of around 25 Tg CH_4 in 16 years. This is not the case for North Asian emissions, which increase again sharply after 1999, possibly caused by the rise of Chinese economy, as noticed previously. The negative trend in European emissions is in qualitative agreement with country-based bottom-up inventories (see Sect. 7.2) but further validation is needed. The processes causing such a decrease are difficult to assess using sparse and noncontinuous atmospheric surface measurements as significant error correlations are found between processes within a region, or between regions, for the same process. However, one can say that most of the negative trend of European emissions is found to be due to anthropogenic emissions (fossil fuels, landfills, and ruminant animals), which represent ~82% of European emissions in our study.

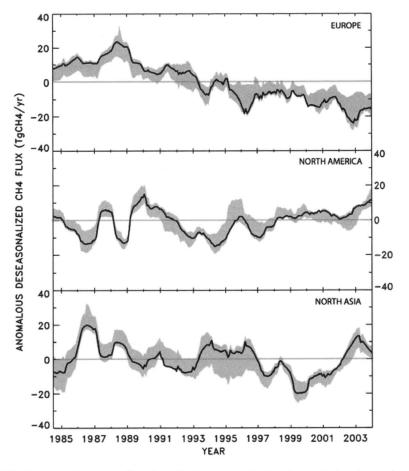

Fig. 7.9 Interannual flux anomalies of the Northern regions broken down into three large subcontinental regions for the control inversion (in Tg CH$_4$/year): Europe, North America, and North Asia. Surface emissions are plotted as deseasonnalized anomalies (12-month running mean subtracted), with a gray area representing the spread of an ensemble of 18 inversions with various settings.

7.5   Conclusions

The presented studies demonstrated recent advances in our understanding of the global and regional CH$_4$ budget.

7.5.1   European Scale: Top-Down Estimates for Individual Countries

Inverse modeling of emissions on the national scale is important for verification of emissions reported under the UNFCCC (Bergamaschi et al. 2004; IPCC 2000). The presented study (Bergamaschi et al. 2005) demonstrates the principal feasibility to derive top-down emissions on the European national scale. The use of the two-way nested zoom model TM5 ensures that European top-down estimates (based on high resolution simulations over Europe) are consistent with global inverse simulations. In particular, recent high-frequency observations from several European sites put significant constraints on emissions from different European countries.

For further improvement of national and regional top-down estimates, a further extension of the measurement network is crucial. Furthermore, detailed intercomparisons of different inverse model systems should be performed in order to better estimate and better understand systematic errors of top-down estimates.

7.5.2 Global Scale: Now Comprehensive View on Global Distribution on CH_4 Sources

SCIAMACHY now provides exciting views on the global CH_4 distribution. The study of Bergamaschi et al. (2007) showed that inverse model simulations optimized with high-accuracy surface measurements can be used to evaluate these new satellite products over remote regions.

First coupled inversions, using both surface and satellite data suggest in particular higher tropical emission than assumed in present bottom-up inventories. Beside the potential role of CH_4 emissions from plants, CH_4 emissions from tropical wetlands remain very uncertain. Spatial distributions of different CH_4 wetland inventories differ substantially and need further validation. The improvement of bottom-up inventories (in particular the description of spatio-temporal emission patterns) of wetlands would be very useful for further improvement of top-down estimates.

7.5.3 Interannual Variations and Trends over Past Two Decades

The processes controlling the variations of CH_4 emissions over 1983–2004 have been quantified separately using 20 years of concentration measurements in an inversion model of atmospheric transport and chemistry. Bousquet et al. (2006) explained the slowing atmospheric CH_4 growth rate by a decline of anthropogenic emissions throughout the 1990s. Since 1999, however, anthropogenic emissions of CH_4 have risen up again, especially over North Asia, whereas Europe shows a continuous negative trend in emissions. Their effects on atmospheric growth rate have remained hidden by a coincident dip in wetland emissions associated with prolonged drying. This suggests that the atmospheric CH_4 may increase again in the near future, with significant impact on climate. For short-term variations, wetland emissions were found to have dominated the interannual variability of CH_4 sources, whereas fire emissions played a smaller role, except during the 1997–1998 El Niño event. Both inverted wetland and fire emission changes are in very good agreement with independent estimates based on the remote sensing information and biogeochemical models.

The presented studies demonstrate the importance of global and regional monitoring of atmospheric GHGs. High-accuracy in situ measurements on global remote stations are central to monitor accurately the atmospheric background (e.g., to determine global trends and interannual variability). For monitoring stations closer to source regions, it is very useful to perform these measurements at high frequency (approximately hourly) in order to capture the full synoptic variability. Both for global and regional stations, it is extremely important that these measurements are performed on a long term, and continuously without major gaps in the time series. Additional important information on the global CH_4 distribution is now also available from SCIAMACHY. It is expected that the quality of satellite products will further improve, with further development of retrieval algorithms, with new sensors, and with validation data such as continuous ground sites and Fourier Transform Infrared Spectrometry (FTIR) measurements. At the same time, however, it is obvious that the high-accuracy in situ monitoring will remain indispensable.

References

Baker, D. F., R. M. Law, K. R. Gurney, P. Rayner, P. Peylin, A. S. Denning, L. P. Bousquet, Bruhwiler, Y.-H. Chen, P. Ciais, I. Y. Fung, M. Heimann, J. John, T. Maki, S. Maksyutov, K. Masarie, M. Prather, B. Pak, S. Taguchi, and Z. Zhu, TransCom3 inversion intercomparison: Impact of transport model errors on the interannual variability of regional CO_2 fluxes, 1988–2003, *Global Biogeochem. Cycles*, 20, doi:10.1029/2004GB002439, 2006.

Bergamaschi, P., R. Hein, M. Heimann, and P. J. Crutzen, Inverse modeling of the global CO cycle 1. Inversion of CO mixing ratios, *J. Geophys. Res.*, 105(D2), 1909–1927, 2000a.

Bergamaschi, P., R. Hein, C. A. M. Brenninkmeijer, and P. J. Crutzen, Inverse modeling of the global CO cycle 2. Inversion of $^{13}C/^{12}C$ and $^{18}O/^{16}O$ isotope ratios, *J. Geophys. Res.*, 105(D2), 1929–1945, 2000b.

Bergamaschi, P., H. Behrend, and A. Jol (Eds.), *Inverse modelling of national and EU greenhouse gas emission inventories—report of the workshop "Inverse modelling for potential verification of national and EU bottom-up GHG inventories" under the mandate of the Monitoring Mechanism Committee WG-1, 23–24 October 2003, JRC Ispra*, 144 pp., European Commission Joint Research Centre, Ispra, 2004.

Bergamaschi, P., M. Krol, F. Dentener, A. Vermeulen, F. Meinhardt, R. Graul, M. Ramonet, W. Peters, and E. J. Dlugokencky, Inverse modelling of national and European CH_4 emissions using the atmospheric zoom model TM5, *Atmos. Chem. Phys.*, 5, 2431–2460, 2005.

Bergamaschi, P., F. Dentener, G. Grassi, A. Leip, Z. Somogyi, S. Federici, G. Seufert, and F. Raes, Methane Emissions from Terrestrial Plants—On the discovery of CH_4 emissions from terrestrial plants and its potential implications—Comments on the paper of Keppler et al. (Methane emissions from terrestrial plants under aerobic conditions, *Nature*, 439, 2006), European Commission, DG Joint Research Centre, Institute for Environment and Sustainability, 6 pp., 2006.

Bergamaschi, P., C. Frankenberg, J. F. Meirink, M. Krol, F. Dentener, T. Wagner, U. Platt, J. O. Kaplan, S. Korner, M. Heimann, E. J. Dlugokencky, and A. Goede, Satellite chartography of atmospheric methane from SCIAMACHY on board ENVISAT: 2. Evaluation based on inverse model simulations, *J. Geophys. Res.*, 112(D2), doi:10.1029/2006JD007268, 2007.

Bousquet, P., P. Peylin, P. Ciais, C. Le Quere, P. Friedlingstein, and P. P. Tans, Regional changes in carbon dioxide fluxes of land and oceans since 1980, *Science*, 290(5495), 1342–1346, 2000.

Bousquet, P., D. A. Hauglustaine, P. Peylin, C. Carouge, and P. Ciais, Two decades of OH variability as inferred by an inversion of atmospheric transport and chemistry of methyl chloroform, *Atmos. Chem. Phys.*, 5, 2635–2656, 2005.

Bousquet, P., P. Ciais, J. B. Miller, E. J. Dlugokencky, D. A. Hauglustaine, C. Prigent, G. van der Werf, P. Peylin, E. Brunke, C. Carouge, R. L. Langenfelds, J. Lathiere, P. F., M. Ramonet, M. Schmidt, L. P. Steele, S. C. Tyler, and J. W. C. White, Contribution of anthropogenic and natural sources methane emissions variability, *Nature*, *443*, 439–443, 2006.

Buchwitz, M., R. de Beek, J. P. Burrows, H. Bovensmann, T. Warneke, J. Notholt, J. F. Meirink, A. P. H. Goede, P. Bergamaschi, S. Korner, M. Heimann, and A. Schulz, Atmospheric methane and carbon dioxide from SCIAMACHY satellite data: Initial comparison with chemistry and transport models, *Atmos. Chem. Phys.*, *5*, 941–962, 2005a.

Buchwitz, M., R. de Beek, S. Noel, J. P. Burrows, H. Bovensmann, H. Bremer, P. Bergamaschi, S. Korner, and M. Heimann, Carbon monoxide, methane and carbon dioxide columns retrieved from SCIAMACHY by WFM-DOAS: Year 2003 initial data set, *Atmos. Chem. Phys.*, *5*, 3313–3329, 2005b.

Clerbaux, C., J. Hadji-Lazaro, S. Turquety, G. Mégie, and P.-F. Coheur, Trace gas measurements from infrared satellite for chemistry and climate applications, *Atmos. Chem. Phys.*, *3*, 1495–1508, 2003.

Crutzen, P. J., Photochemical reactions initiated by and influencing ozone in unpolluted tropospheric air, *Tellus*, *26*, 46–57, 1974.

Dentener, F., W. Peters, M. Krol, M. van Weele, P. Bergamaschi, and J. Lelieveld, Interannual variability and trend of CH_4 lifetime as a measure for OH changes in the 1979–1993 time period, *J. Geophys. Res.*, *108*(D15), 4442, doi:4410.1029/2002JD002916, 2003.

Dlugokencky, E. J., K. A. Masaire, P. M. Lang, P. P. Tans, L. P. Steele, and E. G. Nisbet, A dramatic decrease in the growth-rate of atmospheric methane in the Northern-Hemisphere during 1992, *Geophys. Res. Lett.*, *21*(1), 45–48, 1994.

Dlugokencky, E. J., L. P. Steele, P. M. Lang, and K. A. Masarie, Atmospheric methane at Mauna-Loa and barrow observatories—Presentation and analysis of in-situ measurements, *J. Geophys. Res.*, *100*(D11), 23103–23113, 1995.

Dlugokencky, E. J., E. G. Dutton, P. C. Novelli, P. P. Tans, K. A. Masarie, K. O. Lantz, and S. Madronich, Changes in CH_4 and CO growth rates after the eruption of Mt Pinatubo and their link with changes in tropical tropospheric UV flux, *Geophys. Res. Lett.*, *23*(20), 2761–2764, 1996.

Dlugokencky, E. J., B. P. Walter, K. A. Masarie, P. M. Lang, and E. S. Kasischke, Measurements of an anomalous global methane increase during 1998, *Geophys. Res. Lett.*, *28*(3), 499–502, 2001.

Dlugokencky, E. J., S. Houweling, L. Bruhwiler, K. A. Masarie, P. M. Lang, J. B. Miller, and P. P. Tans, Atmospheric methane levels off: Temporary pause or a new steady-state?, *Geophys. Res. Lett.*, *30*(19), 1992, doi:1910.1029/2003GL018126, 2003.

Dueck, T. A., R. de Visser, H. Poorter, S. Persijn, A. A. Gorissen, W. W. de Visser, A. Schapendonk, J. Verhagen, J. Snel, F. J. M. Harren, A. K. Y. Ngai, F. Verstappen, H. Bouwmeester, L. A. C. J. Voesenek, and A. van der Werf, No evidence for substantial aerobic methane emission by terrestrial plants: A ^{13}C-labelling approach, *New Phytol.*, *175*(1), 29–35, doi:10.1111/j.1469-8137.2007.02103.x., 2007.

EEA, Annual European Community greenhouse gas inventory 1990–2001 and inventory report 2003, European Environment Agency, Copenhagen, 2003.

EEA, Annual European Community greenhouse gas inventory 1990–2002 and inventory report 2004, European Environment Agency, Copenhagen, 2004.

Enting, I. G., C. M. Trudinger, and R. J. Francey, A synthesis inversion of the concentration and δ^{13}C of atmospheric CO_2, *Tellus B*, *47*(1–2), 35–52, 1995.

Enting, I. G., Green's function methods of tracer inversion, in *Inverse methods in global biogeochemical cycles*, P. Kasibhatla et al. (eds.), pp. 19–31, American Geophysical Union, Washington, DC, 2000.

Etheridge, D. M., G. I. Pearman, and P. J. Fraser, Changes in tropospheric methane between 1841 and 1978 from a high accumulation-rate Antarctic ice core, *Tellus B*, *44*, 282–294, 1992.

Frankenberg, C., J. F. Meirink, M. van Weele, U. Platt, and T. Wagner, Assessing methane emissions from global space-borne observations, *Science*, *308*(5724), 1010–1014, 2005.

Frankenberg, C., J. F. Meirink, P. Bergamaschi, A. P. H. Goede, M. Heimann, S. Körner, U. Platt, M. van Weele, and T. Wagner, Satellite chartography of atmospheric methane from

SCIAMACHY onboard ENVISAT: Analysis of the years 2003 and 2004, *J. Geophys. Res.*, *111*, D07303, doi:07310.01029/02005JD006235, 2006.

Fung, I., J. John, J. Lerner, E. Matthews, M. Prather, L. P. Steele, and P. J. Fraser, Three-dimensional model synthesis of global methane cycle, *J. Geophys. Res.*, *96*, 13033–13065, 1991.

GLOBALVIEW-CH4, Cooperative Atmospheric Data Integration Project—Methane. CD-ROM, NOAA CMDL (Also available on Internet via anonymous FTP to ftp.cmdl.noaa.gov, Path: ccg/ch4/GLOBALVIEW), 2005.

Gurney, K. R., R. M. Law, A. S. Denning, P. J. Rayner, D. Baker, P. Bousquet, L. Bruhwiler, Y. H. Chen, P. Ciais, S. Fan, I. Y. Fung, M. Gloor, M. Heimann, K. Higuchi, J. John, T. Maki, S. Maksyutov, K. Masarie, P. Peylin, M. Prather, B. C. Pak, J. Randerson, J. Sarmiento, S. Taguchi, T. Takahashi, and C. W. Yuen, Towards robust regional estimates of CO_2 sources and sinks using atmospheric transport models, *Nature*, *415*(6872), 626–630, 2002.

Hauglustaine, D. A., F. Hourdin, L. Jourdain, M. A. Filiberti, S. Walters, J. F. Lamarque, and E. A. Holland, Interactive chemistry in the Laboratoire de Meteorologie Dynamique general circulation model: Description and background tropospheric chemistry evaluation, *J. Geophys. Res.*, *109*(D4), D04314, doi:04310.01029/02003JD003957, 2004.

Heimann, M., and T. Kaminski, Inverse modeling approaches to infer surface trace gas fluxes from observed atmospheric mixing ratios. Approaches to scaling of trace gas fluxes in ecosystems, in *Approaches to scaling of trace gas fluxes in ecosystems*, A. F. Bouwman (ed.), pp. 275–295, Elsevier, Amsterdam, 1999.

Hein, R., P. J. Crutzen, and M. Heimann, An inverse modeling approach to investigate the global atmospheric methane cycle, *Global Biogeochem. Cycles*, *11*(1), 43–76, 1997.

Hoerling, M., and A. Kumar, The perfect ocean for drought, *Science*, *299*(5607), 691–694, 2003.

Houweling, S., T. Kaminski, F. Dentener, J. Lelieveld, and M. Heimann, Inverse modelling of methane sources and sinks using the adjoint of a global transport model, *J. Geophys. Res.*, *104*, 26137–26160, 1999.

Houweling, S., F. Dentener, and J. Lelieveld, Simulation of preindustrial atmospheric methane to constrain the global source strength of natural wetlands, *J. Geophys. Res.*, *105*(D13), 17243–17255, 2000.

IPCC, Good practice guidance and uncertainty management in national greenhouse gas inventories, Institute for Global Environmental Strategies, Japan, 2000.

IPCC, Climate change 2001: The scientific basis. Contribution of working group I to the third assessment report of the Intergovernmental Panel on Climate Change, 881 pp., Cambridge University Press, Cambridge, UK and New York, NY, 2001.

IPCC, Climate change 2007: The physical science basis. Contribution of working group I to the fourth assessment report of the Intergovernmental Panel on Climate Change 996 pp., Cambridge University Press, Cambridge, United Kingdom and New York, NY, USA, 2007.

Keppler, F., J. T. G. Hamilton, M. Brass, and T. Rockmann, Methane emissions from terrestrial plants under aerobic conditions, *Nature*, *439*, 187–191, 2006.

Kirschbaum, M. U. F., D. Bruhn, D. M. Etheridge, J. R. Evans, G. D. Farquhar, R. M. Gifford, K. I. Paul, and A. J. Winters, A comment on the quantitative significance of aerobic methane release by plants, *Funct. Plant Biol.*, *33*, 521–530, 2006.

Krol, M., and J. Lelieveld, Can the variability in tropospheric OH be deduced from measurements of 1,1,1-trichloroethane (methyl chloroform)?, *J. Geophys. Res.*, *108*(D3), 4125, doi:4110.1029/2002JD002423, 2003.

Krol, M., S. Houweling, B. Bregman, M. van den Broek, A. Segers, P. van Velthoven, W. Peters, F. Dentener, and P. Bergamaschi, The two-way nested global chemistry-transport zoom model TM5: Algorithm and applications, *Atmos. Chem. Phys.*, *5*, 417–432, 2005.

Krol, M. C., J. Lelieveld, D. E. Oram, G. A. Sturrock, S. A. Penkett, C. A. M. Brenninkmeijer, V. Gros, J. Williams, and H. A. Scheeren, Continuing emissions of methyl chloroform from Europe, *Nature*, *421*(6919), 131–135, 2003.

Langenfelds, R. L., R. J. Francey, B. C. Pak, L. P. Steele, J. Lloyd, C. M. Trudinger, and C. E. Allison, Interannual growth rate variations of atmospheric CO_2 and its $\delta^{13}C$, H_2, CH_4, and CO between 1992 and 1999 linked to biomass burning, *Global Biogeochem. Cycles*, *16*(3), 1048, doi:1010.1029/2001GB001466, 2002.

Manning, A. J., D. B. Ryall, R. G. Derwent, P. G. Simmonds, and S. O'Doherty, Estimating European emissions of ozone-depleting and greenhouse gases using observations and a modeling back-attribution technique, *J. Geophys. Res.*, *108*(D14), 4405, doi:4410.1029/2002JD002312, 2003.

Manning, M. R., D. C. Lowe, R. C. Moss, G. E. Bodeker, and W. Allan, Short-term variations in the oxidizing power of the atmosphere, *Nature*, *436*, 1001–1004, 2005.

Matthews, E., and I. Fung, Methane emissions from natural wetlands, global distribution, area and environmental characteristics of sources, *Global Biogeochem. Cycles*, *1*, 61–86, 1987.

Matthews, E., and I. Fung, Methane emissions from rice cultivation: Geographic and seasonal distribution of cultivated areas and emissions, *Global Biogeochem. Cycles*, *5*, 3–24, 1991.

Meirink, J. F., H. J. Eskes, and A. P. H. Goede, Sensitivity analysis of methane emissions derived from SCIAMACHY observations through inverse modelling, *Atmos. Chem. Phys.*, *6*, 1275–1292, 2006.

Mikaloff Fletcher, S. E. M., P. P. Tans, L. M. Bruhwiler, J. B. Miller, and M. Heimann, CH_4 sources estimated from atmospheric observations of CH_4 and its $^{13}C/^{12}C$ isotopic ratios: 1. Inverse modeling of source processes, *Global Biogeochem. Cycles*, *18*(4), GB4004, doi:4010.1029/2004GB002223, 2004a.

Mikaloff Fletcher, S. E. M., P. P. Tans, L. M. Bruhwiler, J. B. Miller, and M. Heimann, CH_4 sources estimated from atmospheric observations of CH_4 and its $^{13}C/^{12}C$ isotopic ratios: 2. Inverse modeling of CH_4 fluxes from geographical regions, *Global Biogeochem. Cycles*, *18*(4), GB4005, doi:4010.1029/2004GB002224, 2004b.

Miller, J. B., K. A. Mack, R. Dissly, J. W. C. White, E. J. Dlugokencky, and P. P. Tans, Development of analytical methods and measurements of $^{13}C/^{12}C$ in atmospheric CH_4 from the NOAA Climate Monitoring and Diagnostics Laboratory global air sampling network, *J. Geophys. Res.*, *107*(D13), 107, doi:110.1029/2001JD000630, 2002.

Olivier, J. G. J., and J. J. M. Berdowski, Global emissions sources and sinks, in *The climate system*, J. Berdowski, et al. (eds.), pp. 33–37, 2001.

Page, S. E., F. Siegert, J. O. Rieley, H. D. V. Boehm, A. Jaya, and S. Limin, The amount of carbon released from peat and forest fires in Indonesia during 1997, *Nature*, *420*(6911), 61–65, 2002.

Petit, J., J. Jouzel, D. Raynaud, N. Barkov, J. Barnola, I. Basile, M. Bender, J. Chappellaz, M. Davis, G. Delaygue, M. Delmotte, V. Kotlyakov, M. Legrand, V. Lipenkov, C. Lorius, L. Pepin, C. Ritz, E. Saltzman, and M. Stievenard, Climate and atmospheric history of the past 420,000 years from the Vostok ice core, Antarctica, *Nature*, *399*(6735), 429–436, 1999.

Peylin, P., D. Baker, J. Sarmiento, P. Ciais, and P. Bousquet, Influence of transport uncertainty on annual mean and seasonal inversions of atmospheric CO_2 data, *J. Geophys. Res.*, *107*(D19), 4385, doi:4310.1029/2001JD000857, 2002.

Platt, U., W. Allan, and D. Lowe, Hemispheric average Cl atom concentration from $^{13}C/^{12}C$ ratios in atmospheric methane, *Atmos. Chem. Phys.*, *4*, 2393–2399, 2004.

Prigent, C., E. Matthews, F. Aires, and W. B. Rossow, Remote sensing of global wetland dynamics with multiple satellite data sets, *Geophys. Res. Lett.*, *28*(24), 4631–4634, doi:4610.1029/2001GL013263, 2001.

Prinn, R. G., R. F. Weiss, B. R. Miller, J. Huang, F. N. Alyea, D. M. Cunnold, P. J. Fraser, D. E. Hartley, and P. G. Simmonds, Atmospheric trends and lifetime of CH_3CCl_3 and global OH concentrations, *Science*, *269*(5221), 187–192, 1995.

Prinn, R. G., R. F. Weiss, P. J. Fraser, P. G. Simmonds, D. M. Cunnold, F. N. Alyea, S. O'Doherty, P. Salameh, B. R. Miller, J. Huang, R. H. J. Wang, D. E. Hartley, C. Harth, L. P. Steele, G. Sturrock, P. M. Midgley, and A. McCulloch, A history of chemically and radiatively important gases in air deduced from ALE/GAGE/AGAGE, *J. Geophys. Res.*, *105*(D14), 17751–17792, 2000.

Prinn, R. G., J. Huang, R. F. Weiss, D. M. Cunnold, P. J. Fraser, P. G. Simmonds, A. McCulloch, C. Harth, P. Salameh, S. O'Doherty, R. H. J. Wang, L. Porter, and B. R. Miller, Evidence for substantial variations of atmospheric hydroxyl radicals in the past two decades, *Science*, *292*, 1882–1888, 2001.

Prinn, R. G., J. Huang, R. F. Weiss, D. M. Cunnold, P. J. Fraser, P. G. Simmonds, A. McCulloch, C. Harth, S. Reimann, P. Salameh, S. O'Doherty, R. H. J. Wang, L. W. Porter, B. R. Miller, and

P. B. Krummel, Evidence for variability of atmospheric hydroxyl radicals over the past quarter century, *Geophys. Res. Lett.*, *32*(7), L07809, doi:07810.01029/02004GL022228, 2005.

Quay, P. D., J. L. Stutsman, D. O. Wilbur, A. K. Snover, E. J. Dlugokencky, and T. A. Brown, The isotopic composition of atmospheric methane, *Global Biogeochem. Cycles*, *13*, 445–461, 1999.

Rayner, P. J., I. G. Enting, R. J. Francey, and R. Langenfelds, Reconstructing the recent carbon cycle from atmospheric CO_2, $\delta^{13}C$ and O_2/N_2 observations, *Tellus B*, *51B*(2), 213–232, 1999.

Rodenbeck, C., S. Houweling, M. Gloor, and M. Heimann, CO_2 flux history 1982–2001 inferred from atmospheric data using a global inversion of atmospheric transport, *Atmos. Chem. Phys.*, *3*, 1919–1964, 2003.

Simmonds, P. G., A. J. Manning, R. G. Derwent, P. Ciais, M. Ramonet, V. Kazan, and D. Ryall, A burning question. Can recent growth rate anomalies in the greenhouse gases be attributed to large-scale biomass burning events?, *Atmos. Environ.*, *39*(14), 2513–2517, 2005.

Spahni, R., J. Chappellaz, T. F. Stocker, L. Loulergue, G. Hausammann, K. Kawamura, J. Flückiger, J. Schwander, D. Raynaud, V. Masson-Delmotte, and J. Jouzel, Atmospheric methane and nitrous oxide of the late Pleistocene from Antarctic ice cores, *Science*, *310*, 1317–1321, 2005.

Spivakovsky, C. M., J. A. Logan, S. A. Montzka, Y. J. Balkanski, M. Foreman-Fowler, D. B. A. Jones, L. W. Horowitz, A. C. Fusco, C. A. M. Brenninkmeijer, M. J. Prather, S. C. Wofsy, and M. B. McElroy, Three-dimensional climatological distribution of tropospheric OH: Update and evaluation, *J. Geophys. Res.*, *105*(D7), 8931–8980, 2000.

Turquety, S., J. Hadji-Lazaro, C. Clerbaux, D. A. Hauglustaine, S. A. Clough, V. Casse, P. Schlüssel, and G. Megie, Operational trace gas retrieval algorithm for the Infrared Atmospheric Sounding Interferometer, *J. Geophys. Res.*, *109*, doi:10.1029/2004JD004821, 2004.

Tyler, S. C., H. O. Ajie, M. L. Gupta, R. J. Cicerone, D. R. Blake, and E. J. Dlugokencky, Stable carbon isotopic composition of atmospheric methane: A comparison of surface level and free tropospheric air, *J. Geophys. Res.*, *104*(D11), 13895–13910, 1999.

Uppala, S. M., P. W. Koallberg, A. J. Simmons, U. Andrae, V. da Costa Bechtold, M. Fiorino, J. K. Gibson, J. Haseler, A. Hernandez, G. Kelly, X. Li, K. Onogi, S. Saarinen, N. Sokka, R. P. Allan, E. Andersson, K. Arpe, M. A. Balmaseda, A. C. M. Beljaars, L. van de Berg, J. Bidlot, N. Bormann, S. Caires, F. Chevallier, D. A, M. Dragosavac, M. Fisher, M. Fuentes, S. Hagemann, E. Holm, B. J. Hoskins, L. Isaksen, P. A. E. M. Janssen, R. Jenne, A. P. A. McNally, J.-F. Mahfouf, J.-J. Morcrette, N. A. Rayner, R. W. Saunders, P. Simon, A. Sterl, K. E. Trenberth, A. Untch, D. Vasiljevic, P. Viterbo, and J. Woollen, The ERA-40 Reanalysis, *J. Roy. Met. Soc.*, *131*, 2961–3012, 2005.

Van der Werf, G. R., J. T. Randerson, G. J. Collatz, and L. Giglio, Carbon emissions from fires in tropical and subtropical ecosystems, *Global Change Biol.*, *9*(4), 547–562, 2003.

Van der Werf, G. R., J. T. Randerson, G. J. Collatz, L. Giglio, P. S. Kasibhatla, A. F. Arellano, S. C. Olsen, and E. S. Kasischke, Continental-scale partitioning of fire emissions during the 1997 to 2001 El Nino/La Nina period, *Science*, *303*(5654), 73–76, 2004.

Vermeulen, A., R. Eisma, A. Hensen, and J. Slanina, Transport calculations of NW-European methane emissions, *Env. Sci. & Policy*, *2*, 315–324, 1999.

Wahlen, M., N. Tanaka, R. Henry, and T. Yoshinari, ^{13}C, D and ^{14}C in Methane, *Eos*, *68, No. 44*, 1220, 1987.

Walter, B. P., M. Heimann, and E. Matthews, Modeling modern methane emissions from natural wetlands 2. Interannual variations 1982–1993, *J. Geophys. Res.*, *106*(D24), 34207–34219, 2001.

Wang, J. S., J. A. Logan, M. B. McElroy, B. N. Duncan, I. A. Megretskaia, and R. M. Yantosca, A 3-D model analysis of the slowdown and interannual variability in the methane growth rate from 1988 to 1997, *Global Biogeochem. Cycles*, *18*(3), 3011, doi:3010.1029/2003GB002180, 2004.

Warwick, N. J., S. Bekki, K. S. Law, E. G. Nisbet, and J. A. Pyle, The impact of meteorology on the interannual growth rate of atmospheric methane, *Geophys. Res. Lett.*, *29*(20), 1947, doi: 1910.1029/2002GLO15282, 2002.

WMO, Global Atmospheric Watch World Data Centre for Greenhouse Gases, 92 pp., Japan Meteorological Agency in co-operation with World Meteorological Organisation, 2003.

Chapter 8
Designing an Observation Strategy for N$_2$O

Annette Freibauer

8.1   Introduction

Nitrous oxide (N$_2$O) is a powerful greenhouse gas (GHG) with a global warming potential (GWP) of 296 in relation to CO$_2$ (IPCC 2001). In the context of the United Nations Framework Convention on Climate Change (UNFCCC), a GWP value of 310 is used (IPCC 1996). For easy comparison, we express N$_2$O fluxes in units of nitrogen (N) and in CO$_2$-carbon equivalents (Ceq), applying the politically relevant GWP of 310 (Eq. 8.1). N$_2$O is also involved in the depletion of the stratospheric ozone layer.

$$1 \, \text{kg N}_2\text{O-N} = 44/28 \times 310 \times 12/44 \, \text{kg Ceq} = 133 \, \text{kg Ceq} \qquad (8.1)$$

where 44/28 is the conversion from N$_2$O-N to N$_2$O, the GWP of 1 kg N$_2$O = 310 kg CO$_2$ equivalents and 12/44 is the conversion from CO$_2$ equivalents to C equivalents.

The atmospheric concentration of N$_2$O has increased from 270 ppb to more than 320 ppb today (IPCC 2001). The lifetime of N$_2$O in the atmosphere is about 114 years (IPCC 2001). Globally, the microbial processes of nitrification and denitrification in soils constitute the largest source of N$_2$O. Global anthropogenic N$_2$O emissions are estimated to be 6.9 Tg N year^{-1} (0.87 Gt Ceq year^{-1}, average in 1990s), more than half of which result from fertilizer application to soils. Human activities contribute about 40% to the total global N$_2$O emissions. N$_2$O accounts for 6% of the global anthropogenic GHG emissions (IPCC 2001).

N$_2$O is photochemically destroyed in the stratosphere through reaction with excited oxygen atoms. There is an evidence that a large proportion of the current N$_2$O emissions is removed in the stratosphere, but the global N$_2$O sources have not yet been fully identified (IPCC 2001).

This chapter will give an overview of the present N$_2$O emissions in Europe, and then focus on the major source, the soilborne N$_2$O emissions. The spatio-temporal variation of soil N$_2$O fluxes and underlying biogeochemical processes will be described, as well as approaches to observe and quantify N$_2$O fluxes, associated uncertainties and ways towards a systematic operational N$_2$O monitoring. In the latter part, N$_2$O emissions from other sources than soils will be included.

8.2 N_2O Emissions in Europe

According to the UNFCCC (http://ghg.unfccc.int/index.html), the European Community (EU-27) emitted 0.86 Tg N as anthropogenic N_2O in 2005 (Table 8.1). Emissions tended to decrease during the past 15 years. The N_2O emission in 2005 was equivalent to 13% of the global anthropogenic N_2O emissions and 7% of Europe's direct GHG emissions. The national communications contain data on anthropogenic N_2O emissions: industry, road traffic, waste, animal husbandry, fertilizer application and so-called "indirect N_2O emissions" due to atmospheric N deposition. The most important sources of N_2O in Europe are agricultural soils, followed by industrial nylon and nitric acid production (Table 8.1). The anthropogenic emissions of N_2O from soils are caused by fertilizer use, N fixation by agricultural crops and atmospheric N deposition (Nevison and Holland 1997). Soils also emit N_2O naturally without human intervention. This is often called "natural background". Soils contribute by far the largest uncertainty to anthropogenic N_2O emission estimates (Table 8.1). In total, N_2O flux rates are associated with much higher uncertainty than CO_2 exchange in terrestrial ecosystems.

Table 8.1 Estimated budget of anthropogenic N_2O for the European Community (ETC, 2007). Emissions are derived from the national communications (http://ghg.unfccc.int/index.html) and uncertainties are based on IPCC (2000) and Olivier et al. (1998)

EU-27 in 2005	N_2O emissions Gg N year[-1]	Uncertainty Gg N year[-1]	%
Agricultural soils, "natural background" (Bouwman, 1996)	EU-27: 200 EU-15: 130	160 104	80
Anthropogenic N_2O emissions from national communications			
Soils anthropogenic total	EU-27: 499 EU-15: 346	400 277	80
Agricultural soils direct	267	214	80
Agricultural soils indirect	169	135	80
Manure management	60	45	75
LULUCF	5	4	80
Industrial processes	128	13	10
Transport	52	26	50
Other energy	74	22	30
Waste	27	27	100
Other sources	9	9	100
Anthropogenic N_2O (Gg N year[-1])	EU-27: 861 EU-15: 687	551 440	64 64
Anthropogenic N_2O in EU-27 (Tg Ceq year[-1])	91	58	
Soilborne N_2O emissions in EU-15 from models Agricultural soils in EU-15	442[a]	116[a]	26
Forest soils in EU-15	78–87[b]		
Soil N_2O emissions in EU-15 [Gg N year[-1]]	525		

LULUCF Land-Use, Land-Use Change and Forestry
[a]Freibauer (2003)
[b]Kesik et al. (2005)

Soilborne N$_2$O emissions are separated into agricultural N fertilization (Table 8.1, "agricultural soils direct") and additional N$_2$O emissions from soil that can be directly attributed to human N input, including enhanced N$_2$O emissions in natural ecosystems because of atmospheric N deposition (Table 8.1, "agricultural soils indirect"). The so-called "natural background N$_2$O emissions" from agricultural soils are excluded from these anthropogenic N$_2$O emission calculations. Assuming the old default value of 1 kg N$_2$O-N ha^{-1} year^{-1} as background (Bouwman 1996), an estimate of these "background emissions" is given in Table 8.1 as well. Total soilborne N$_2$O emissions have also been calculated by empirical and biogeochemical models (Table 8.1). The reported N$_2$O emissions agree with these independent model-based estimates of the N$_2$O emissions from agricultural soils and forests when the "background emissions" are added (Table 8.1), but uncertainties in both estimates are high.

8.3   Regulation of N$_2$O Emissions in the Terrestrial Biosphere

8.3.1   N$_2$O Formation in Soils

N$_2$O is a by-product of nitrification, an aerobic microbial process converting ammonium to nitrate, and of denitrification, an anaerobic microbial process converting nitrate in two steps to N$_2$O, and then N$_2$O to N$_2$. At the micro-scale, process level control of N$_2$O production by nitrification and denitrification is well known (Mosier et al. 1996; Skiba et al. 1998; Smith et al. 1998). Both processes are highly sensitive to temperature, substrate availability, pH and oxygen, the latter being controlled by soil moisture status and microbial activity. In agricultural soils, N$_2$O emissions are additionally controlled by management factors, among which crop type, fertilizer type and N application rates seem to be most relevant (Stehfest and Bouwman 2006). Figure 8.1 shows the dependence of N$_2$O production on soil temperature and moisture in a schematic way. It is obvious that the relationships are highly non-linear and complex when considering that temperature and soil moisture interact with each other and that nitrification and denitrification can occur simultaneously. Environmental factors are in most cases unknown at the micro-scale, but only measured at the site scale. It is also the ecosystem scale at which model input parameters are determined (Mosier et al. 1996). Temperature, precipitation/soil moisture and management activities regulate N$_2$O emissions in a complex non-linear way at the daily timescale (Freibauer and Kaltschmitt 2003; Smith et al. 1998). However, at annual timescales, the explanatory power of each of the environmental and management factors at the ecosystem scale is usually low (Freibauer and Kaltschmitt 2003).

Natural abundance of ^{15}N-N$_2$O and ^{18}O-N$_2$O may be a suitable proxy for attribution of tropospheric N$_2$O to agricultural soils (Kaiser et al. 2004; Kim and Craig 1993; Perez et al. 2001; Röckmann et al. 2003) although the signal is partly disturbed by ocean and in situ atmospheric processes, and the level of depletion of N$_2$O from agricultural soils in ^{15}N and ^{18}O is not constant (Tilsner et al. 2003;

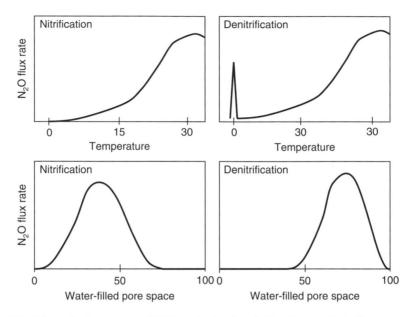

Fig. 8.1 Schematic dependence of N$_2$O production by nitrification and denitrification on soil temperature and moisture. Graphs for soil moisture are redrawn from Firestone and Davidson (1989).

Yamulki et al. 2001). At ecosystem level, N$_2$O forming processes of nitrification and denitrification can be distinguished by isotopic studies of natural abundance ^{15}N-N$_2$O and ^{18}O-N$_2$O or labelling experiments in conjunction with laboratory incubation studies (Tilsner et al. 2003; Yamulki et al. 2001). This distinction between processes provides valuable insights for validation of process-based models.

8.3.2 Spatio-Temporal Emission Patterns of Soilborne N$_2$O Emissions

Brumme et al. (1999) introduce a valuable concept of controls on seasonal and annual N$_2$O fluxes for forest ecosystems, which can be expanded to other land use types (Fig. 8.2).

8.3.2.1 Long-Term Controls of N$_2$O Fluxes (Ecosystem and Climatic Properties)

Background emission patterns show no correlation with long-term controls of N$_2$O fluxes and are characterized by low fluxes throughout the year. Seasonal emission patterns are explained by the combined effect of high annual precipitation,

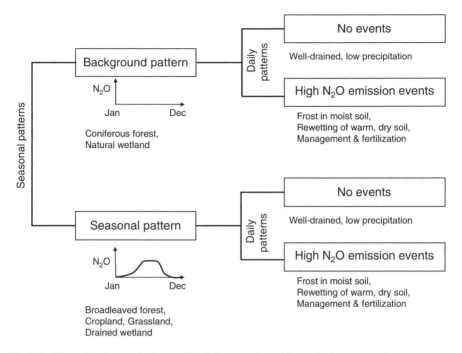

Fig. 8.2 Hierarchical control of annual N₂O fluxes and resulting emission patterns in ecosystems, generalized and extended from Brumme et al. (1999).

broadleaved trees, amount and structure of soil litter layer, high mineral bulk densities and plant community (Brumme et al. 1999). They roughly follow the seasonal cycle of temperature.

8.3.2.2   Short-Term Controls of N₂O Fluxes (Weather, Management)

Short-term bursts of N₂O occur on top of these seasonal patterns. Unlike CO_2, N₂O emissions from soils can have a strong episodic nature. Extreme events lasting a few hours to days can contribute up to 80% of the annual N₂O emissions.

N₂O emissions typically exhibit a high degree of spatial variability (75–100%; reviewed by Freibauer and Kaltschmitt 2003). Managed, in particular fertilized, soils tend to show a wider site-to-site variation in the magnitude of annual N₂O emissions than unmanaged ones. In contrast, "indirect" N₂O emissions from diffuse pollution of aquifers is probably smaller than originally expected (Nevison 2000).

Effects of fertilizer type, timing and quantity, and of crop species and crop rotation have been commonly studied by factorial designs. They have provided an extensive dataset for deriving emission factors, parametrization and validation of process-based models and the attribution of emissions to recent human interference versus natural/land use history-based emissions (Bouwman et al. 2002a; Stehfest and Bouwman 2006). Long-term studies with factorial design are the key source of information for N_2O monitoring.

Adequate information of key drivers of N_2O emissions at ecosystem scale is essential for bottom–up flux estimates. Whilst general information about vegetation type, tree species and agricultural crops is available in reasonable quality, major gaps in knowledge still exist with regard to spatially explicit soil characteristics (organic C content, moisture) and agricultural management. The latter can be overcome by coupling farm-economic with ecological models (Neufeldt et al. 2004) and better exploitation of agricultural census data.

8.4   Current Monitoring Capabilities

N_2O poses much greater challenges than CO_2 to an adequate observational design and protocol at ecosystem level, and to quantitative modelling and upscaling. However, considerable progress has been made in the bottom–up quantification of N_2O emissions over the last decade.

8.4.1   Flux Measurements at Ecosystem Level

Annual or longer N_2O flux measurements have been performed at more than 100 sites in Europe, covering a wide range of ecosystems. On agricultural sites, measurements often encompass various fertilizer treatments or crop types. Most researchers have applied manual static chambers combined with gas chromatography and electron capture detector or infrared gas analysis and measure in time steps of days to weeks, so the necessary temporal interpolation between measurements introduces an additional uncertainty to annual N_2O flux estimates. A comparison with continuous measurements demonstrated that a combination of weekly with event-driven sampling can produce robust annual N_2O flux estimates with an accuracy of better than 25% (Ruser 1999; Butterbach-Bahl personal communication). Chamber techniques are well-established, cheap and easily applicable but labour intensive unless sampling is automated. Automated chambers are increasingly used. Automated systems often turned out too expensive to allow a sufficient number of replicates so that they are used in combination with manual chambers to address spatial variability. Chambers are useful in experiments with different treatments (fertilizer, crop type) as in agricultural systems. For monitoring, an adequate number of replicates and a minimum surface area of $1\,m^2$ covered by the chambers, depending on the spatial variability of a site, are recommended (Heinemeyer et al. 1995). Tunable diode lasers and most recently quantum cascade lasers offer a new methodology for N_2O flux measurements by eddy covariance (Kroon et al. 2007). Chamber and micrometeorological techniques were shown to produce comparable results (Christensen et al. 1996; Smith et al. 1994). Well-aerated soils have been widely covered in experiments, but there are clear gaps in knowledge about soils with high organic contents, gleyic and stagnic properties and under alternating-wet conditions, as well as in Mediterranean and boreal climate. The European research projects, NOFRETETE and NitroEurope, are providing the first multi-year, quasi-continuous N_2O flux measurements in ecosystems across Europe with automated static chambers.

8.4.2 Observations at Regional Level

Micrometeorological measurements have successfully been applied in campaign mode using flux gradient methods, eddy covariance, mass balance (outgoing vs. incoming winds), and nocturnal boundary layer approaches (Denmead et al. 2000; Griffith et al. 2002; Hensen et al. 1999). Whilst results generally confirm the plot-scale measurements, operability of FTIR and tunable diode lasers sensors and sometimes inadequate sensitivity of the approaches to detect small fluxes still limit their use for monitoring purposes.

8.4.3 Atmospheric Observations at Continental to Global Level

The first global atmospheric N_2O monitoring networks started around 1980 (ALE/ GAGE/AGAGE,[1] NOAA-CMDL[2]). AGAGE operates continuous N_2O measurements in the lower troposphere at ground-level stations in unpolluted areas. NOAA-CMDL operates a discontinuous flask measurement programme. In Europe, further national stations came into operation in the mid-1990s. In the frame of CarboEurope, N_2O was included in the emerging multi-species observation network in Europe. N_2O concentration measurements are, however, less standardized than those of CO_2. The CarboEurope network comprises 9 continuous stations (2 ground monitoring stations, 7 tall towers), 21 flask sampling sites and 5 (+3 in Russia) vertical aircraft profile sites at which N_2O concentrations have been or are measured on a voluntary basis. Because of the small concentration gradients of N_2O, detecting an atmospheric signal remains difficult (Schmidt et al. 2001). In contrast, the information contained in existing N_2O concentration measurements has not been systematically exploited. A major limitation in these data was the lack of an international primary scale standard for N_2O that would allow inter-calibrating the various existing N_2O observational sites and networks. Such a scale has only very recently been established (Hall et al. 2007). The small concentration gradients are, therefore, further perturbed by likely calibration offsets. The isotopic information in atmospheric N_2O has rarely been

---

[1] ALE, Atmospheric Lifetime Experiment; GAGE, Global Atmospheric Gases Experiment; AGAGE, Advanced Global Atmospheric Gases Experiment.

[2] N_2O and Halocarbons Group, Climate Monitoring and Diagnostics Laboratory, NOAA.

measured, and the observed long-term trends in Antarctica (Röckmann and Levin 2005) have only been used for source/sink partitioning of the changes in the global atmospheric N_2O burden not to attribute regional N_2O sources. Chedin et al. (2002) propose retrieving night-time atmospheric N_2O concentrations from existing satellite observations in a latitudinal band relevant for Europe. It is unclear whether this satellite product will become operational and useful for top–down modelling.

8.4.4   Bottom–Up Modelling

Literature provides numerous plot-scale studies, from which empirical emission factors or models for inventory estimates were derived. Examples for global sectoral inventories are Bouwman et al. (2002b) and those summarized in Table 4.4 of IPCC (2001). Inventories apparently give a surprisingly robust estimate of fluxes when aggregated over larger areas (national to continental) within large uncertainties, but fail to depict the variation of N_2O emissions with inter-annual variability of climate and crop rotations, or regional patterns. Empirical models depict some of the regional patterns of soil and climate with large uncertainties, but still have a limited capacity to explain the variation in local N_2O fluxes. In Europe and EU member states, inventories of N_2O agricultural emissions were calculated by regional emission factors based on IPCC (national communications; Boeckx and Van Cleemput 2001; Brown et al. 2001), based on measurements (Lilly et al. 2003), by empirical regressions (Bouwman et al. 2002b; Conen et al. 2000; Freibauer and Kaltschmitt 2003; Sozanska et al. 2002; Stehfest and Bouwman 2006), regionalization of farm types and agro-pedological regions (Boeckx et al. 2001).

Mechanistic process models have so far been applied at local, national and regional levels only. They provide good diagnostics of temporal patterns of N_2O fluxes. The models are sensitive to soil temperature, moisture, texture and organic carbon (C) content. N_2O flux maps have been presented for agriculture and/or forest in selected regions of Europe based on DNDC (agriculture; Brown et al. 2002; Butterbach-Bahl et al. 2001), pNET-DNDC and regional versions of Expert-N (forest; Butterbach-Bahl et al. 2001; Butterbach-Bahl et al. 2004; Kesik et al. 2005; Schulte-Bisping et al. 2003). The modelled spatial patterns of N_2O fluxes seem to adequately represent the expected magnitude of N_2O emissions in standard situations of well-aerated soils with low to medium organic C content, but uncertain in under-sampled areas with likely high emissions (gleyic, stagnic soil properties and high soil organic C). Nevertheless, the explanatory power of the models for site-level N_2O fluxes is often low. Obviously, not only the magnitude of drivers, for example, fertilizer application rates, but also the intensity of gradients at small spatial scales matter. This goes beyond the feasible level of detail in existing models.

The comparison between approaches cited above reveals differences in national N_2O budgets of up to a factor of 2, which can be partly explained by different scope or omissions of sources in older inventories. At sub-continental aggregation, a better agreement is achieved (e.g. Freibauer 2003; Li et al. 2001).

8.4.5 Top–Down Modelling

N_2O fluxes have some principal differences to CO_2 that complicate observations and spatial scaling of fluxes. The earth surface almost entirely acts as a source of N_2O; small sinks were only found in some wet situations for a few days. N_2O sources are very homogeneously spread in space, but extreme N_2O emission events, for example, during freeze–thaw periods, after fertilization or when rewetting dry soils, make them very variable at synoptical timescales in a presently unpredictable manner. This feature, together with the long atmospheric residence time of N_2O, produces complex atmospheric N_2O concentration patterns and very small atmospheric N_2O concentration gradient signals. There are also N_2O sources in the atmosphere, which complicate the top–down verification, for example, review in Bouwman et al. (2000).

Nevertheless, the global N_2O emissions have been estimated by forward atmospheric transport modelling with very coarse resolution, but uncertainties are large (Bouwman et al. 2000) as a consequence of limited concentration data and small gradients. Using a ^{222}Rn tracer approach, Biraud et al. (2002) and Schmidt et al. (2001) demonstrated that high-precision continuous multi-species monitoring can derive top–down estimates of N_2O fluxes at landscape to continental scale. Top–down estimates are particularly useful to constrain regional budgets of the large-area, episodic N_2O sources from soils.

Hirsch et al. (2006) were the first to use the NOAA-CMDL flask data to inversely model N_2O source/sink patterns for four to six regions of the globe with an atmospheric transport model. They demonstrate the feasibility to calculate robust long-term average N_2O fluxes for large global regions. Uncertainties, however, remain large. A major constraint is the unknown temporal and spatial variation in stratosphere–troposphere exchange of N_2O.

8.5 Strategy Towards a Pan-European (Continental Scale) GHG Monitoring Network

This section sets forth a framework for integrated European N_2O monitoring based on observation-driven modelling. This co-ordinated system has three main objectives:

1. To provide the long-term observations and census information required to improve understanding of the present state and future behaviour of European N_2O emissions, particularly the interaction between human and natural factors responsible for major sources.
2. To integrate observations and models of the terrestrial C-, N- and water cycle.
3. To monitor and assess the effectiveness of N_2O emission reduction activities in Europe, including attribution of sources by region and sector.

The system will meet those objectives by routinely quantifying the regional distribution of N_2O fluxes exchanged between the European land surface and the atmosphere via measurements and models, along with observations and census information of drivers of nitrification and denitrification. Table 8.1 identifies soils as the dominant source of N_2O, and of uncertainty. Therefore, we will focus the strategy on the N_2O exchange between soils and the atmosphere.

N_2O can in principle be addressed by a monitoring system which has many analogies with CO_2 monitoring, that is, a multiple-constraint framework in which measurements at local scale feed into mechanistic process-oriented soil models with high time-space resolution. The results will be tested against atmospheric inverse model estimates based on atmospheric N_2O concentration measurements and, if feasible, remotely sensed N_2O column densities. As the correlation between environmental and management factors with observed N_2O fluxes is often weak mechanistic models may need some more development than for CO_2. Even more than in the case of CO_2, observation-driven, empirical models can provide equally accurate means of calculating contemporate N_2O fluxes as biogeochemical models. It is expected that it will take longer to develop fully operational observational and data assimilation schemes for N_2O than for CO_2. Methodological synergies with CO_2 and CH_4 monitoring will facilitate progress.

The long-term vision is to fully couple the C-, N- and water cycle in ecosystem models. This will allow to predict the response of the full GHG balance of ecosystems to climate variability and changing human management, and to simultaneously trace the fate of C and N in ecosystems. N_2O is a small constituent (typically ~1%) of the terrestrial N cycle. Consequently, even large errors and uncertainties in modelled N_2O only have a small effect on the overall predictive capacity of the models for other elements of the N cycle, but significantly affect the capacity of models for quantifying the feedback between climate change, human activities and ecosystem GHG fluxes.

8.6   Future Directions for Enhanced Observational Strategies

Systematic N_2O observations are still in their infancy as compared with CO_2. Future observational strategies need to test various designs to find out what is needed as minimum and what is feasible. Measurements will need to balance between process level information needed for model calibration and validation and spatial representativeness. N_2O monitoring can learn from CO_2 monitoring and should be closely linked with other trace gas observations. A continental N_2O observation network needs to address the high temporal and spatial variability of N_2O, in particular the irregular extreme N_2O emission events. Monitoring therefore has to be continuous and has to consider the various human factors of land management.

8.6.1 Directions for Enhanced Ecosystem Scale Measurements

To maximize synergy with existing observations of ecosystem GHG fluxes, systematic long-term in situ N_2O observations at ecosystem level should best be linked with the existing CO_2 observation network. The N_2O observation network should be balanced to cover all ecosystem types according to the emission patterns of Fig. 8.1, with a higher density in ecosystems that contribute most to European emissions, that is, croplands, grasslands, and N-rich wetlands. Measurements should aim at filling observational gaps in hotspots like (partially) moist ecosystems and compacted, N-rich farm areas and tractor lines that are potentially high N_2O emitters. Because of the strong influence of fertilization, fertilizer timing and other agricultural management practices, a thorough measurement design needs to rely on test farms where all representative crops of a regional crop rotation are measured in parallel as well as small, but critical hotspot areas. Continuous chamber measurements have proven to be a reliable operational means for these measurements.

Chamber measurements need to be performed in automatic, quasi-continuous mode, in particular in periods of likely high emission events. Measurement designs covering adequate surface area (>1 m^2) and adequate number of replicates (~8) have proven robust. Collocated data on meteorology, soil moisture, soil C and N, detailed management activities and, if possible, mineral N in soil, are required.

N_2O chamber measurements should be performed in a factorial design to cover the full range of possible emission events, local hotspots and different intensity levels of test farms and landscapes [e.g. variation in crop type, fertilizer type, tree type (coniferous/broadleaved), management, soil type, soil water status, compacted areas, areas where livestock preferably rests]. It is practically not feasible to have high-frequency measurements at all target locations with an adequate number of replicates. A good alternative is therefore to combine continuous chamber measurements at high sampling frequency with continuous measurements in a cumulative way as weekly or monthly average N_2O fluxes. The European NitroEurope project (www.nitroeurope.eu) explores cumulative approaches based on the well-established chamber technique.

For this purpose, cost-effective, low maintenance, in situ sensors for terrestrial ecosystem N_2O fluxes need to be developed. Such sensors could be constructed based on the knowledge for N_2O monitoring systems in hospitals but require a higher precision and sensitivity for N_2O at ambient concentration.

Unlike in the case of ecosystem C fluxes, N_2O measurements have not yet systematically been collected in a central manner. An international initiative to enhance data harmonization and inter-comparability, archiving and distribution of compiled observational data to the modelling community is needed. Existing international programmes in the field of reactive N could integrate N_2O into their activities, such as the International Nitrogen Initiative (INI), the COST Action 729 and the ESF programme NiNE.

8.6.2 Directions for Enhanced Regional Scale Measurements

New observational approaches should be tested to integrate N_2O emissions at farm to regional scale and to verify the highly uncertain upscaled local fluxes. Combinations of chamber measurements with mobile plume measurements appear promising (Hensen et al. 2006) as well as night-time accumulation of N_2O in the tropospheric inversion layer.

Isotope tracer methods should be further investigated to attribute atmospheric N_2O to different sources. Isotopic tracers give valuable additional information that helps interpreting the small atmospheric N_2O concentration gradients. In a similar way, the existing long-term continuous atmospheric CO_2 monitoring stations should be extended by high-precision, inter-calibrated N_2O and ^{222}Rn to multi-species sampling by flasks and also best in continuous mode. Further inner-continental stations are needed to reduce uncertainties over Eastern Europe where in situ measurements and information about drivers are particularly scarce. The inter-calibration of N_2O concentration measurements and exhaustive use of existing and upcoming N_2O concentration data from multi-species monitoring stations and flasks is essential to formulate the exact requirements for atmospheric N_2O observations. The use of remotely sensed N_2O column data should be further investigated. A pan-European towards systematic long-term observations of N_2O is needed. Unfortunately, N_2O was not considered as target species in the recent initiative towards an Integrated Carbon Observing System (ICOS, Chap. 18).

8.6.3 Requirements for Ancillary Information for Scaling

The major drivers for N_2O emissions have been identified and need to be monitored together with N_2O fluxes: Soil temperature and soil moisture status and their variability, as well as land use and management, in particular N fertilization.

Data sources for meteorological information are operational. Soil moisture can be calculated by the models. However, it would be worth testing to what extent remote sensing products for topsoil moisture agree with these calculations and with direct observations, and whether they provide additional constraints. Remote sensing products for soil moisture only exist in research mode.

Valuable census data at farm level are presently collected as a prerequisite for obtaining subsidies under the European Common Agricultural Policy, but these data are not public. At least in the context of the European network of test farms, plot-level and farm-gate C- and N-balances should be made available to the research and monitoring community. Regional C- and N-models require georeferenced information about the distribution of agricultural crop types and management dates. In this field, remote sensing methods may provide useful information. Crop types and field sizes are already being monitored from space in a statistical sampling mode in order to verify the farmers' submissions for subsidies under the Common Agricultural

Policy. Developing operational remote sensing products for agricultural crops and land management dates (harvest, ploughing) would significantly reduce the uncertainty in bottom–up modelling of N$_2$O emissions.

For top–down estimates, it remains to be tested whether continental scale tropospheric N$_2$O concentration gradients and vertical columns can be measured in adequate precision and resolution to constrain the terrestrial N$_2$O budget at this scale.

8.6.4 Stepwise Implementation of the Modelling Strategy

At present, bottom–up modelling of N$_2$O fluxes is immature as compared to CO$_2$. Because of the difficulty in modelling the temporal N$_2$O emission patterns and the large uncertainty in annual N$_2$O fluxes, a robust N$_2$O modelling strategy must rely on a suite of different models. Before N$_2$O modelling can become operational, several steps need to be taken.

An obvious first step is the choice of promising models for continental scale N$_2$O modelling. Existing process-oriented and empirical models need to be validated against independent observations from a wide range of ecosystems and further improved. Modules for important N$_2$O sources other than soil need to be added. Sensitivity tests shall identify the most important input parameters that need to be constrained by observations.

One major source of uncertainty is the inadequate spatio-temporal representation of soil moisture and of land management. It must be a priority to produce maps of farming patterns, fertilization rates and crop types. It has been proven very useful to couple N$_2$O models to farm-economic models, which allow spatializing the management patterns and farming intensity (Neufeldt et al. 2006). This approach needs to be further developed and expanded to European and global level.

Ultimately, as for CO$_2$, a suite of different biogeochemical models with fully coupled C–N–water cycles should be developed that can deliver routine diagnostics of the European N$_2$O emissions from soils. This effort should include bottom–up data assimilation schemes for N species.

For the top–down estimates, the suitability of various atmospheric models and approaches should be tested to constrain regional and continental N$_2$O budgets.

8.7 Conclusion

The key to improving N$_2$O emission estimates lies in the soilborne fluxes. This requires the scientific development of cheap and routine operational flux measurement techniques to improve the density of observations in critical ecosystems and seasons prone to high N$_2$O emissions, as well as improved process models. The ecosystem level N$_2$O observation network in Europe has emerged in an opportunistic way around interested research groups. The NitroEurope project is starting to co-ordinate and harmonize the ecosystem level observations. However, important potential emission hotspots are likely to be yet uncovered by observations. These can only be tackled by regional N_2O concentration measurements in the troposphere in conjunction with vertical atmospheric profiles to constrain the stratosphere–troposphere exchange of N_2O. Such observations are, however, not yet co-ordinated in Europe, but proved more demanding than those of CO_2. The continental co-ordination of observations and a systematic strategy for addressing the challenges of temporal and spatial scale transitions are still in their infancy.

References

Biraud, S. et al. 2002. Quantification of carbon dioxide, methane, nitrous oxide and chloroform emissions over Ireland from atmospheric observations at Mace Head. Tellus Series B-Chemical & Physical Meteorology, 54(1): 41–60.

Boeckx, P. and Van Cleemput, O. 2001. Estimates of N_2O and CH_4 fluxes from agricultural lands in various regions in Europe. Nutrient Cycling in Agroecosystems, 60(1–3): 35–47.

Boeckx, P., Van Moortel, E. and Van Cleemput, O. 2001. Spatial and sectorial disaggregation of N_2O emissions from agriculture in Belgium. Nutrient Cycling in Agroecosystems, 60(1–3): 197–208.

Bouwman, A.F. 1996. Direct emission of nitrous oxide from agricultural soils. Nutrient Cycling in Agroecosystems, 46: 53–70.

Bouwman, A.F., Boumans, L.J.M. and Batjes, N.H. 2002a. Emissions of N_2O and NO from fertilized fields: Summary of available measurement data. Global Biogeochemical Cycles, 16(18 October 2002).

Bouwman, A.F., Boumans, L.J.M. and Batjes, N.H. 2002b. Modeling global annual N_2O and NO emissions from fertilized fields. Global Biogeochemical Cycles, 16(4): 1080, doi:10.1029/2001BG001812.

Bouwman, A.F., Taylor, J.A. and Kroeze, C. 2000. Testing hypotheses on global emissions of nitrous oxide using atmospheric models. Chemosphere—Global Change Science, 2(3–4): 475–492.

Brown, L. et al. 2001. An inventory of nitrous oxide emissions from agriculture in the UK using the IPCC methodology: Emission estimate, uncertainty and sensitivity analysis. Atmospheric Environment, 35(8): 1439–1449.

Brown, L. et al. 2002. Development and application of a mechanistic model to estimate emission of nitrous oxide from UK agriculture. Atmospheric Environment, 36(6): 917–928.

Brumme, R., Borken, W. and Finke, S. 1999. Hierarchical control on nitrous oxide emission in forest ecosystems. Global Biogeochemical Cycles, 13(4): 1137–1148.

Butterbach-Bahl, K., Kesik, M., Miehle, P., Papen, H. and Li, C. 2004. Quantifying the regional source strength of N-trace gases across agricultural and forest ecosystems with process based models. Plant & Soil, 260(1–2): 311–329.

Butterbach-Bahl, K., Stange, F., Papen, H. and Li, C.S. 2001. Regional inventory of nitric oxide and nitrous oxide emissions for forest soils of southeast Germany using the biogeochemical model PnET-N-DNDC. Journal of Geophysical Research-Atmospheres, 106(D24): 34155–34166, doi 2000JD000173.

Chedin, A. et al. 2002. Annual and seasonal variations of atmospheric CO_2, N_2O and CO concentrations retrieved from NOAA/TOVS satellite observations. Geophysical Research Letters, 29(8): doi 10.1029/2001BL014082.

Christensen, S. et al. 1996. Nitrous oxide emission from an agricultural field—Comparison between measurements by flux chamber and micrometerological techniques. Atmospheric Environment, 30(24): 4183–4190.

Conen, F., Dobbie, K.E. and Smith, K.A. 2000. Predicting N_2O emissions from agricultural land through related soil parameters. Global Change Biology, 6(4): 417–426.

Denmead, O.T., Leuning, R., Jamie, I. and Griffith, D.W.T. 2000. Nitrous oxide emissions from grazed pastures: Measurements at different scales. Chemosphere—Global Change Science, 2(3–4): 301–312, doi:10.1016/S1465-9972(00)00035-0.

ETC. 2007. National greenhouse gases inventories (IPCC Common Reporting Format sector classification). The European Topic Centre on Air and Climate Change (Database version 2.3, 12 June 2007), national submissions to UNFCCC or to EU Monitoring Mechanism of CO_2 and other greenhouse emissions. http://dataservice.eea.europa.eu/dataservice/metadetails. asp?id=971, downloaded on 10 April 2007.

Firestone, M.K. and Davidson, E.A. 1989. Microbiological basis of NO and N_2O production and consumption in soil. In: M.O. Andreae and D.S. Schimel (Eds.), Exchange of Trace Gases between Terrestrial Ecosystems and the Atmosphere. Dahlem Konferenzen. John Wiley & Sons, Chichester, pp. 7–21.

Freibauer, A. 2003. Regionalised inventory of biogenic greenhouse gas emissions from European agriculture. European Journal of Agronomy, 19(2): 135–160.

Freibauer, A. and Kaltschmitt, M. 2003. Controls and models for estimating direct nitrous oxide emissions from temperate and boreal agricultural mineral soils in Europe. Biogeochemistry, 63(1): 93–115.

Griffith, D.W.T., Leuning, R., Denmead, O.T. and Jamie, I.M. 2002. Air-land exchanges of CO_2, CH_4 and N_2O measured by FTIR spectrometry and micrometeorological techniques. Atmospheric Environment, 36(11): 1833–1842.

Hall, B., Dutton, G.S. and Elkins, J.W. 2007. The NOAA Nitrous oxide standard scale for atmospheric observations. Journal of Geophysical Research, 112, D09305, doi:10.1029/2006JD007954.

Heinemeyer, O., Munch, J.C. and Kaiser, E.-A. 1995. Variabilität von N_2O-Emissionen—Bedeutung der Gasauffangsysteme. Mitt Dtsch Bodenkundl Ges, 76: 543–546.

Hensen, A., Dieguez Villar, A. and Vermeulen, A.T. 1999. Emission estimates based on ambient N_2O concentrations measured at a 200 m tower in the Netherlands 1995–1997. In: A. Freibauer and M. Kaltschmitt (Eds.), Approaches to Greenhouse Gas Inventories of Biogenic Sources in Agriculture, Stuttgart, pp. 215–228.

Hensen, A. et al. 2006. Dairy farm CH_4 and N_2O emissions, from one square metre to the full farm scale. Agriculture Ecosystems & Environment, 112(2–3 Special Issue SI): 146–152.

IPCC. 1996. Global Change 1995—The Science of Climate Change. Contribution of Working Group I to the Second Assessment Report of the Intergovernmental Panel on Cliamte Change, Intergovernmental Panel on Climate Change.

Hirsch, A. et al. 2006. Inverse estimates of the global nitrous oxide surface flux from 1998–2001. Global Biogeochem. Cycles 20, GB1008, doi:10.1029/2004GB002443.

IPCC. 2000. Good Practice Guidance and Uncertainty Management in National Greenhouse Gas Inventories. IPCC National Greenhouse Gas Inventories Programme. Published for the IPCC by the Institute for Global Environmental Strategies, Japan.

IPCC. 2001. Climate Change 2001: The Scientific Basis. IPCC Third Assessment Report. http://www.ipcc.ch/.

Kaiser, J., Röckmann, T. and Brenninkmeijer, C.A.M. 2004. Contribution of mass-dependent fractionation to the oxygen isotope anomaly of atmospheric nitrous oxide. Journal of Geophysical Research-Atmospheres, 109(D3): art. no. D03305.

Kesik, M. et al. 2005. Inventories of N_2O and NO emissions from European forest soils. Biogeosciences, 2(4): 353–375.

Kim, K.R. and Craig, H. 1993. Nitrogen-15 and oxygen-18 characteristics of nitrous oxide—A global perspective. Science, 262(5141): 1855–1857.

Kroon, P.S. Hensen, A., Jonker, H.J.J., Zahniser, M.S., van't Veen, W.H. and Vermeulen, A.T. 2007. Suitability of quantum cascade laser spectrometry for CH_4 and N_2O eddy covariance measurements. Biogeosciences Discussions, 4: 1137–1165.

Li, C.S. et al. 2001. Comparing a process-based agro-ecosystem model to the IPCC methodology for developing a national inventory of N_2O emissions from arable lands in China. Nutrient Cycling in Agroecosystems, 60(1–3): 159–175.

Lilly, A., Ball, B.C., McTaggart, I.P. and Horne, P.L. 2003. Spatial and temporal scaling of nitrous oxide emissions from the field to the regional scale in Scotland. Nutrient Cycling in Agroecosystems, 66(3): 241–257.

Mosier, A.R., Duxbury, J.M., Freney, J.R., Heinemeyer, O. and Minami, K. 1996. Nitrous oxide emissions from agricultural fields—Assessment, measurement and mitigation. Plant & Soil, 181(1): 95–108.

Neufeldt, H., Schäfer, M. and Angenendt, E. 2004. Modelling regional GHG emissions from farm production and agricultural soils with EFEM-DNDC. In: L. Institute for Energy and Environment (Editor), International Conference Greenhouse Gas Emissions from Agriculture—Mitigation Options and Strategies. Institute for Energy and Environment, Leipzig, Leipzig, Germany, pp. 165–171.

Neufeldt, H. et al. 2006. Disaggregated greenhouse gas emission inventories from agriculture via a coupled economic-ecosystem model. Agriculture Ecosystems & Environment, 112(2–3 Special Issue SI): 233–240.

Nevison, C. 2000. Review of the IPCC methodology for estimating nitrous oxide emissions associated with agricultural leaching and runoff. Chemosphere—Global Change Science, 2(3–4): 493–500.

Nevison, C. and Holland, E. 1997. A reexamination of the impact of anthropogenically fixed nitrogen on atmospheric N_2O and the stratospheric ozone layer. Journal of Geophysical Research, 102: 25519–25536.

Olivier, J.G.J., Bouwman, A.F., Van der Hoek, K.W. and Berdowski, J.J.M. 1998. Global air emission inventories for anthropogenic sources of NO_x, NH_3 and N_2O in 1990. Environmental Pollution, 102(Suppl. 1): 135–148.

Perez, T. et al. 2001. Identifying the agricultural imprint on the global N_2O budget using stable isotopes. Journal of Geophysical Research-Atmospheres, 106(D9): 9869–9878, doi 2000JD900809.

Röckmann, T., Kaiser, J. and Brenninkmeijer, C.A.M. 2003. The isotopic fingerprint of the pre-industrial and the anthropogenic N_2O source. Atmospheric Chemistry and Physics, 3: 315–323.

Röckmann, T. and Levin, I. 2005. High-precision determination of the changing isotopic composition of atmospheric N2O from 1990 to 2002. Journal of Geophysical Research, 110, D21304, doi:10.1029/2005JD006066.

Ruser, R. 1999. Freisetzung und Verbrauch der klimarelevanten Spurengase N_2O und CH_4 eines landwirtschaftlich genutzten Bodens in Abhängigkeit von Kultur und N-Duengung, unter besonderer Berücksichtigung des Kartoffelbaus. FAM-Bericht 36, Hieronymus-Verlag, München.

Schmidt, M., Glatzel-Mattheier, H., Sartorius, H., Worthy, D. and Levin, I. 2001. Western European N_2O emissions: A top-down approach based on atmospheric observations. Journal of Geophysical Research-Atmospheres, 106(D6): 5507–5516.

Schulte-Bisping, H., Brumme, R. and Priesack, E. 2003. Nitrous oxide emission inventory of German forest soils. Journal of Geophysical Research-Atmospheres, 108(D4): doi 10.1029/2002JD002292.

Skiba, U.M., Sheppard, L.J., Macdonald, J. and Fowler, D. 1998. Some key environmental variables controlling nitrous oxide emissions from agricultural and semi-natural soils in Scotland. Atmospheric Environment, 32(19): 3311–3320.

Smith, K.A. et al. 1994. Micrometeorological and chamber methods for measurement of nitrous oxide fluxes between soils and the atmosphere: Overview and conclusions. Journal of Geophysical Research, 99(D8): 16541–16548.

Smith, K.A., Thomson, P.E., Clayton, H., Mctaggart, I.P. and Conen, F. 1998. Effects of temperature, water content and nitrogen fertilisation on emissions of nitrous oxide by soils. Atmospheric Environment, 32(19): 3301–3309.

Sozanska, M., Skiba, U. and Metcalfe, S. 2002. Developing an inventory of N$_2$O emissions from British soils. Atmospheric Environment, 36(6): 987–998.

Stehfest, E. and Bouwman, L. 2006. N$_2$O and NO emission from agricultural fields and soils under natural vegetation: Summarizing available measurement data and modeling of global annual emissions. Nutrient Cycling in Agroecosystems, 74(3): 207–228.

Tilsner, J., Wrage, N., Lauf, J. and Gebauer, G. 2003. Emission of gaseous nitrogen oxides from an extensively managed grassland in NE Bavaria, Germany. Biogeochemistry, 63(3): 249–267.

Yamulki, S. et al. 2001. Diurnal fluxes and the isotopomer ratios of N$_2$O in a temperate grassland following urine amendment. Rapid Communications in Mass Spectrometry, 15(15): 1263–1269.

Chapter 9
Monitoring Carbon Stock Changes in European Soils: Process Understanding and Sampling Strategies

**Marion Schrumpf, Jens Schumacher, Ingo Schöning,
and Ernst-Detlef Schulze**

9.1 Introduction

Soils are the main reservoir for carbon (C) in terrestrial ecosystems. On a global average, they contain about 2–3 times as much organic carbon (OC) as the atmosphere or standing biomass, namely about 1500–2000 Gt (Janzen 2005).

It is well established that this reservoir is not inert, but in a dynamic stage of accumulation or decomposition. These processes are influenced by human activities. A major anthropogenic disturbance of soils is a land-use change from forest or natural grassland to agricultural soils (Johnson and Curtis 2001; Guo and Gifford 2002), and the permanent mechanical disturbance by ploughing. Prairie soils, for instance, lost 50% of their original soil C after 50 years of cultivation, and at a rate of about $70 g m^{-2} y^{-2}$ (Matson et al. 1997). The remaining 50% are temporary stabilized against decomposition by various mechanisms (Gleixner et al. 2001), but also this C can be mobilized under changing conditions, although at a much slower rate. Apparently, there is no organic matter in soils which is totally protected against microbial attack. But accumulation or degradation of soil C are influenced by environmental conditions and management.

The total amount of soil C is so large that an activation of this reservoir could result in a significant increase in CO_2 emissions, relevant for the Earth's climate. In a recent study, Bellamy et al. (2005) demonstrated a C loss across England over the past 20 years which is equivalent to the fossil fuel reduction of England since 1990 (Schulze and Freibauer 2005). On the contrary, studies in forests suggest that there is a long-term accumulation of C despite forest management (Schulze et al. 1996), but this sink is small and becomes most likely only measurable if management comes to a halt (Mund and Schulze 2006). Thus, soils are another example for a 'slow in, rapid out' behavior as defined by Körner (2003). Nevertheless, soils have accumulated C across Europe since the pleistocene retreat of glaciers, and contain C which is more than 1000 years old. This net C accumulation from the past could turn into a net C source in the future because of human activity and climate change.

Since soils are very dynamic systems, it appears necessary to take changes in soils into account when estimating a global or national C balance. Such a soil monitoring program will include the determination of C pools at national or continental scale and the detection of changes by repeated re-sampling after defined time intervals. Both aims, however, will be complicated by the inherent variability of soils at all scales, which makes high numbers of samples necessary. In order to generate a practicable soil monitoring system, we have to optimize sampling schemes and generate the database for the application and improvement of soil models. For those models, process studies of soil C dynamics in soils are the prerequisite and additional long-term monitoring sites are important for model calibration.

Thus, the following section will discuss (1) the processes that lead to C stabilization or mobilization, (2) approaches of soil C assessment, (3) existing approaches for the determination of soil C changes and (4) optimization strategies for soil sampling at the European scale.

9.2   Soil Processes That Lead to Stabilization and Mobilization

9.2.1   Determinants of Carbon Contents in Soils

Process understanding is not only important for soil models but also for optimizing sampling schemes for monitoring systems for which we have to understand the factors influencing the variability and the spatial distribution of soil C stocks and stock changes. Overall, soil organic matter (SOM) is a balance of processes that add C to the soil and processes that cause C losses. Changes in soil C can thus be described by the following equation:

$$\Delta SOM = \frac{dC}{dt} = \frac{(C_{in} - C_{out})}{(t_1 - t_0)} \tag{9.1}$$

The details of this equation are complicated (see Chapin et al. 2006). C_{in} would consist of a number of C fluxes (F) which consist mainly of litter input from plants (leaves and roots including root exudates), input from dead soil fauna and from bacteria (F_{litter}). Recent analyses indicate that CO_2 can be directly fixed by microbes in the soil, contributing to SOM (Miltner et al. 2005). Furthermore, advective inputs from the surrounding ecosystems can add C to the soil ($F_{advection}$). Among them are dissolved organic (DOC) or advective transport of litter (e.g. by wind), soil erosion by water and wind, and exogenous C inputs from land management (e.g. manure, slash, compost and sewage sludge).

$$C_{in} = F_{litter} + F_{advection} \tag{9.2}$$

C_{out} would result from a large number of processes. Mainly it is decomposition and the associated emission of CO_2 (F_{CO_2}). However, decomposition releases also intermediate products, such as CH_4 and CO (F_{CH_4}, CO) (Conrad 1996). Carbon is also lost by hydrological processes, namely water which percolates through the profile and which takes dissolvable organic and inorganic C (F_{DOC}, F_{DIC}) or particles (F_p)

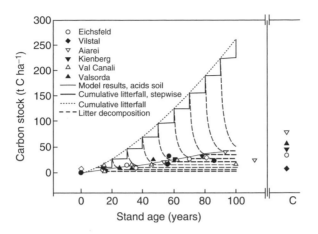

Fig. 9.1 Carbon stocks in the organic layer of a forest chronosequence. Cumulative litterfall and litter decomposition were calculated according to measured data. A model was used for later stages of litter decomposition. Black symbols are for forest stands on acid bedrocks, white symbols for stands on calcareous bedrocks. Right side: C in continuously forested control plots (taken from Thuille et al. 2006).

into deeper soil layers or via lateral flow into the next ecosystem. Organic acids and mineral acids (SO_4^{2-}) can dissolve soil minerals, for example, carbonates, and release dissolved inorganic C (DIC) or CO_2. Carbon is also lost by other forms of advection, for example, soil erosion ($F_{advection}$).

$$C_{out} = F_{CO_2} + F_{CH_4}, CO + F_{DOC} + F_p + F_{DIC} + F_{advection} \qquad (9.3)$$

Effects by humans through harvest and by animal activity through grazing of the food chain further complicate the soil C budget. Given the diversity of biological and physical processes that cause C inputs and outputs, it is not to be expected that soils ever reach a long-term equilibrium of litterfall and decomposition. The magnitude of these processes becomes apparent from Fig. 9.1 which shows the cumulative flux of litter, C losses by decomposition and changes in SOM in an afforestation of grasslands by spruce. After 100 years, the cumulative input of leaf litter was about 250 t C ha^{-1}. The aggregated increase in SOM and C in the organic layer amounted about 4–5 t C ha^{-1}, which is 1–2% of the leaf litter input. Figure 9.1 underestimates the total C-input by neglecting inputs from roots turnover and inputs from the food chain. Total C-input might be exceeding leaf litter input by a factor of 2, reducing the rate of C-accumulation in soils to 0.5–1% of the total input.

9.2.2 Carbon Stabilization

A reduction in C mineralization or an immobilization of DOC decreases C losses and thus increases C contents in soils. Some of the most relevant C stabilization mechanisms are summarized in reviews by Sollins et al. (1996), Hedges and Oades

(1997), Six et al. (2002), and v. Lützow et al. (2006). The activity of microorganisms, which determines mineralization rates, depends on environmental conditions, the amount and quality of organic substrates and their accessibility. Decomposition of organic matter is usually raised by an increase in soil temperature and moisture, but only as long as soil moisture is below field capacity, while anaerobic soil conditions reduce microbial activity (Prescott 2005). Besides rainfall amounts, also the relief position and the permeability of the soil and the underlying parent material determine soil moisture conditions.

A first hindrance to the rapid decay of organic matter in the soil presents the relative chemical recalcitrance of some organic compounds, which can be either plant derived or produced during decomposition. The breakdown of complex substances with aromatic or aliphatic structures as lignin or waxes is limited because of the low energy output for the microbes and often only performed by specialized species. Those substances have hence lower turnover rates in laboratory incubation experiments as compared to easier decomposable and more favourable substances like glucose or amino acids (Gleixner et al. 2001; Sollins et al. 1996). Also, secondary C products like black C, which has a complex, highly condensed aromatic structure produced by fire, contribute to the stable C pool of soils at sites affected by vegetation fires or the deposition of particles from fossil fuel combustion (Schmidt and Noack 2000). However, it turns out that some of the complex chemical plant compounds, such as lignin, can be rapidly decomposed in agricultural soils, while carbohydrates and proteins can be long lasting (Schöning 2005a; Gleixner et al. 2002). This indicates that—in addition to chemical recalcitrance—other mechanisms are also important with respect to the stabilization of organic matter.

Several studies have demonstrated that the accessibility of organic compounds to microorganisms is important for their protection against microbial attack. Biodegradation requires direct contact between microbes and substrate in the case of small molecules or between extra-cellular enzymes and substrates in the case of macromolecules. In contrast to fungi which can actively grow towards new resources, like freshly fallen litter, it is assumed that microbes have to wait until they are either reached by or carried to a new resource. But studies by Kohlmeier et al. (2005) indicate that some motile bacteria might be able to use water films surrounding fungal hyphae as 'fungal highways' in order to move within the soil matrix. Still, organic substrates could be protected against microbial access in fine soil pores which cannot be entered by microbes. Between 15% and 52% of soil pores are inaccessible to microbes because the pore diameter is smaller than 0.2 μm (Chenu and Stotzky 2002). However, Eusterhues et al. (2005) found no indications for physical protection of organic matter in micropores of sandy acid forest subsoils. Another mechanism which separates substrates and microorganisms is the occlusion of soil organic matter into microaggregates (<250 μm) and macroaggregates (>250 μm). A positive relation between soil aggregation and soil organic carbon (SOC) contents was often observed (Six et al. 2002). Organic material occluded into aggregates is protected against decomposition by reduced access of a number of microbes which are unable to penetrate into the aggregates because of small pore sizes, or by the reduced diffusion of oxygen (Six et al. 2002) and of exo-enzymes into the aggregates. Despite this, organic matter enclosed in aggregates still undergoes chemical transformation, exhibiting an increase in recalcitrant alkyl-C and a decrease in easy decomposable O-alkyl-C as compared to free organic matter (Six et al. 2002). The microbial decomposition of organic matter in macroaggregates is also an important process for aggregate stabilization and the formation of stable microaggregates, since the binding agents produced by the microbes help organic fragments to become encrusted with clay particles (Six et al. 2004). Aggregate formation in the soil is promoted by a number of biotic and abiotic processes including the activity of soil fauna, especially earthworms, bacteria and fungi, plant roots and mycorrhizal hyphae, and inorganic binding agents, such as Fe- and Al-oxides or calcium (Six et al. 2004). Arbuscular mycorrhizae, for example, contain a glyco-protein, glomalin, which increases aggregate formation and stability by acting as glue on hyphal surfaces and by sealing aggregate surfaces (Rillig 2005).

The turnover of organic matter associated with aggregates is closely connected to the life time of the aggregates. As estimated by Plante et al. (2002), the mean residence time of macroaggregates can be quite small with an average of 27 days. Generally, organic material is better protected in microaggregates as compared to macroaggregates as can be seen from the longer turnover times of microaggregate-associated C (412 years vs 140 years in macroaggregates; Six et al. 2002).

Another important process for C stabilization in the soil is the formation of organo-mineral and metal–organic complexes (v. Lützow et al. 2006). The sorption of organic molecules to mineral surfaces depends on the composition of organic substances in solutes, the pH, the presence of polyvalent cations and the mineral types (Kalbitz et al. 2005; Dontsova and Bigham 2005). While the high capacity of poorly crystalline minerals (allophone, ferrihydrate, amorphous Fe and Al hydroxides) to form organic associations was often emphasized, sorption of organic molecules also occurs on surfaces of crystalline Fe and Al oxides like goethite and gibbsite and on layer silicates like smectite, kaolinite and illite (Kleber et al. 2005; Schöning et al. 2005b, Kaiser and Guggenberger 2003; Baldock and Skjemstad 2000). There are indications that the stability of the mineral–organic bonding depends on the already present organic coverage of mineral surfaces. Kaiser and Guggenberger (2003) showed that at low occupancy of the mineral surface, organic matter preferentially binds at the edges of micropores as those sites facilitate more stable multiple bindings with the mineral surface. Thus, the binding capacity of mineral surfaces is not unlimited and with the increasing accumulation of organic matter (OM) on particle surfaces, binding strength can be reduced and microbial activity can be enhanced because of the high density of available substrate. One hypothesis for the resistance of OM sorbed to mineral surfaces is that multiple binding to mineral surfaces changes the conformation and the electron distribution of organic molecules, thus making them unavailable for enzymatic degradation (Chenu and Stotzky 2002; Khanna et al. 1998). Similar mechanisms were expected to be responsible for the stabilization of metal–organic complexes, but also metal toxicity can be involved in the reduction of microbial activity (Baldock and Skjemstad 2000; Marschner and Kalbitz 2003). Reduced mineralization rates were reported for complexes between organic molecules and Ca, Fe, Al and heavy metals

(Renella et al. 2004; Schwesig et al. 2003; Baldock and Skjemstad 2000; Francis et al. 1992), but there are also indications that metal ions have no or even a promoting effect on decomposition of dissolved organic matter (Marschner and Kalbitz 2003).

Besides gaseous losses during decomposition (mainly as CO_2), another way by which C can leave the ecosystem is via leaching of DOC. About 1–10 g OC m^{-2} are assumed to leave forest soils annually as DOC (Guggenberger and Kaiser 2003). The magnitude of DOC losses depends on the mobilization of DOC in the organic layer and the mineral soil and on factors removing DOC from the soil solution. Since C is not easily assessed by microbes while dissolved in the soil solution, association of DOC to soil particles is often assumed to be the main process for C removal from the soil solution. This process increases the stability of DOC if it occurs on fresh mineral surfaces, and sorbed OM is added to the soil C pool (Kaiser and Guggenberger 2003). In soil compartments, where mineral surfaces are already covered by a biofilm (mixture of bacterial cells growing within an organic matrix of extracellular polymeric substances), as along preferential flow paths or at the surface of microaggregates, a further sorption to these hot spots of biological activity can increase the decomposition of additionally sorbed OC (Guggenberger and Kaiser 2003). DOC may also be protected in micropores of aggregates, but there is no study on the significance of this process yet (Marschner and Kalbitz 2003). Furthermore, DOC can be precipitated in soils at high metal concentrations of Al or Fe and by increasing soil pH as in the B horizons of Podzols (Jansen et al. 2005; Schwesig et al. 2003), thus increasing the SOC content of those layers.

A process that recently gained more attention with respect to SOM stabilization is the transfer of C in the soil food chain. Microorganisms use the majority of the consumed organic material as energy source or electron acceptor, producing CO_2 or CH_4. But microorganisms also utilize a part of the assimilated organic compounds for the building of new cell material and incorporate it into proteins or lipids for the construction of, for example, new membranes (Gleixner et al. 2005). Some substances, like pyridine, which are vital for life, cannot be produced by all soil organisms and thus persist-in the living biomass being transferred from one organism to the next. Thus, old C is carried over within the food chain, a process which is basically unlimited in time. This also explains why some microbial-derived substances have a much higher ^{14}C age than the life span of microbes. Old and modern C can also become mixed within the same molecule through biological, that is, enzymatic processes. When interpreting ^{14}C ages for turnover rates, this process should be considered since some substances, such as carbohydrates, which were considered to be very labile in soils, might have a fairly long life time, being transferred between food chains.

In summary, all the described processes can act together in the soil to prevent rapid C losses. For some processes, like the absorption of C to mineral surfaces, the capacity of a given soil compartment for C storage seems to be limited and further C accumulation via this mechanism can only occur by exploring new mineral surfaces, for example, at deeper soil layers or by exposure of minerals formerly hidden, for example, in microaggregates. Transport of organic matter from upper to lower soil horizons via bioturbation or DOC can thus increase the C storage capacity of soils. The relevance of individual processes for C stabilization will depend on soil types and environmental conditions. In mountainous regions, for example, low temperatures and often high rainfall amounts can determine soil C stocks by limiting microbial activity, thus leading to a pronounced accumulation of organic matter in otherwise less developed Regosols. In soils of volcanic origin, like Andosols, organo-mineral complexes are usually responsible for the accumulation of high amounts of SOM. The relevance of individual mechanisms for C storage will also affect the reaction of soil C contents to changes in environmental conditions, like climate change. But for most soils, data on the quantitative importance of individual stabilization mechanisms under different environmental conditions are not available.

9.2.3   Processes of Carbon Mobilization

Generally, processes that decrease the efficiency of C retention mechanisms described above will lead to C mobilization. Mobilization may be initiated by a number of changes in the soil environment, like changes in soil pH, the redox potential, or the cation composition of the soil solution, which can affect the bonding strength of organo-mineral complexes or lead to the destruction of soil aggregation (Sollins et al. 1996). Several factors that promote microbial activity like increasing temperature, a more favourable moisture regime or higher nutrient inputs can also increase C losses (Sollins et al. 1996). The temperature sensitivity of SOC was recently subject of some debate. Soil warming experiments have demonstrated that respiration rates increase only for one to three years and then decline again to previous rates (Melillo et al. 2002). Thus, in the longer run, microbial activity in the soil was reduced again by acclimation or a decrease of the labile C pool in the soil (Eliasson et al. 2005). If all ecosystem components are considered, the effect of global warming is complicated by interactions with other environmental factors, for example, warmer temperatures can increase the growing period of the vegetation and thus enhance litter inputs in some regions while in other areas, higher temperatures can come along with drier conditions reducing both, plant growth and microbial activity. Therefore, future soil respiration models should not only focus on direct effects of temperature on the mineralization of organic substances but also take into account all the various processes and interactions involved (Davidson et al. 2005).

Other important factors that lead to C-mobilization are agriculture and land-use changes. The conversion of natural forest or grassland into agricultural land usually leads to a decline in soil C stocks (e.g. Matson et al. 1997; Johnson and Curtis 2001; Guo and Gifford 2002). Especially tillage has been shown to increase C losses from soils (Sollins et al. 1996). Land-use changes that are commonly considered to increase C contents in the soil, such as afforestation, can also cause initial losses of soil C. Thuille and Schulze (2006) observed that while C stocks in the organic layer increased following afforestation of grasslands in Thuringia and the Alps, C stocks in the mineral soil decreased as fine litter inputs to the topsoil were lower in the forest as compared to the meadows. Since it took about 80 years until the mineral soil regained its original C contents, long rotation times are necessary to account for initial C losses after afforestation of grasslands.

The relevance of soil erosion for C losses depends on the scale of interest. While C-exports can be relevant at the plot scale, the process might be less significant if larger areas, including the deposition zones, are considered. If eroded soil material is deposited again and buries former organic rich topsoils, erosion can also lead to C-stabilization in deposition zones. Losses by soil erosion depend on environmental factors like the topography (e.g. slopes promote erosion by water) or the frequency of rainstorms and wind speed. But also land-use changes affect erosion by disaggregating the soil particles, and the reduction of the vegetation coverage during some time of the year. For more information on soil erosion see Valentin et al. (2005), Boix-Fayos et al. (2005) or Nordstrom and Hotta (2004).

Summarizingly, processes that affect soil C-contents and their changes include (1) factors that influence the amount and quality of organic matter inputs and (2) factors that determine microbial activity and their access to the substrate. Overall, land use, climate and clay content and composition seem to be among the factors most important for C-storage and distribution.

9.3   Spatial Variability of Soils

9.3.1   Variability of Soils—A Question of Scale and Disturbance

Soil properties vary at spatial scales ranging from a few millimetres to the continental scale. Figures 9.2 and 9.3 present examples for soil C-distributions at the regional and continental scale. The regional map of soil C stocks in Saxony-Anhalt in Germany shows a high variability with C stocks ranging from 7 to 12.8 kg C m^{-2} (0–100 cm soil depth, Grabe et al. 2003). At the European scale, the variability of C concentrations (0–30 cm soil depth) is reflected, for example, on the map produced by Jones et al. (2005a). Depending on the scale of interest, different factors are controlling the variability of soil C contents. On a plot scale, for example, the heterogeneity of the C content in the soil will depend on small-scale variations in C inputs (accumulation of litter in depressions, variations in the root distribution), on bioturbation, on the depth to a stagnic horizon, on the clay content, or the content of iron oxides. Differences in geology or topography affect the C content at the regional level, while at even larger scales, climate could be the most important driver. Nevertheless, also anthropogenic effects have to be considered since land use and land-use changes have significantly altered soil properties and led to a higher variability of C stocks in space and time.

A number of studies were done to characterize spatial heterogeneity of SOC on the plot scale. Klironomos et al. (1999), who analyzed 40 soil samples from a 30-m^2 plot covered by chaparral, observed a coefficient of variation of 58% for the organic matter content of the upper 15 cm of the soil. Also, Conant et al. (2003) examined the spatial variability of soil C in microplots on a cultivated and a forest site in Tennessee and

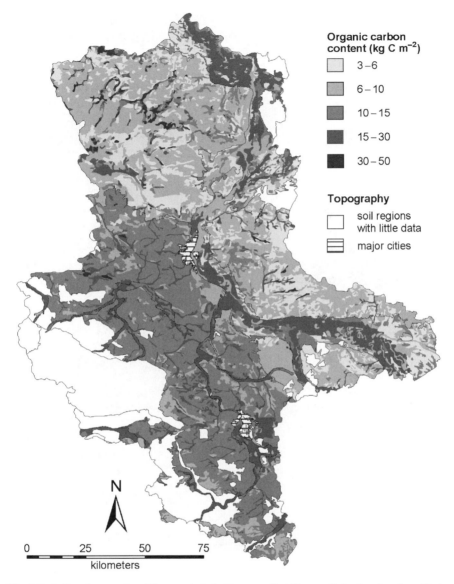

Fig. 9.2 Soil carbon stocks of Saxony-Anhalt, Germany (0–100 cm soil depth, taken from Grabe et al. 2003).

on an old growth forest and a secondary forest site in Washington. Three micro-plots were analyzed per site, each of which consisted of six coring points with a distance of about 2 m. The coefficient of variation of C stocks within the microplots ranged between 5% and 118% depending on the site conditions (Table 9.1). Highest variability occurred in the old growth forest, which was probably induced by the irregular occurrence of buried decomposing logs (Conant et al. 2003). In order to simulate

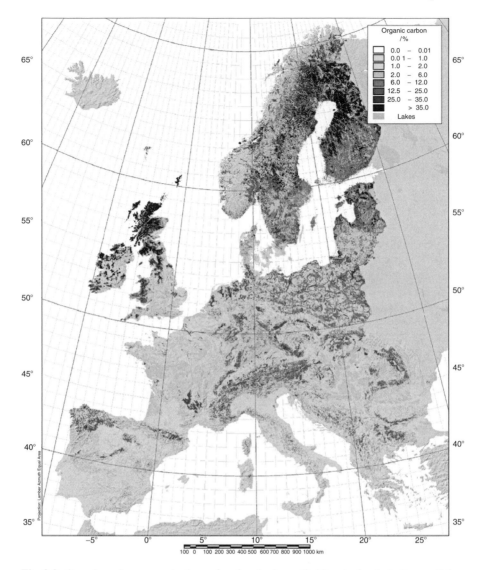

Fig. 9.3 Organic carbon concentrations of surface horizons (0–30 cm) of soils in Europe (taken from Jones et al. 2005a).

re-sampling, microplots were sampled a second time 1 m apart from the first sampling points. The results show that while the difference between the re-sampled microplots was only 2–8% on the cultivated site, it was up to 27% and 40% on old growth and secondary forests, respectively, again underlining the huge small-scale variability on those sites. Therefore, the small-scale variability within a few metres has to be accounted for in inventories of both, small- and larger-scale soil C inventories.

Table 9.1 Example for small-scale variability at two sites in Tennessee and two sites in Washington. Three microplots of 10m² were installed per plot and six soil samples were taken within each microplot for two times in order to simulate an immediate paired resampling of the microplots. Means, standard deviations (SD) and coefficients of variation (CV) are given for each microplot, as well as means and variability of microplots within plots (Data taken from Conant et al. 2003)

	Micro-plot	"1st Sampling" mean (n = 6) Mg C ha⁻¹	CV %	"2nd Sampling" mean (n = 6) Mg C ha⁻¹	CV %	Difference Mg C ha⁻¹
Tennessee						
Cultivated	1	18.5 (1.8)	9	20.0 (4.1)	21	-1.5
Cultivated	2	19.0 (1.2)	7	19.3 (1.4)	7	-0.3
Cultivated	3	17.3 (2.2)	12	17.8 (1.5)	8	-0.5
Mean between microplots		18.2 (0.8)		19.0 (1.1)		-0.8 (0.6)
Forest	1	30.3 (7.1)	23	28.8 (3.3)	11	1.5
Forest	2	29.5 (1.4)	5	26.6 (3.8)	15	2.9
Forest	3	29.4 (3.6)	12	34.7 (7.3)	21	-5.3
Mean between microplots		29.8 (0.5)		30.0 (4.2)		-0.2 (4.4)
Washington						
Old growth forest	1	24.8 (7.2)	29	27.4 (2.6)	10	-2.6
Old growth forest	2	142 (118)	83	140 (125)	89	1.5
Old growth forest	3	54.7 (33.1)	61	39.9 (10.7)	27	14.8
Mean between microplots		73.7 (60.6)		69.1 (61.8)		4.6 (9.1)
Second growth forest	1	78.0 (42.9)	55	46.8 (22.1)	47	31.2
Second growth forest	2	51.1 (30.6)	60	45.0 (43.1)	96	6.1
Second growth forest	3	38.0 (10.3)	27	29.8 (6.3)	20	8.2
Mean between microplots		55.7 (20.4)		40.5 (9.3)		15.2 (14.0)

The variability between microplots on a site was lowest for the cultivated site with a coefficient of variation of only 5% and again highest for the old growth forest site where it was more than 80% due to one microplot having more than threefold higher C contents than the other two (Table 9.1).

Only a few studies analyzed soil C variability at wider range of spatial scales. Palmer et al. (2002), for example, divided the total variance of C stocks in the state of Georgia in three different levels. On average, around 21% of the total variance of C stocks within the upper 20cm of the soil was within subplots with a sampling distance of less than 1m, 38% was between subplots of 0.017ha within plots of 1ha, and 41% of the variance was between plots distributed over the state. In a study on the spatial variability of C concentrations in US grasslands, Conant and Paustian (2002) observed an increase in the coefficient of variation from the country (Dundy County, coefficient of variation of 39%) to the national scale (USA, 63%). By comparing studies on the variability of soil C conducted at different spatial scales, Conen et al. (2005) found a logarithmic relation between relative plot sizes and relative variances in C stocks, indicating that a reduction in plot size over six orders of magnitude (e.g. from 100ha to 1m²) would be necessary to halve the coefficient of variation.

9.3.2 Implications for Monitoring Soil Carbon

The variability of soils at various scales makes their investigation difficult and the detection of change a real challenge, especially since the expected changes are small compared to the high C contents already present in the soils. Thus, a high number of samples will be necessary to detect changes. A way to improve the sampling design for monitoring is the usage of paired sampling, where exactly the same spots are re-sampled. The advantage of a paired design is that only the variance of the differences between very close sampling points is relevant but not the total variability of C contents within the area under consideration. Since the immediate surroundings of a sampling point can be affected by sampling, a distance of about 1 m between sampling and re-sampling seems to be necessary. Thus, the small-scale variability within 1 m distance as described above cannot be fully avoided also in a paired sampling scheme, but larger-scale variances in C distribution would not be relevant for the detection of changes.

The study of Conant et al. (2003), described above (Sect. 5.3.1), can be used as an example for the effect of a paired design on the minimum detectable difference in soil C stocks using the mean of the microplots as sampling points. To simulate an unpaired, or random sampling design, the standard deviation of the C stocks between microplots within sites (Table 9.1) was used as basis for a two group t-test, assuming that the variance between microplots will be the same for sampling and re-sampling. For the paired design, the standard deviation of the differences between the two samplings of the same microplots (Table 9.1) was used in a paired t-test, assuming that an increase or decrease in C stocks between the first and second sampling will not increase the variability of the differences.

Overall, smallest changes will be detectable on the cultivated site, which had on average the lowest standard deviation for C stocks (Fig. 9.4). Since the standard

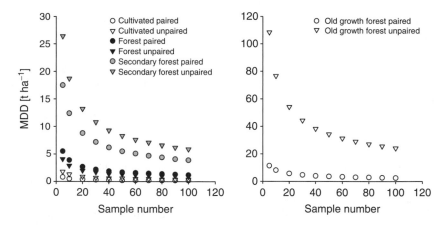

Fig. 9.4 Relation between minimum detectable difference (MDD, t-test) and sample size using the results of Conant et al. (2003) for a paired and unpaired plot design on a cultivated and a forest in Tennessee and a secondary forest and an old growth forest in Washington.

deviation between the microplots was not much different from the standard deviation of the difference between the two samplings, the paired design also only resulted in a slight improvement of the minimum detectable difference. On the old growth forest site on the other hand, a very high number of samples would be necessary in an unpaired design to detect changes below 20t ha^{-1} because of the high variability among microplots. Since the differences between two samplings of the same microplots resulted in a much lower standard deviation, the minimum detectable difference was much smaller using a paired design (Fig. 9.4). Only in the case of the forest site in Tennessee, the paired design did not result in an improvement of the change detection. In this case, the variability of the differences between the first and second sampling was higher than the variability between sampling sites. This is most probably because of the low sample size of only three replicates in this study, since it can be assumed that for most situations the small-scale variability between sampling points in close neighbourhood will be smaller than the variability between sampling points of greater distances. This should be especially the case if not only the plot scale is considered as in this example but also the continental scale. Therefore, a paired design is strongly recommended for a soil monitoring system in Europe.

Furthermore, an appropriate design for the distribution of sampling points over Europe, which should be representative but using a minimum number of samples, has to be identified. Therefore, in the following section, (1) a comparison of the most common sampling schemes for inventories is made, (2) existing soil inventories at national and European scale are presented and (3) the approach of the European project CarboEuropeIP for monitoring European soil C stocks is introduced.

9.4 Existing Soil Carbon Inventories and Monitoring Systems at Different Scales

9.4.1 Common Sampling Strategies

The most commonly used sampling strategies for soil inventories are based on complete random sampling, regular sampling (e.g. according to a grid) or stratified sampling designs (Fig. 9.5). An alternative to the pre-defined sampling sites in stratified or random sampling schemes is the judgement-based sampling design, where sampling sites are chosen only on the basis of expert knowledge. This has the disadvantages that it is not only objective but also relies on the expertise, the ambition and expectations of individual persons. Furthermore, there is a risk for bias, since the person conducting the study might tend to chose similar sites, thus possibly also underestimating the real variability of a region.

Random and regular sampling schemes on the other hand are both hardly influenceable from outside, and they are different from stratified sampling because they do not require any prior knowledge about the study site. Nevertheless, random sampling also bears the risk of not representatively covering the area and for both,

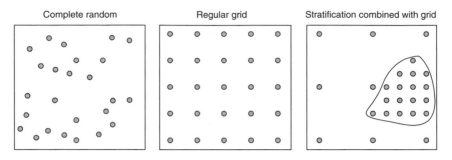

Fig. 9.5 Commonly applied sampling schemes for inventories (grey dots: sampling points).

the random and the regular sampling, it is possible that hot spots of minor coverage may be ignored. For practical purposes in monitoring, grid designs are often easier to handle than random sampling, since the points are easier to locate in the field, for the first inventory as well as for the re-sampling.

The aim of stratified sampling, where some regions are more intensively analyzed than others, is to reduce the variance within strata as compared to the total variance in order to minimize sampling efforts and to get the best results. Stratified sampling schemes can be combined with both, random and regular sampling schemes. But it requires some information about the study site and the knowledge about the dependency of, for example, soil C stocks, their changes and variability on environmental parameters. Some information might be available as maps of soil types, land use, topography or climate. On the basis of this knowledge, sensitive areas can be defined, which will be sampled at a higher resolution than others. Still, evenly distributed regular systems are more flexible if the topic of interest is changing during long-term monitoring and other environmental problems and parameters are getting more importance, for which the monitoring was not optimized. For more details on different sampling schemes for inventories and monitoring programs, see de Gruijter et al. (2006).

9.4.2 Existing Soil Inventories and Soil Monitoring Systems in Europe

The first step towards a European monitoring network for soil C is the inventory of the existing soil C stocks at the beginning of a monitoring phase. Several attempts have been made to get this standard at the national as well as the European level.

9.4.2.1 National Level

Although a number of national soil inventories and soil monitoring programs were initiated during the last decades (Jones et al. 2005b), national estimates of total soil

C stocks are only for some countries available by now, including France (Arrouays et al. 2001), Denmark (Krogh et al. 2003), Belgium (Lettens et al. 2004) and Great Britain (Milne and Brown 1997). According to those studies, mean C density is about 6.0 ± 5.7 kg m^{-2} for the upper 30 cm of the soil in France, 5.1–7.0 kg m^{-2} (depending on land-use type) for the top 30 cm in Belgium, 16 kg m^{-2} for the upper 100 cm in Denmark, and in England and Wales, and Scotland, surveys of whole profiles resulted in a mean C density of 21 ± 5 and 41 ± 10 kg m^{-2}, respectively.

For many countries, no single national database exists which includes all necessary soil information for the determination of C stocks (C concentration, bulk density and stone content) obtained in a harmonized way to a sufficient soil depths and which representatively covers the country (Table 9.2). Most existing soil surveys were conducted for special purposes, like soil mapping, to study the effects of air pollution on forests or for soil fertility analyses in agricultural sites. Thus, individual parameters necessary for the determination of soil C stocks, like stone content or bulk density, were often not determined and frequently, studies were restricted to the topsoil or the plow-layer. As a consequence, various models, including pedo-transfer functions and multiple regression analyses, had to be used to complete the datasets. Liebens and VanMolle (2003) tested different models to estimate the soil bulk density against each other, which led to a difference of 6% for the total estimate of soil C stocks in Flanders. Furthermore, available soil inventories were often restricted to certain regions or land-use types. The example of Belgium in Table 9.2 shows that in some countries also the federal structure complicates harmonized countrywide inventories, since each administrative region has the authority for environmental, forestry and agricultural policies so that most inventories are restricted to those regions (e.g. Wallonia or Flanders, Lettens et al. 2005a). Overall, there seems to be a need for complete spatial databases of high quality to reduce the uncertainty of countrywide C-stock estimates (Liebens and VanMolle, 2003).

If national soil surveys are to be used as a basis for a European monitoring, the low level of standardization within and between national databases is problematic. One restriction is that soil data within and between various databases are usually not obtained in a harmonized way with respect to sampling procedures and analytical methods. Another shortcoming of the usage of various soil databases as a basis for monitoring is that they were usually not created during one standard year, which could then be used as a baseline, but databases were usually obtained over much longer time periods (see Table 9.2). There is also no standard available for combining data of different databases and for upscaling to the national or even continental level. At the moment each country uses different definitions of soil series and land-use types for scaling up point data to the region, making it difficult to combine results across country boarders. However, Liebens and VanMolle (2003) tested different variations of scaling-up methods for Flanders and concluded that with less than 2%, the effect on country estimates of C stocks was comparatively small.

However, since soil, vegetation and land-use maps are frequently used for upscaling in soil C mapping and total stock estimates, the quality of the obtained results also depends on accuracy and scale of the maps used. Usually, soil information is not available at a similar high spatial resolution than land-cover data.

Table 9.2 Examples for databases used to assess regional or national soil carbon contents

	Land use	Sampling scheme	Sampling time	Sampling procedure	Sampling depth	Bulk density	Stone content	Sample number
England and Wales[1] National Soil Inventory	all	grid 5 × 5 km²	1978–1983	composite of 25 soil cores, by depth	0–15 cm			5,662
Ireland[2] National survey	grassland (pasture)	two sites per 10 × 10 km²	1964	by depth	0–10 cm	n.i.	n.i.	678
Regional survey	grassland (pasture)	grid 10 × 10 km²	1995–1996	composite of 15 samples, by depth	0–10 cm	n.i.	n.i.	220
France[3] ICP forest level 1 and 2 plots	forest	grid 16 × 16 km² for level 1	1993–1994	soil profiles, by depth	0–60 cm	measured	n.i.	540 + 102
Soil survey staff, Soil mapping	forest, cropland, range soils	n.i.	n.i.	soil profiles, by horizon	whole profiles	only few records	n.i.	n.i.
n.i.	cropland	irregular	n.i.	composite of 15 soil cores, by depth	0–30 cm	no	n.i.	5,000 fields
Denmark[4] Soil Profile Database (SPD)	all	for 837 profiles: 7 × 7 km² grid	1987–1991	soil profiles, by horizon	whole profiles	measured	n.i.	2,000
Danish Soil Classification System (DSC)	cropland, grassland	n.i.	1975–1978	composite of 25 samples, by depth	0–20 cm	no	n.i.	36,000
Belgium, Flanders[5] Aardewerk	all	n.i.	1950–1970	soil profiles, by horizon	whole profiles	no	measured	13,033

Belgium, Flanders[6]	Belgian Soil Service	cropland, grassland	n.i.	1989–1991 and 1998–2000	by depth	0–23 cm crops, 0–6 cm grass	estimated	no	1,674 means for various units
	Institute for forestry and Game management	forest	subsample of grid 1000 × 500 m²	1997–2002	soil profiles, by horizon	0–20 cm	measured	soil map	290
Belgium, Wallonia[6]	Centre d'Information Agricole a) Luxembourg b) Liège c) Hainaut	cropland, grassland for a) also forest	n.i.	1989–1991 and 1999–2001	by depth	0–15 cm, forest 0–20 cm	estimated	no	12,283 points for a), 367 averages for b) and c)

n.i. no information, [1]Bellamy et al. 2005, [2]Zhang and McGrath 2004, [3]Arrouays et al. 2001, [4]Krogh et al. 2003, [5]Lettens et al. 2004, [6]Lettens et al. 2005a

In response to the creation of the European Soil Database, almost all European countries have prepared national soil maps at scales of 1:1,000,000 or 1:500,000, but only 20% of the EU and bordering countries have a complete coverage of soil maps at scales of 1:50,000 or larger (Jones et al. 2005b). Thus, in many countries, detailed information on the spatial distribution of soil types is lacking or restricted to certain regions. This will be especially problematic in heterogeneous regions, where the variability of soil types and characteristics within mapping units can be considerable. General problems that can be associated with the usage of mapped soil types for upscaling will be discussed below. Thus, overall, there seems to be a need for a standardized European monitoring program in addition to national efforts.

9.4.2.2 European Wide

There are currently three international initiatives to create a common and harmonized soil database for entire Europe.

Geochemical Baseline Mapping Program (FOREGS)

A European wide soil inventory was performed within the Geochemical Baseline Mapping Program (FOREGS) which started in 1997. A total of 925 sample sites located in 26 countries covering an area of 4,250,000 km^2 were visited to obtain a sample density of one site per 5,000 km^2. The FOREGS sampling grid was based on GTN grid cells (Global Terrestrial Network, 160 × 160 km^2) developed for the purpose of Global Geochemical Baseline mapping (Darnley et al. 1995). Fieldwork was carried out between 1997 and 2001 at five randomly selected points in each grid cell. The sampling includes the O-layer, the topsoil (0–25 cm) and the C-horizon. For standardization, all equipment for sampling was bought in Finland. The samples were prepared in only one laboratory in Slovakia and then analyzed in different European survey laboratories. Finally, maps were produced and published in the Geochemical Atlas of Europe (Salminen 2005). Carbon maps present C concentrations of the topsoil and the subsoil. The disadvantages of this study were that it had the lowest spatial resolution of the three European inventories presented and no bulk densities were determined, so that only information on concentrations, but no stocks are available.

The European Soil Database

The European Soil Database consists of four databases, three of which can be interesting for soil C monitoring. The Soil Geographical Database of Europe (SGDBE) at scale 1:1,000,000 (1:250,000 is planned and currently applied in test areas) is a digitized soil map and part of the European Soil Information System (EUSIS). The database contains a list of 'soil typological units' (STU), which includes information on soil classification [FAO 1974, 1990 and World Reference Base (WRB), 1997], parent material, surface and subsurface textural class, depth to textural change, soil water regime, land use, slope, elevation, obstacles to roots and limitations to agricultural use. Thus, parameters important to calculate soil C stocks like OC concentration, bulk density or stone content are not directly available. Since it is not possible to present all STUs in a map, they were aggregated to 'soil mapping units' (SMUs) to form soil associations. Detailed information on this process is available in the European Commission publication reference: EUR 20422 EN. To increase the application of the database and its interpretability for ecological questions, the Pedotransfer Rules Database (PTRDB) was created (Van Ranst et al. 1995). Those pedotransfer rules help to include soil properties in the database, which are not measured but can be estimated from a combination of other data taken from the STU and some external sources (Daroussin and King 1997). Among the outputs of the PTRDB are the profile mineralogy and the cation exchange capacity, but also the topsoil OC content, the topsoil packing density and the volume of stones. The third database is the Soil Profile Analytical Database of Europe (SPADBE). This database contains two categories of soil profile information with physical and chemical analyses. One category consists of geo-referenced, measured profile information, which includes SOM content, bulk density and stone content. Analytical methods are not harmonized but have to be reported. The second category consists of measured or estimated typical soil profile characteristics for each dominant STU. Still, for studying soil C in Europe, the database is not comprehensive enough yet (Rusco et al. 2001).

Jones et al. (2005a) used the Soil Geographical Database to create a European map of soil C concentrations but not C stocks of the upper 30 cm of the soil. Modified pedotransfer rules of the PTRDB were used to estimate C concentration ranges. As input parameters, the full FAO soil code (first, second and third character in item soil in the database), the dominant surface textural class (obtained from the database) and land-cover information (European land-cover data) were used. A separate correction to account for the influence of temperature on the SOM content using a heuristic function and an external database on average annual cumulative temperatures was implemented. The resulting map gives an appropriate overview of the distribution of C concentrations over Europe (Fig. 9.3). However, it relies on the precision of the input data and that of the empirically generated pedotransfer rules. For the detection of changes, besides C concentrations, also C stocks are important. Those could as well be generated by pedotransfer functions estimating the stone content and the bulk density, but the more pedotransfer functions are used to estimate integrated values, the higher gets the associated uncertainty of the estimate. The authors stress the applicability of the mapping procedure to estimate effects of different future scenarios by using different land-use input data or climate situations as a basis for the pedotransfer rules. However, because of the wide range of C concentrations used for each mapping unit (0–0.01%, 0.01–1%, 1–2%, 2–6%, 6–12% ...), the map itself in its current state appears not to be useful as a basis for the detection of changes since those changes are probably much smaller than the ranges of the C mapping units for most areas.

International Co-operative Programme on Assessment and Monitoring
of Air Pollution Effects on Forests (ICP Forests)

The ICP Forest Level I database was initiated to monitor the effects of air pollution
on forests and soils. Thus, it is restricted to forested areas. The Level I network consists of more than 5,000 plots which are systematically arranged along a $16 \times 16 \text{km}^2$
grid covering Europe. A first soil survey was executed in 28 countries between 1990
and 1995. The collection of the organic layer and the upper 0–10 and 10–20 cm of the
mineral soil was mandatory. A composite sample of at least five subsamples had to
be collected per plot. The determination of the OC content was mandatory for the
upper 20 cm of the soil. Although the bulk density had to be reported for the upper
10 cm of the soil, it could be obtained from pedotransfer functions and did not have
to be measured directly. The same was true for the stone content that was often only
obtained by visual estimation. Wirth et al. (2004) tested the estimates of bulk density
and stone contents collected during the German national soil inventory (BZE, 8 ×
8 km² grid, integrated in the ICP forest grid) in Thuringia, and concluded that the bulk
density was on average overestimated by 35% and the stone content by 40%. Since
overestimation of bulk densities occurred primarily in C-rich horizons, this resulted
in a net overestimation of C stocks of 18 t ha⁻¹. These results emphasize the need for
direct measurements of all parameters which are necessary for the determination of
C stocks. Another problem is the incomplete reporting of mandatory parameters.
While data on OC concentrations are available for more than 80% of the sites, it is
only 43% for the bulk density and 73% for the stone content (Baritz et al. 2005).
Thus, again the dataset has to be completed with various models.

The ICP soil database was used by the EU-CarboInvent project as part of the
CarboEurope cluster. It aims at quantifying C stocks and C stock changes in
European forests, including soils. But as soil depths were limited to 20 cm in the
soil inventory, also the C stocks obtained from the database are limited to that
depths although they can be substantially higher for deep soils. Overall, the total
amount of C in the organic layers of the studied forests averages at about 10 t ha⁻¹,
and exceeds 20 t ha⁻¹ only in Gleysols. In the mineral soil, about 50 t ha⁻¹ are found
on average in the top 20 cm.

Figure 9.6 demonstrates the measured soil C stocks in the organic layer and the
mineral soil of different soil types separated in seven defined climatic regions
(Baritz et al. 2005). Although organic layers often contain only a small fraction of
the total soil C pool, in some soils, such as Podzols, Gleysols and Arenosols, the
organic layer contains as much C as the mineral soil, but the associated variation is
huge. If C stocks in the mineral soil are compared between soil types, differences
are hard to obtain since the variation within soil types is much larger than the difference in C stocks between soil types. Only the Arenosols, which represent very
sandy soils, seemed to have lower C stocks on average.

Overall, these figures are not very promising if soil types shall be used as criteria
for stratification or for upscaling in soil C maps, since the variability of C stocks
within the soil groups are very huge while differences between soil types are small.
But maybe the soil groups as they are currently defined in WRB are no meaningful

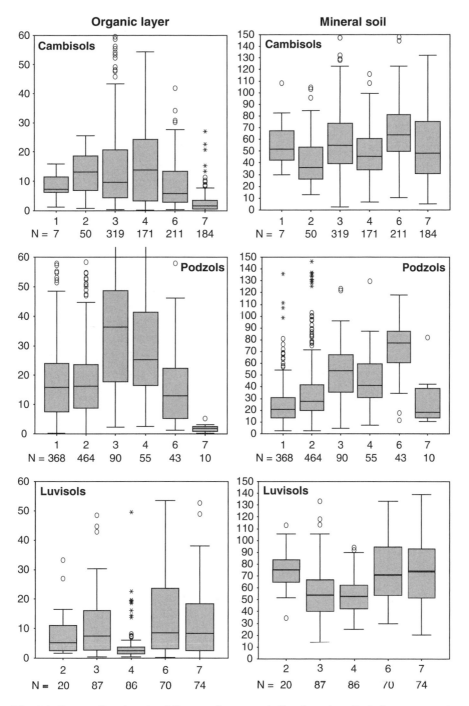

Fig. 9.6 Forest soil carbon for different soil types and climatic regions. Left diagram: organic layer, right diagram: mineral horizon (0–20 cm, Germany 0–30 cm). Climatic regions: (1) subboreal - boreal, (2) boreal - temperate, (3) temperate-oceanic – suboceanic, (4) temperate-suboceanic – subcontinental, (5) temperate - continental, (6) temperate - mountainous, (7) mediterranean (Baritz et al. 2005).

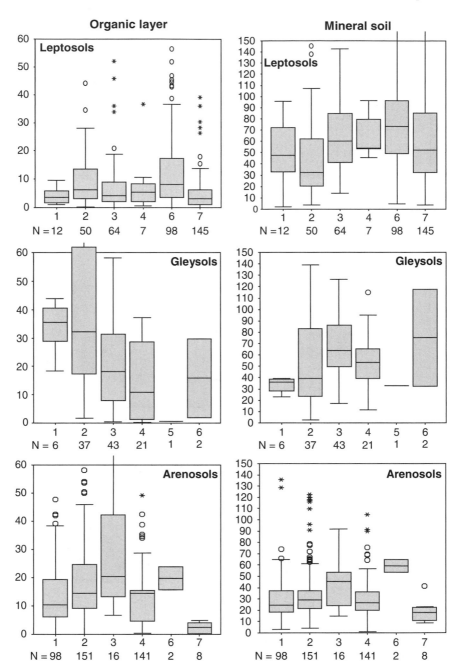

Fig. 9.6 (continued)

predictors for soil C stocks. Krogh and Greve (1999) tested the highest level of WRB soil groups for consistency of pH and silt and clay content within groups for Danish soils and found that cultivation led to a considerable increase in within-group variability of soil properties, since land use affects topsoil characteristics which are relevant for soil classification. Since land use also has a strong impact on soil C contents, it probably also contributes to the high variability of C stocks within soil groups. Kern et al. (1998) and Davis et al. (2004) state that lower taxonomic categories are more reliable predictors for soil C than higher ones and recommend the usage of the great groups of the US Soil Taxonomy, since they include information, like drainage, climate and texture, which are believed to affect soil C. Still, Davis et al. (2004) found substantial differences between regional and national averages for the same great groups indicating that there is still huge large-scale variability within those groups.

In the European-wide study by Baritz et al. (2005), also an effect of climate on C stocks is not clearly visible for all soil types (Fig. 9.6). With the exception of Luvisols and Gleysols, the amount of C stored in the organic layer in the Mediterranean region (7) was overall lower than in the other climatic regions. Gleysols tended to have higher C stocks in the organic layer of the boreal region (1 and 2) than in the temperate zone (3, 4 and 5), while for Podzols, it was the other way round. Maybe a further differentiation of the C stocks in the organic layer between forest types, for example, into deciduous and coniferous forests, which tend to accumulate different amounts of organic matter in the organic layer (Ulrich and Puhe 1994), would have helped to reveal a better soil type or climate-specific effect by reducing the overall variance.

In the mineral soil, Podzols had lowest soil C stocks under boreal climate, higher stocks under temperate climate and highest stocks in the mountain regions (6). Otherwise, with the possible exception of Gleysols and Arenosols, no general trends for higher C stocks in mountainous as compared to lowland temperate climate regions could be found in the mineral soil. If only German soils are considered, Ulrich and Puhe (1994) observed an increase in mineral soil C at elevations above 800 masl, as a result of decreasing temperature and increasing rainfall amounts. Also Lal (2005) stresses the influence of the relation between rainfall and potential evapotranspiration on the C content of forest soils and Kane et al. (2005) observed a temperature effect on forest soil C contents in Alaska. Similarly, Verheijen et al. (2005) and Liebens and VanMolle (2003) observed an effect of soil moisture regime (rainfall and soil drainage) on soil C contents in England, Wales and Flanders. Other factors influencing soil C contents were clay content and land use.

9.5 Recent Evidence for Changes in Soil Carbon

9.5.1 Examples for Measured Soil C Changes

In the following section, we will discuss some cases where soil C changes have recently been reported. In particular, the study of Bellamy et al. (2005) attracted some attention since they observed a drastic decline of on average $64\,g\ C\ kg^{-1}\ y^{-1}$

in the upper 15 cm of the mineral soil in England and Wales between 1978 and 2003. Besides agricultural soils, also non-arable lands lost detectable amounts of C. This indicates that processes which affect the whole country independent of land coverage, like climate change, have to be considered as possible reasons for the overall declining C concentrations.

A comparison of soil C stocks in Belgium between 1960 and 2000 reveals that C stocks in forests and grasslands increased during the last 40 years (27 and 9 t C ha^{-1} over 40 years, respectively) while C stocks in croplands exhibited only a small, but significant decline (-1 t C ha^{-1}). But these trends were not linear, since highest amounts of C in grasslands and croplands were obtained in 1990 (for forests, no data was available for 1990). Carbon stocks in grasslands, for example, were 70 t C ha^{-1} in 1960, 84 t C ha^{-1} in 1990 and 79 t C ha^{-1} in 2000 (Lettens et al. 2005b). Thus, grasslands lost C between 1990 and 2000, but C stocks of 2000 were still higher than those of 1960. Regional varieties in land management led to differences in soil C trends between northern (Flanders) and southern (Wallonia) Belgium. SOC under croplands in Flanders, for example, increased by 14% between 1960 and 1990 because of an increase in farmyard application as a result of a growing livestock density. In Wallonia, more mineral fertilizer, including lime, was applied which increases mineralization rates, and also higher C losses by erosion were assumed. Therefore, C stocks in croplands of Wallonia decreased by 16% during the same time, resulting in only a low net change in soil C for the whole country (Lettens et al. 2005b). Overall, these results demonstrate that (1) while it is necessary to study changes in soil C over longer time periods because otherwise no changes are detectable, the time periods should not be chosen too long otherwise minor fluctuations will be overlooked or obscure the result, and (2) spatial differences in soil C trends have to be considered in monitoring systems and their interpretation.

Regional differences in soil C trends were also observed by Zhang and McGrath (2004) for grasslands in southeastern Ireland. Because of differences in the frequency of pasture improving plowing, coastal areas exhibited an overall increase in SOC concentrations (+10% to + 30%) between 1964 and 1996, while inland areas exhibited an overall decrease (-20% to -50%). Again only a regional stratification reveals land use-related changes in soil, while country-wide averages only indicate minor, insignificant changes.

9.5.2 Discussion of Methods Applied

While in all those studies, significant changes in soil C stocks could be detected, different approaches had to be used for comparing two, often rather unequal, datasets of different study years (Fig. 9.7). Overall, the study by Bellamy et al. (2005) had the most appropriate dataset for the detection of changes, since it was based on a regular sampling according to a dense grid of 5×5 km^2, a total of 25 subsamples per grid point reduced the influence of small-scale variability, samples were collected and treated in a harmonized way, and the re-sampling was

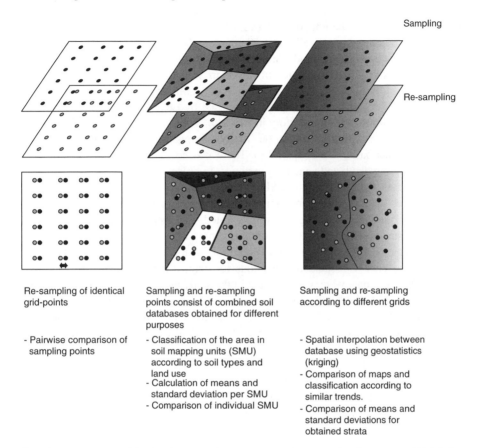

Fig. 9.7 Comparison of the data bases and interpretation techniques of Bellamy et al. (2005, left), Lettens et al. (2005a,b middle) and Zhang and McGrath (2004, right).

conducted at the same places as the first survey. Although re-sampling was not conducted in the same year for all land-use types, this design enabled a paired comparison of sampling points, which reduces the variance and increases the chance to detect changes.

Also the Irish study by Zhang and McGrath (2004) was grid-based, but different grids and sampling points were used for the first sampling in 1964 (up to two randomly chosen points per $10 \times 10 \, km^2$) and in 1996 (sampling at the nodes of a $10 \times 10 \, km^2$ grid). Again, between 15 and 20 subsamples of the upper 10 cm of permanent grassland soils were taken to reduce small-scale variability, but no paired comparison of sampling points from the different surveys was feasible. Maps showing the spatial patterns of SOC concentrations were created for both years via spatial interpolation by kriging, using information of preliminary calculated semivariograms. For comparison, the relative difference in SOC concentration between each pixel on both maps was calculated and plotted. The kriging variance was used to calculate the uncertainty level of the differences at each pixel point of the combined maps (for details, see Zhang and McGrath 2004). This example shows that geostatistical methods and geographic information system (GIS) can be used to identify regions with different trends in SOC contents that can afterwards be treated separately for statistical evaluation of significant differences.

The most heterogeneous dataset had to be used by Lettens et al. (2005a,b). Since there was no nationwide soil inventory in Belgium, a dataset of a number of individual soil surveys which were done for various reasons during different time periods had to be used as a data basis (see Chap. 4.2.1). There was no standardization for sampling strategy, sampling depth and analytical methods. In order to be able to aggregate these different datasets, landscape units (LSU) were used based on land-cover data (Corine land-cover geo-dataset, European Commission 1993) and a digitized soil map. A total of 289 LSU (with 72,908 polygons) were created. For each LSU, the average SOC content and standard deviation is then obtained from all soil profiles within this LSU. For some regions, datasets do not consist of individual soil profile analyses, but only average C contents with standard deviations for defined land units (usually land-use units) exist. These data were also included in the study and their contribution to the C contents of the LSU used in this study was area weighted. A special statistical procedure described in Lettens et al. (2005a,b) was then used to estimate the standard deviation of the C contents of those LSU. SOC stocks of individual LSU are compared pairwise among the different years to obtain C fluxes between 1960, 1990 and 2000. Since geometrically identical LSU had to be used for each study year, effects of land-use changes cannot be analyzed by this approach.

In all studies, stratification was used to increase the interpretability of the results. Bellamy et al. (2005) used stratification according to land-use types to reduce the sampling effort for re-sampling and also for the interpretation of the results. In the Belgium study, Lettens et al. (2005a,b) used stratification according to land use and soil types by inventing the LSU, thus decreasing the variance within each LSU compared to the total variance and increasing the chance to detect changes. Finally, Zhang and McGrath (2004) used geostatistics for a stratification of the study area according to the sign of changes. Limitations of two of the studies in England and Ireland were that results were restricted to the topsoil and only present changes in C concentrations, due to lack of data on bulk density. In the Irish study the bogs were excluded. Not only in Ireland, but also in the UK bogs contain considerable amounts of total C stocks, but changes can probably not be detected by simple concentration measurements, since the bogs can have risen or collapsed during the studied time period without changing C concentrations in the top soil.

9.6   The Approach of CarboEurope-IP

The integrated European project CarboEurope started in 2004 and aims at understanding, quantifying and predicting the terrestrial C balance of Europe at various scales. Soils are considered at the ecosystem and the continental scale.

At ecosystem level, the aims are: (1) to provide soil information for the interpretation of eddy flux measurements, (2) to monitor soil C stocks to obtain an independent estimate of the net biome productivity (NBP) on selected main eddy tower sites, (3) to improve process understanding of C storage in soils. Thus, the focus is on footprint areas of main eddy tower sites in CarboEurope, which vary in size between 1 ha and 10 ha depending on land-use type and wind situations. For all 52 main tower sites, soil maps of the footprint area are to be produced to get information about the homogeneity of those sites and current C stocks of major soil types are determined.

On 12 special so-called verification sites, a huge effort has been made to detect changes in soil C stocks within 5 years—the commitment period of the Kyoto Protocol. Among the sites chosen out of the total 52 main tower sites are three deciduous forests, three coniferous forests, three grasslands and three agricultural sites distributed over Europe (Fig. 9.8). Preliminary studies have shown that a sample size of 100 samples per site might be a good compromise between practicability and a fair chance to detect changes—if they do occur on the chosen sites. Thus, a total of 100 soil cores were taken along a regular grid in the main footprint area of the eddy covariance towers using a hydraulic corer system by Eijkelkamp (core diameter: 8 cm) to a depth of 70 cm or until hard bedrock was reached. Before coring, the organic layer was removed and collected at each point using a square frame with 25-cm side length. Soil cores were pedologically described and afterwards separated into segments of 0–5, 5–10, 10–20, 20–30 … cm. At two sites, where coring was not possible because of the high stone content of the soil, an excavation method of soil monolithes covering an area of 30×30 cm² was applied with a reduced number of samples. In total, about 9,000 soil samples were collected in 2004.

For all soil samples, stone and root contents, bulk density and C and N contents will be analyzed and C stocks calculated from the results. The results provide detailed baseline information on C stocks and their variance in the footprints of eddy covariance stations in the year 2004. Statistics can then be used to calculate the minimum detectable difference of C concentrations and C stocks at individual sites, dependent on present C amounts and their variability. The first re-sampling is planned in 2009, which may afterwards be repeated every 5 years according to the Kyoto cycles. A paired design will be used, so that the next samples will be taken 1 m apart from the original sampling points, which are well documented and marked in the field. After the second soil survey, we will then be able to state with a certain statistical confidence that changes of a certain magnitude occurred or that they were below a certain magnitude. Together with results of a continuous monitoring of biomass C, an independent assessment of changes in ecosystem C stocks can be made to verify the long-term eddy covariance measurements.

On 10 randomly selected soil cores out of the 100 cores per site, further analyses will be conducted to get more information on the processes behind C gains and losses in the soils. Mineralization experiments, aggregate and density fractionation, ^{13}C, ^{14}C and ^{15}N analyses as well as particle size and mineralogical observations will

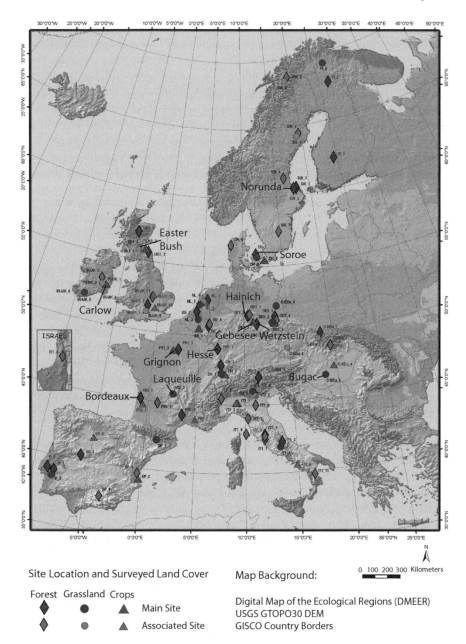

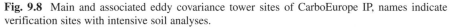

Fig. 9.8 Main and associated eddy covariance tower sites of CarboEurope IP, names indicate verification sites with intensive soil analyses.

be done at different European institutes on the same samples. The advantage of this strategy is that the results have a high accuracy per site because of the high number of replicates and new information on the medium-scale spatial variability of various soil properties will be obtained. Furthermore, results of individual analyses can be directly compared and correlated.

Although these studies will provide very detailed insights into the soil C budgets of individual ecosystems, they are still restricted to a limited area and not all possible combinations of soil and land-use types could be included in this study. Thus, the results cannot be directly used for upscaling changes in soil C to the whole European continent. But they are valuable as test sites to calibrate, verify and improve soil models, which can be applied to larger areas. To facilitate upscaling, the continental integration group within CarboEurope will additionally provide georeferenced inventories of soils and biomass covering entire Europe as a basis for terrestrial ecosystem models. This project part is closely linked to the activities of the EU project CarboInvent (see Chap. 4.2.2.) and will produce a soil C map of Europe including not only forested areas but also grasslands and croplands.

9.7 Towards an Improved Sampling Scheme for Europe

Two main questions need to be answered on the European scale:

(1) How big is the current European soil C stock?
(2) Are there changes in European soil C stocks and where and why do they occur?

Currently, most monitoring systems use a grid design for soil sampling. The advantage of a sampling design with a uniform distribution of samples is that it has some flexibility for new research questions, but that goes along with a huge sampling effort. Thus, for analyzing existing soil C stocks, stratification would be useful according to parameters that reduce the variance within the strata as compared to the total area. In the past, soil type, land use and climate were used as parameters for stratification. The soil analyses of Baritz et al. (2005), which were restricted to the land-use type of forest showed only minor differences in C stocks and variance between soil types. Thus the question arises, whether soil types according to the current soil classification by FAO, WRB or Soil Taxonomy define suitable strata for precise estimates of C stocks and their changes at high taxonomic levels. Maybe a more operational grouping of soils would be useful according to differences in their C accumulation potential. On the assumption that soil C contents depend on texture, mineralogy and litter input, and that microbial activity is reduced by anaerobic conditions in the soil, Intergovernmental Panel on Climate Change (IPCC, 1996) invented six major soil groups which include (1) high clay activity mineral soils, (2) low clay activity mineral soils, (3) sandy soils, (4) volcanic soils, (5) aquatic soils (soils influenced by stagnating or ground water) and (6) organic soils. A detailed and extensive sensitivity analysis should be done to find the most suitable parameters for stratification. A database for such an analysis should include soil samples of all naturally occurring combinations of soil types, climate regions and land use, as those are the most probable factors affecting soil C.

The situation gets more difficult, if the aim is not to reduce the variance of the total estimate of C stocks, but the variance of changes would be necessary for the detection of changes using a paired design. Besides C stocks, also changes can vary at different spatial scales. Some changes can be driven by more local factors, such as land-use change, while impacts of climate change would cover larger areas. Still, at a European scale, climate change will probably not affect all areas in the same way. Again soil properties might affect the magnitude of those effects by different abilities to stabilize C which is either already present in the soil or entering it as a new C. Overall, greatest changes might be expected in areas which are intensively used, where land-use changes occur or climate change is most pronounced, while in areas with constant land cover, changes are probably rather small. Also the direction of changes should be considered and it might be useful to stratify between sites with expected decreases (intensively used agricultural sites) or increases in C stocks (forest sites, sites where land-use practise has changed towards conservative tillage or higher inputs of organic matter).

Unfortunately, the available database for changes in soil C, its spatial distribution and dependencies is rather weak. An alternative to real data is the use of soil models to predict soil C changes under different scenarios. Smith et al. (2005, 2006) analyzed changes in European mineral soil C between 1990 and 2080 for different climate scenarios, separately for croplands, grasslands and forests. They identified climate change as a key driver for SOC dynamics during the next century, but its effects are overlaid by influences of changes in biomass production and advances in technology for grass- and croplands and changes in age structures for forests. Including all those parameters in the analyses, on average all European land-use systems seem to accumulate soil C during the first half of the twenty-first century. But there are regional differences (Fig. 9.9). Although an overall increase in cropland C contents is expected because of advances in technology, some areas in northern Europe are still expected to loose C. For grasslands, only the very eastern part of Europe in Finland and Greece seems to loose some C, the coastal areas of southern Europe remain more or less unchanged, while the remaining part of Europe seems to accumulate C. Forested areas are expected to accumulate C until 2050 because of an increase in litter production as a consequence of the forest age structure. Afterwards, scenarios vary with some predicting-C losses after 2050. A strong effect of land-use changes for the entire European budget was also predicted in these studies. Overall, the results indicate that stratification between land-use types and different climatic regions might be useful, together with the identification of areas where land-use changes occur. The use of different scenarios as input parameters for soil models as adopted by Smith et al. (2005, 2006) will be a useful tool to identify sensitive areas for soil C changes and give useful information for a stratification of sampling to verify the model results. Peltoniemi et al. (2007) show how soil models can be used to improve the stratification of soil sampling

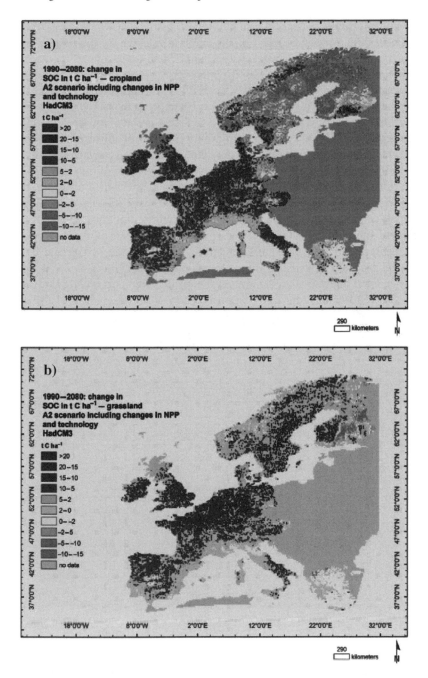

Fig. 9.9 Maps showing model results of difference in mean SOC stocks in croplands (a) and grasslands (b) between 1990 and 2080 for a climate change scenario (A2), including changes in NPP (net primary productivity), advances in technology and regional differences in SOC stocks (taken from Smith et al. 2005).

for monitoring soil C changes. In their study of changes in soil C across England and Wales, Bellamy et al. (2005) based the stratification for the re-sampling of sites on a combination of land-use types and initially measured C content. The definition of strata according to initially measured values is tempting because the resulting stratification automatically reduces within-strata variance. One has, however, to bear in mind that the initial values are subject to measurement errors (as a result of the variance in the field as well as of measurement errors in the lab), so that extreme values in the first sampling period will probably have less extreme values in the re-sampling phase. This may lead to a spurious negative relationship between initial C content and rate of change. The magnitude of this effect depends on the ratio between within-site variability and total variability in the first sampling period and has to be considered in a proper statistical analysis. To facilitate such analyses, information about within-site variability should be obtained at least for a small subset of the sampling sites.

For analyses of the ecosystem C budget, a harmonization of soil and biomass inventories is required (see Chap. 10 for forest inventories). Only if soil and biomass analyses were conducted at the same sites, the results of both inventories are directly comparable and enable studies to improve process understanding of the whole

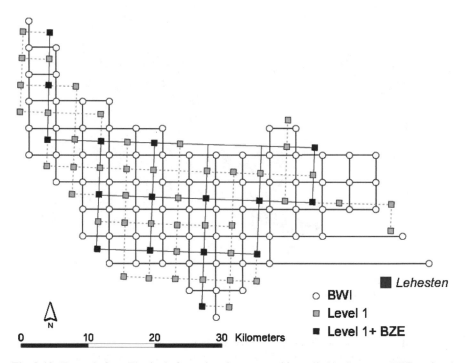

Fig. 9.10 Example from Thuringia for various inventory grids applied in Germany. BWI: national forest inventory, BZE: national soil inventory, Level 1 and Level 2: sampling sites of the European ICP forest inventory.

ecosystem. But according to a FSCC report (2003, http://www.ibw.vlaanderen.be/publicaties/rapporten/fscc/grid_study.pdf), only 64% of the Forest Level I sites in the forest soil conditions database are sites represented in the 2001 Crown Condition database as well. Also long traditions and political reasons often impede the harmonization of inventories. In Thuringia, Germany, for example, the ICP Level 1 grid size was reduced to $4 \times 4 \, km^2$ for monitoring forest crown damage and to $8 \times 8 \, km^2$ for soil analyses (BZE). The grid used for forest inventories in Germany (BWI) is $4 \times 4 \, km^2$, also, but it has a different orientation, thus impeding direct comparisons between sampling points (Fig. 9.10). Thus, if an attempt will be made to monitor soil C across Europe, also existing or upcoming biomass monitoring systems should be considered in order to improve the output of the overall monitoring efforts.

9.8   Conclusions

In summary, we recommend the use of the available databases and modelling results to identify parameters for an effective stratification of soil sampling, either optimized for C stocks or for stock changes. Using estimated variances of those obtained strata, an analysis of the minimum detectable difference should be made and sample sizes chosen according to the desired accuracy. In the optimal case, grid schemes of already existing monitoring systems could be reused and a random subset of grid points would be selected according to the chosen stratification. King and Montanarella (2002) made a representativity analyses for several grid sizes, and concluded that a grid of $16 \times 16 \, km^2$ would be needed to account for all soil- and land-cover types in Europe. For EU25 with a total area of $3,972,868 \, km^2$, that would result in 16,000 sampling points. Since the same grid size was also used for the ICP forest monitoring, it might be a good basis for the analyses. Also, existing national inventory or monitoring grids could serve as a basis, but the hindrance for integrating them into international systems is the lack of harmonization in sampling and analytical methods. Usually, countries do not want to change their original procedures because the new surveys would then not be comparable to the old ones and using both methods is often considered too laborious, although it would help to develop conversion factors for the different methods applied. A paired design should be used for monitoring with re-sampling being performed as close as possible to the original sampling. In order to reduce effects of small-scale variability, not only one sample should be taken per chosen grid point but rather the mean of 9–25 individual samples. Geostatistical methods, like kriging, are recommended to get additional information on the spatial distribution of C stocks and changes over Europe as well as on the representativity of the sampling for different scales. The obtained maps can be used to identify regional differences in trends in C stock changes, which can be used for a further stratification for analyses of significant differences. Besides grid analyses on a continental scale, additionally more detailed ecosystem analyses are required to improve process understanding and calibrate ecosystem models.

References

Arrouays D, Deslais W, Badeau V (2001) The carbon content of topsoil and its geographical distribution in France. Soil Use and Management 17:7–11.

Baldock JA, Skjemstad JO (2000) Role of soil matrix and minerals in protecting natural organic material against biological attack. Organic Geochemistry 31:697–710.

Baritz R, Van Ranst E, Seufert G (2005) Soil carbon default values relevant for evaluation of the carbon status of forest soils in Europe. CarboInvent-WP3-D3.2-RUG, Joanneum Research, Austria.

Bellamy PH, Loveland PJ, Bradley RI, Lark RM, Kirk GJD (2005) Carbon losses from all soils across England and Wales 1978–2003. Nature 437:245–248.

Boix-Fayos C, Martinez-Mena M, Calvo-Cases A, Castillo V, Albaladejo J (2005) Concise review of interrill erosion studies in SE Spain (Alicante and Murcia): Erosion rates and progress of knowledge from the 1980s. Land Degradation & Development 16:517–528.

Chapin III FS, Woodwell GM, Randerson JT, Rastetter EB, Lovett GM, Baldocchi DD, Clark DA, Harmon ME, Schimel DS, Valentini R, Wirth C, Aber JD, Cole JJ, Goulden ML, Harden JW, Heimann M, Howarth RW, Matson PA, McGuire AD, Melillo JM, Mooney HA, Neff JC, Houghton RA, Pace ML, Ryan MG, Running SW, Sala OE, Schlesinger WH, Schulze E-D (2006) Reconciling carbon-cycle concepts, terminology, and methods. Ecosystems 9:1041–1050.

Chenu C, Stotzky G (2002) Interactions between microorganisms and soil particles: An overview. In: Huang PM, Bollag J-M, Senesi N (eds) Interactions Between Soil Particles and Microorganisms. John Wiley & Sons Ltd., Chichester, pp 3–40.

Conen F, Zerva A, Arrouays D, Jolivet C, Jarvis PG, Grace J, Mencuccini M (2005) The carbon balance of forest soils: Detectability of changes in soil carbon stocks in temperate and boreal forests. In: Griffiths H, Jarvis PG (eds) The Carbon Balance of Forest Biomes. Taylor and Francis, London, pp 233–247.

Conant RT, Smith GR, Paustian K (2003) Spatial variability of soil carbon in forested and cultivated sites: Implications for change detection. Journal of Environmental Quality 32:278–286.

Conant RT, Paustian K (2002) Spatial variability of soil organic carbon in grasslands: Implications for detecting change at different scales. Environmental Pollution 116:127–135.

Conrad R (1996) Soil microorganisms as controllers of atmospheric trace gases (H_2, CO, CH_4, OCS, N_2O, and NO). Microbiological Reviews 60:608–640.

Darnley AG, Björklund A, Bölviken B, Gustavsson N, Koval PV, Plant JA, Steenfelt A, Tauchid M, Xie X (1995) A global geochemical database for environmental and resource management. Final report of IGCP Project 259. Earth Sciences 19, UNESCO Publishing, Paris, 122 pp.

Daroussin J, King D (1997) A pedortransfer rules database to interpret the soil geographical database of Europe for environmental purposes. In: Bruand A, Duval O, Wosten H, Lilly A (eds) The Use of Pedotransfer Functions in Soil Hydrology Research in Europe. European Soil Bureaux Research Report No 3, EUR 17307 EN, INRA, Orleans, pp 25–40.

Davidson EA, Janssens IA, Luo Y (2005) On the variability of respiration in terrestrial ecosystems: Moving beyond Q_{10}. Global Change Biology 12:154–164.

Davis AA, Stolt MH, Compton JE (2004) Spatial distribution of soil carbon in southern new England hardwood forest landscapes. Soil Science Society of America Journal 68:895–903.

de Gruijter J, Brus D, Bierkens M, Knotters M (2006) Sampling for natural resource monitoring. Springer, Heidelberg, 332 pp.

Dontsova KM, Bigham JM (2005) Anionic polysaccharide sorption by clay minerals. Soil Science Society of America Journal 69:1026–1035.

Eliasson PE, McMurtrie RE, Pepper DA, Strömgren M, Linder S, Ågren GI (2005) The response of heterotrophic CO_2 flux to soil warming. Global Change Biology 11:167–181.

Eusterhues K, Rumpel C, Kögel-Knabner I (2005) Organo-mineral associations in sandy acid forest soils: Importance of specific surface area, iron oxides and micropores. European Journal of Soil Science 56:753–763.

Francis AJ, Dodge CJ, Gillow JB (1992) Biodegradation of metal citrate complexes and implication of toxic-metal mobility. Nature 356:140–142.

Gleixner G, Czimczik C, Kramer C, Lühker BM, Schmidt MWI (2001) Plant compounds and their turnover and stability as soil organic matter. In: Schulze ED, Heimann M, Harrison SP, Holland EA, Lloyd J, Prentice IC, Schimel DS (eds) Global Biogeochemical Cycles in the Climate System. Academic Press, San Diego, pp 201–216.

Gleixner G, Poirier N, Bol R, Balesdent J (2002) Molecular dynamics of organic matter in cultivated soil. Organic Geochemistry 33:357–366.

Gleixner G, Kramer C, Hahn V, Sachse D (2005) The effect of biodiversity on carbon storage in soils. Ecological Studies 176:165–184.

Grabe M, Kleber M, Hartmann KJ, Jahn R (2003) Preparing a soil carbon inventory of Saxony-Anhalt, Central Germany using GIS and the state soil data base SABO_P. Journal of Plant Nutrition and Soil Science 166:642–648.

Guggenberger G, Kaiser K (2003) Dissolved organic matter in soil: Challenging the paradigm of sorptive preservation. Geoderma 113:293–310.

Guo LB, Gifford RM (2002) Soil carbon stocks and land use change: A meta analysis. Global Change Biology 8:345–360.

Hedges JI, Oades JM (1997) Comparative organic geochemistries of soils and marine sediments. Organic Geochemistry 27:319–361.

Jansen B, Nierop KGJ, Verstraten JM (2005) Mechanisms controlling the mobility of dissolved organic matter, aluminium and iron in podzol B horizons. European Journal of Soil Science 56:537–550.

Janzen HH (2005) Soil carbon: A measure of ecosystem response in a changing world? Canadian Journal of Soil Science 85:467–480.

Johnson CE, Curtis PS (2001) Effects of forest management on soil C and N storage: Meta analysis. Forest Ecology Management 140:277–238.

Jones, RJA, Hiederer, R, Rusco, E, Montanarella, L (2005a) Estimating organic carbon in the soils of Europe for policy support. European Journal of Soil Science 56:655–671.

Jones RJA, Houskováá B, Bullok P, Montanarella L (2005b) European Soil Bureaux Research Report No 9 EUR 20559 EN, Office for Official Publications of the European Communities, Luxembourg.

Khanna M, Yoder M, Calamai L, Stotzky G (1998) X-ray diffractometry and electron microscopy of DNA from *Bacillus subtilis* bound on clay minerals. Sciences of Soils 3:1–10.

Kaiser K, Guggenberger G (2003) Mineral surfaces and soil organic matter. European Journal of Soil Science 54:219–236.

Kalbitz K, Schwesig D, Rethemeyer J, Matzner E (2005) Stabilization of dissolved organic matter by sorption to the mineral soil. Soil Biology and Biochemistry 37:1319–1331.

Kane ES, Valentine DW, Schuur EAG, Dutta K (2005) Soil carbon stabilization along climate and stand productivity gradients in black spruce forests of interior Alaska. Canadian Journal of Forest Research 35:2118–2129.

Kern JS, Turner, DP, Dodson, RF (1998) Spatial patterns in soil organic carbon pool size in the Northwestern United States. In: Lal R, Kimble JM, Follett RF, Stewart BA (eds) Soil Processes and the Carbon Cycle. CRC Press, Boston, pp 29–44.

King D, Montanarella L (2002) Inventaire et surveillance des sols en Europe. Étude et Gestion des Sols 9:137–148.

Kleber M, Mikutta R, Torn MS, Jahn R (2005) Poorly crystalline mineral phases protect organic matter in acid subsoil horizons. European Journal of Soil Science 56:717–725.

Klironomos JN, Rillig MC, Allen MF (1999) Designing belowground field experiments with the help of semi-variance and power analyses. Applied Soil Ecology 12:227–238.

Körner C (2003) Slow in, raip out—Carbon flux studies and Kyoto targets. Science 300:1242.

Kohlmeier S, Smits THM, Ford RM, Keel C, Harms H, Wick LY (2005) Taking the fungal highway: Mobilization of pollutant-degrading bacteria and fungi. Environmental Science & Technology 39:4640–4646.

Krogh L, Greve MH (1999) Evaluation of world reference base for soil resources and FAO soil map of the world using nationwide grid soil data from Denmark. Soil Use and Management 15:157–166.

Krogh L, Noergaard A, Hermansen M, Greve MH, Balstroem T, Breuning-Madsen H (2003) Preliminary estimates of contemporary soil organic carbon stocks in Denmark using multiple datasets and four scaling-up methods. Agriculture, Ecosystems and Environment 96:19–28.

Lal R (2005) Forest soils and carbon sequestration. Forest Ecology and Management 220:242–258.

Lettens S, Van Orshoven J, van Wesemael B, Muys, B (2004) Soil organic and inorganic carbon contents of landscape units in Belgium derived using data from 1950 to 1970. Soil Use and Management 20:40–47.

Lettens, S, Van Orshovena, J, van Wesemael, B, De Vos, B, Muys, B (2005a) Stocks and fluxes of soil organic carbon for landscape units in Belgium derived from heterogeneous data sets for 1990 and 2000. Geoderma 127:11–23.

Lettens, S van Orshoven, J, van Wesemael, B, Muys, B, Perrin, D (2005b) Soil organic carbon changes in landscape units of Belgium between 1960 and 2000 with reference to 1990. Global Change Biology 11:2128–2140.

Liebens J, VanMolle M (2003) Influence of estimation procedure on soil organic carbon stock assessment in Flanders, Belgium. Soil Use and Management 19:364–371.

v. Lützow M, Kögel-Knabner I, Ekschmitt K, Matzner E, Guggenberger G, Marschner B, Flessa H (2006) Stabilization of organic matter in temperate soils: Mechanisms and their relevance under different soil conditions—A review. European Journal of Soil Science 57:426–445.

Marschner B, Kalbitz K (2003) Controls of bioavailability and biodegradability of dissolved organic matter in soils. Geoderma 113:211–235.

Matson PA, Parton WJ, Power WJ, Swift MJ (1997) Agricultural intensification and ecosystem properties. Science 277:504–508.

Melillo JM, Steudker PA, Aber JD, Newkirk K, Lux H, Bowles FP, Catricala C, Magill A, Ahrens T, Morrisseau S (2002) Soil warming and carbon-cycle feedbacks to the climate system. Science 298:2173–2176.

Milne, R, Brown, TA (1997) Carbon in the vegetation and soils of Great Britain. Journal of Environmental Management 49:413–433.

Miltner A, Kopinke FD, Kindler R, Selesi DE, Hartmann A, Kastner M (2005) Non-phototrophic CO_2 fixation by soil microorganisms. Plant and Soil 269:193–203.

Mund M, Schulze ED (2006) Impacts of silvicultural practices on the carbon budget of European beech forests. Allgemeine Forst- und Jagdzeitung 177:47–63.

Nordstrom KF, Hotta S (2004) Wind erosion from cropland solutions in the USA: A review of problems, and prospects. Geoderma 121:157–167.

Palmer CJ, Smith WD, Conkling BL (2002) Development of a protocol for monitoring status and trends in forest soil carbon at a national level. Environmental Pollution 116:209–219.

Peltoniemi M, Heikkinen J, Mäkipää R (2007) Stratification of regional sampling by model-predicted changes of carbon stocks in forested mineral soils. Silva Fennica (September, in press).

Plante AF, Feng Y, McGill WB (2002) A modeling approach to quantifying soil macroaggregate dynamics. Canadian Journal of Soil Science 82:181–190.

Prescott CE (2005) Decomposition and mineralization of nutrients from litter and humus. In: BassiriRad H (ed) Nutrient Acquisition by Plants. An Ecological Perspective. Springer, Ecological Studies 181, Berlin, pp 15–41.

Renella G, Landi L, Nannipieri P (2004) Degradation of low molecular weight organic acids complexed with heavy metals in soil. Geoderma 122:311–315.

Rillig MC (2005) Polymers and microorganisms. In D Hillil (ed) Encyclopedia of soils in the environment. Elsevier, Oxford, pp 287–294.

Rusco E, Jones R, Bidoglio G (2001) Organic matter in the soils of Europe: Present status, and future trends. European Soil Bureaux Research, EUR 20556 EN, JRC IES, Ispra.

Salminen R (2005) Geochemical Atlas of Europe. Part 1—Background Information, Methodology and Maps. ISBN 951–690–913–2 (electronic version).

Schmidt MWI, Noack AG (2000) Black carbon in soils and sediments: Analysis, distribution, implications and current challenges. Global Biogeochemical Cycles 14:777–793.

Schöning I, Morgenroth G, Kögel-Knabner I (2005a) O/N-alkyl and alkyl C are stabilised in fine particle size fractions of forest soils. Biogeochemistry 73:475–497.

Schöning I, Knicker H, Kögel-Knabner I (2005b) Intimate association between O/N-alkyl carbon and iron oxides in clay fractions of forest soils. Organic Geochemistry 36:1378–1390.

Schulze ED, Freibauer A (2005) Carbon unlocked from soils. Nature 437:205–206.

Schulze E-D, Kelliher FM, Körner C, Lloyd J, Hollinger DJ, Vygodskaya NN (1996) The role of vegetation in controlling carbon dioxide and water exchange between land surface and the atmosphere. In: Walker B, Steffen W (eds) Global Change and Terrestrial Ecosystems. IGBP-Series 2, Cambridge University Press, Cambridge, pp 77–92.

Schwesig D, Kalbitz K, Matzner E (2003) Effects of aluminium on the mineralization of dissolved organic carbon derived from forest floors. European Journal of Soil Science 54: 311–322.

Six J, Conant RT, Paul EA, Paustian K (2002) Stabilization mechanisms of soil organic matter: Implications for C-saturation of soils. Plant and Soil 241:155–176.

Six J, Bossuyt H, Degryze S, Denef K (2004) A history of research on the link between (micro)aggregates, soil biota, and soil organic matter dynamics. Soil and Tillage Research 79:7–31.

Smith J, Smith P, Wattenbach M, Zaehle S, Hiederer R, Jones RJA, Montanarella L, Rounsevell MDA, Reginster I, Ewert F (2005) Projected changes in mineral soil carbon of European croplands and grasslands, 1990–2080. Global Change Biology 11:2141–2152.

Smith P, Smith J, Wattenbach M, Meyer J, Lindner M, Zaehle S, Hiederer R, Jones RJA, Montanarella L, Rounsevell MDA, Reginster I, Kankaanpää S (2006) Projected changes in mineral soil carbon of European forests, 1990–2100. Canadian Journal of Soil Science 86:159–169.

Sollins P, Homann P, Caldwell BA (1996) Stabilization and destabilization of soil organic matter: Mechanisms and controls. Geoderma 74:65–105.

Thuille A, Schulze ED (2006) Carbon dynamics in successional and afforested spruce stands in Thuringia and the Alps. Global Change Biology 12:325–342.

Ulrich B, Puhe J (1994) Auswirkungen der zukünftigen Klimaveränderung auf mitteleuropäische Waldökosysteme und deren Rückkopplungen auf den Treibhauseffekt. Enquete-Kommission "Schutz der Erdatmosphäre" des deutschen Bundestages (eds) Studienprogramm Bd. 2: Wälder, Studie B (208 S.). Bonn, Economica Verlag.

Valentin C, Poesen J, Yong Li (2005) Gully erosion: Impacts, factors and control. Catena 63: 132–153.

Van Ranst, E, Thomasson, AJ, Daroussin, J, Hollins JM, Jones RJA, Jamagne M (1995) Elaboration of an extended knowledge database to interpret the 1:1000000 EU Soil Map for environmental purposes. In: King D, Jones RJA, Thomasson AJ (eds) European Land Information Systems for Agro-Environmental Monitoring. EUR 16232 EN, Office for Official Publications of the European Communities, Luxembourg, pp 71–84.

Verheijen FGA, Bellamy PH, Kibblewhite MG, Gaunt JL (2005) Organic carbon ranges in arable soils of England and Wales. Soil Use and Management 21:2–9.

Wirth C, Schulze ED, Schwalbe G, Tomczyk S, Weber G, Weller E (2004) Dynamik der Kohlenstoffvorräte in den Wäldern Thüringens. Mittelungen der Thüringer Landesanstalt für Wald, Jagd und Fischerei 23/2004:308 pp.

Zhang CS, McGrath D (2004) Geostatistical and GIS analyses on soil organic carbon concentrations in grassland of southeastern Ireland from two different periods. Geoderma 119:261–275.

Chapter 10
Monitoring Carbon Stock Changes in European Forests Using Forest Inventory Data

Raisa Mäkipää, Aleksi Lehtonen, and Mikko Peltoniemi

10.1 Introduction

The forest carbon stock in Europe is large and changes in it may contribute notably to atmospheric CO_2 concentration. The area of forested land in Europe is about 1,000 million ha, which is about 47% of the land area (e.g. MCPFE 2003). The largest part, more than 800 million ha, of the forests in Europe is located in the Russian Federation, whereas the forested area of EU15 is 137 million ha. The percentage of forested land varies considerably between counties ranging from 68% in Finland and Sweden to 1% in Iceland. In the forests of Europe (excluding Russia), the carbon stock of the vegetation was estimated to be about 8,000 Tg (Nabuurs et al. 1997; Goodale et al. 2002). Estimates of the soil carbon stocks range from 5,000 to 14,000 Tg and are evidently more uncertain than the estimated carbon stocks of the vegetation (Goodale et al. 2002; Liski et al. 2002; Nabuurs et al. 2003). Current estimates of the changes in the carbon stock of vegetation in Europe (excluding Russia) range from 50 to 100 Tg C year^{-1}, and the changes in the soil range from 13 to 61 Tg C year^{-1}, depending on the methods applied as well as on the reference area and period (Goodale et al. 2002; Liski et al. 2002; Karjalainen et al. 2003; Nabuurs et al. 2003). In general, large carbon stocks of peatlands soils are not fully accounted in these studies, because of the lack of representative data and/or models.

The changes in the carbon stock of forested areas are partly due to annual variation in net primary production (NPP) and soil respiration, both resulting from climatic variation (Ciais et al. 2005). Furthermore, most of the annual changes in the forest carbon stock are the result of variation in commercial harvests and natural disturbances (Kauppi et al. 1992; UNECE 2000; Nabuurs et al. 2003). In general, the role of natural disturbance has been marginal in comparison to magnitude of commercial harvests in Europe. Exceptionally severe storms in 1999 reduced the carbon sink of biomass in that year (by 52 Tg C), but the increasing trend in the net carbon sink in Europe was only briefly interrupted by these large-scale disturbances (Nabuurs et al. 2003). Under future climatic conditions, the frequency of large-scale disturbances may, however, increase remarkably and they may also play a larger role in Europe's forests.

In addition to the annual variation, long-term trends can also be found in the forest carbon sequestration of Europe. In general, the forest carbon stock of a country tends to increase as a result of forest management and increased growing stock (UNECE 2000; Nabuurs et al. 2003; Lehtonen 2005; Liski et al. 2006). In Northern Europe, the increase in forest carbon stock may be partly a recovery from the period of extensive slash-and-burn cultivation and from overuse of timber resources in the 1800s and early 1900s, especially in Finland. In addition, the growth and biomass stock of trees have increased by drainage and amelioration of peatland forests (Hökkä et al. 2002; Minkkinen et al. 2002). In central Europe, increase in forest carbon stock is partly the result of increase in forested area (MCPFE 2003). The increase in forested area, as a result of afforestation projects and abandonment of agricultural fields, has been remarkable and has played a major role in forest carbon sequestration, for example, in the United Kingdom (Cannell and Dewar 1995). In Scandinavia, as much as 68% of the land area is forested and afforestation has played smaller role in carbon sequestration (Mäkipää and Tomppo 1998; IPCC 2000; UNECE 2000). However, change from other wooded land has increased the area of forested land, especially in Finland, where drainage of peatland has resulted in improved productivity and transformation of less productive peatlands to forested land. This transformation has evidently increased the carbon stock of trees, but the response of soil greenhouse gas (GHG) balance to drainage is more uncertain (Minkkinen et al. 2002).

On a European scale, monitoring of changes in the carbon stock of forest vegetation and soil is challenging because design of the monitoring system should be capable of detecting all these major trends as well as the annual variation that contributes to the carbon balance of forests. Because the factors that control forest carbon balance operate at different spatial and temporal scales, it is necessary to complement existing monitoring networks such as national forest inventories (NFIs) with statistical and dynamic modelling to establish a proper carbon-monitoring system for the forests of Europe.

10.2 Forest Inventories as a Source of Information for Assessment of Carbon

National forest inventories that are designed for monitoring of forest resources at national scales provide a firm basis for large-scale carbon assessments. Forest inventories that cover the entire country have been conducted in Scandinavia since the 1920s, while in France and in Germany the first NFIs were conducted in 1960 and in 1981, respectively (Laitat et al. 2000) (Table 10.1). Time span, sampling methods and quantity (and quality) of the information collected by NFIs vary among countries according to local conditions and information needs (UN-ECE/FAO 1985; UNECE 2000) (Table 10.1), but recent efforts of the research community are also aiming to harmonise inventory methods (COST E43 project, http://www.metla.fi/eu/cost/e43/). In Western European countries with substantial forest industries,

Table 10.1 National Forest Inventories as Data Sources (adopted and modified from Laitat et al. 2000). Country codes according to ISO 3166-1 Aptha-2 code elements, AT = Austria, BE = Belgium, CH = Switzerland, DE = Germany, DK = Denmark, ES = Spain, FI = Finland, FR = France, GB = United Kingdom, GR = Greece, IR = Ireland, IS = Iceland, NL = The Netherlands, NO = Norway, and SE = Sweden

Country	AT	BE	CH	DE	DK	ES	FI	FR	GB	GR	HU	IR	IS	NL	NO	SE
Start	1961	1980 1994[h]	1983	1986	1981	1964	1921	1960	1924	1963	1950	1998	1972	1940	1919	1923
Inventory type[a]	SB	SB	SB	SB	ST	SB	SB	SB	SA	SA	ST	–	–	SA	SA	SA
Periodicity	5	10	10	15	10	10–20	10	10	15–20	30	10	–	–	–	–	–
Forest area[b] (1,000 ha)	3,862	667	1,221	11,076	500	17,915	22,500	15,554	2,845	3,752	1,976	669	46	365	9,387	27,528
Grid	3.9×3.9	1×1	1.4×1.4	4×4	–	1×1	6×6	various	various	1×0.5	–	–	–	–	3×3	various
Plots	11,000	10,600	6,500	12,580	–	84,203	70,000	133,500	–	2,744	–	–	–	3,400	10,500	18,000
Percentage S.E. in area	1.2	0.42	0.3	1.1	–	–	0.48	0.71	–	0.2	–	1	–	–	0.96	0.5
Percentage S.E. in volume	1.6	5.1	1	0.8	–	1.13	0.57	0.54	–	2.6	–	–	–	5	1.36	0.6
Percentage S.E. in volume growth	–	–	0.9	–	–	–	0.8	0.59	–	–	–	–	–	–	1.4	0.4
Biomass, above ground[c,d]	E	E	E	E,F	E	E	E	–	O	E	E	E	O[e]	F	F	F
Biomass, below ground[c,d]	E	E	E	E,F	E	E	E	–	O	E	–	–	–	F	F	F
Origin of biomass functions/ expansion factors[c,d]	CS	CN	D	CS, D	CN	CS	CS	–	CS	D, CS	CS	CS	–	CS	CN	CS

(continued)

Table 10.1 (continued)

Country	AT	BE	CH	DE	DK	ES	FI	FR	GB	GR	HU	IR	IS	NL	NO	SE
Soil estimate in NIR	–	–	–	–	–	–	–	–	YES[f]	–	–	–	–	–	YES[f]	–
DOM estimate in–NIR	–	–	–	–	–	–	–	–	YES[f]	YES	–	–	–	YES	YES[f]	–
Nationwide soil survey existing[g]	–	YES	–	–	YES	–	–	YES	YES	–	–	–	–	YES	–	YES
Repeated sampling in soil survey	–	–	–	–	–	–	–	–	YES	–	–	–	–	–	–	YES

[a] SB refers to sample-based inventory, while ST refers to stand-wise inventory (based on forest compartments)

[b] According to FAO forest definition, source of data: www.fao.org/forestry/

[c] E Biomass expansion factor, F Biomass function (tree-level accounting), O Other method, CS Country-specific model, CN model adopted from neighbouring countries, D IPCC default

[d] Information based on Annex I Party GHG Inventory Submissions, National Inventory Reports (NIR) 2005 from http://unfccc.int

[e] Only afforested land in calculation

[f] Modelled estimate

[g] Based on Jones et al. 2004

[h] Respectively for the Walloon and Flemish regions

systematic samples of forest sites have been collected several times, while Eastern European countries have conducted their first rounds of systematic NFI only recently. Before the current NFIs were established, forest resources were quantified based on regional data compilations that often under- or overestimated target variables, such as growing stock on a national level (UN-ECE/FAO 1985; UNECE 2000). Traditionally, an NFI covering a large country required almost a decade to carry out, while smaller countries were able to accomplish their NFIs in a few years. Because the most important target variables of the NFIs have been forested area, growing stock and increment, the inventories are designed and optimised for quantifying these variables. However, due to the long rotation times of NFIs, the annual estimates of increment were not directly available. Therefore, estimation of annual carbon stock changes is based on interpolation or on extrapolation.

10.3 Carbon Inventories Under the Climate Convention are Guided by IPCC

The United Nations Framework Convention on Climate Change (UNFCCC 1992), aims at stabilising the GHG concentrations in the atmosphere at a non-dangerous level and the Kyoto Protocol (UNFCCC 1997) under it, setting legally binding commitments to reduce emissions in the industrialised countries. In the Kyoto Protocol, forest carbon sinks were accepted as one of the mitigation options of climate change. The protocol allows that a limited part of the forest carbon sinks can compensate for the emission reductions of a country. Because the credited sinks may compensate for internationally agreed emission reductions, the need for more accurate sink estimates and more transparent reporting became evident.

Requirements concerning GHG reporting were already defined in the Climate Convention and in the Kyoto Protocol, but more detailed reporting rules were set in subsequent meetings (e.g. in Bonn and Marrakesh 2001). At the request of the Convention, the Intergovernmental Panel on Climate Change (IPCC) has developed guidance for estimation and reporting of GHG emissions and removals of land use, land-use change and forestry sectors including all different carbon pools (above- and belowground biomass, deadwood, litter and soil organic carbon) (IPCC 2003, 2006).

IPCC guidance (2003, 2006) stated that forest carbon inventories should be based on representative nationwide information on forest resources, such as forested area, area of other land-use classes and exchanges between them, estimate of growing stock, annual growth, commercial harvests and other losses. In general, this information is collected by the NFIs (UNECE 2000). Furthermore, relevant information of nationwide soil surveys or forest soil monitoring programmes can be used in the carbon inventories (IPCC 2003).

Verification of reported forest carbon stock change estimates must be conducted with independent methods and material. Verification of the forest carbon inventory can be based on remote sensing, ecosystem modelling or combinations of these (IPCC 2003). Furthermore, inverse modelling that calculates fluxes from the concentration measurements and atmospheric transport models can be used for verification on a continental scale and direct measurements of ecosystem fluxes on a local scale (IPCC 2003) (Chap. 17).

10.4   Estimation of Forest Carbon Balance Based on Inventory Data and Modelling of Litter and Soil

10.4.1   Biomass

On a national scale, changes in the biomass and carbon stocks of trees can be estimated based on NFI data. The change can be estimated based on the difference between two consecutive inventories or on estimated increment and losses (drain). These two approaches are also recommended by the IPCC (2003). The default method of the IPCC, in which estimates of both annual losses and growth are needed, assumes that countries are able to quantify all components of the drain (losses), such as natural mortality, fuelwood gathering and loggings, as well as growth on an annual basis. For example, in Finland, data on commercial drain and domestic fuel wood use are aggregated by species and gathered at the national level by questionnaires sent to wood users (both forest industry and households), while estimates of natural mortality have been derived from permanent sample plots of the NFI. In the case of Finland, the forest industry is obliged to report its wood use and therefore the estimate of the annual drain resulting from commercial harvests is considered and tested to be reliable (Kuusela 1979). In Sweden, the carbon stock change is based on consecutive measurements from permanent sample plots of NFI (Ståhl et al. 2004). The use of permanent sample plots gives a more reliable estimate of stock change than the use of temporary sample plots (with the same number of plots), since the high covariance between the following measurements increases the accuracy of the change estimate.

Countries have annually reported changes in the biomass carbon stock based on the NFI to the United Nations Framework Convention on Climate Change (UNFCCC) according to the guidelines developed by the IPCC. Usually the net changes in carbon stock have been obtained by deducting losses from the increment of growing stock, and thereafter this net change of volume has been converted to biomass by biomass expansion factors (BEFs). These BEFs were developed to obtain whole-tree biomass, including canopy and roots (e.g. IPCC 2003; Lehtonen et al. 2004; Levy et al. 2004; Lehtonen et al. 2007) or the biomass of roots can be estimated separately with shoot–root ratios available from the literature (IPCC 2003; Levy et al. 2004). The values of these BEFs and the methodology developed vary considerably among countries (Löwe et al. 2000; Somogyi et al. 2007). The main disadvantages of current BEFs are that often the uncertainty estimates are lacking or that even the origin of the estimates is unknown. Furthermore, most of the BEFs are constant and developed for conversion from stem volume to whole-tree biomass, but they are often applied for increment and losses.

Biomass equations are applied at the tree-level and tree-wise biomasses are predicted as a function of tree dimensions (such as diameter and height) and other predictors. Most of the existing biomass equations are based on a couple of sites with only a few felled sample trees (Zianis et al. 2005; Muukkonen and Mäkipää 2006a). Therefore, the main limitation of biomass equations is poor representativeness in national-scale inventories. However, a few exceptions exist, for example Marklund (1988) in which uniform sampling was conducted for the whole of Sweden. Recently there have also been other works that compiled data from various studies and derived new biomass equations, as Wirth et al. (2004) did for Norway spruce in central Europe.

In addition to the biomass equations and constant BEFs, there are other options for regional biomass estimation. Fang et al. (2001) modelled BEFs as a function of stem volume based on direct field measurements in China, while Jenkins et al. (2003) grouped biomass equations and developed generalised functions for species groups in the USA and Muukkonen (2007) developed generalised functions for the major tree species in Europe. Levy et al. (2004) modelled BEFs as a function of tree height in the UK, while Lehtonen et al. (2004) modelled BEFs in Finland as a function of stand age. In general, these biomass models use measurements or existing biomass models from single trees for development of more robust biomass estimation methods on regional or national scales.

Liski et al. (2006) estimated changes in biomass and soil carbon stocks for Finland, based on NFI data and modelling. Stem volume estimates reported by NFIs were converted to biomass components (foliage, branches, stem and roots of different sizes) with BEFs. These BEFs with uncertainty estimates were developed by age-class and dominant species (Lehtonen et al. 2004). Tree biomass is the main component of the carbon stock of vegetation, but the understorey vegetation may still play an important role, especially in litter input. The biomass of understorey vegetation was modelled as a function of stand age. The models that were based on nationwide vegetation data were complemented with models on the relationship between biomass and coverage of understorey vegetation (Muukkonen and Mäkipää 2006b; Muukkonen et al. 2006).

10.4.2   Deadwood

The amount of carbon in deadwood is highly variable between stands across the landscape, both in managed and unmanaged forests. Unmanaged or natural forests contain considerably larger quantities of deadwood than managed forests (Siitonen 2001; Jonsson et al. 2005). Coarse woody debris (deadwood) may be a notable carbon stock in forests, especially after major disturbances (Harmon et al. 1986; Krankina et al. 2001). Furthermore, changes in management practices may show wide and relatively rapid influence on the carbon stock of deadwood. Current management practices are already aiming to increase the amount of deadwood because of its value for the maintenance of biodiversity. In addition to the management practices that control the amount of deadwood left on a site, the amount of deadwood is dependent on the time and severity of the disturbances, the rate of natural mortality in general and the decay rate of deadwood (Harmon et al. 1986; Siitonen 2001; Stokland 2001).

In carbon inventories, changes in the carbon stock of deadwood are usually calculated using the stock change method or by calculating the difference between transfer of carbon into deadwood and transfer out of the deadwood pool that is modelled with a decomposition model (IPCC 2003).

The stock change method can be applied by countries in which measurements of the carbon stock of deadwood are included in the forest inventories and where such an inventory is repeated. Measurements of the volume of deadwood are conducted in the NFIs, for example, in Sweden, Norway, Switzerland and Finland, but repeated nationwide inventories of deadwood are rare. Deadwood volume is measured according to decay classes (Ståhl et al. 2001; Stokland 2001), and the volume estimates can be converted to biomass and carbon estimates by decay classes (inventory measure) using species-specific statistical models (Kruys et al. 2002; Mäkinen et al. 2006).

Change in the carbon stock of deadwood can also be calculated based on the difference between input and output of carbon in this pool. Carbon input into the deadwood pool includes harvest residues left at the site and natural mortality (both self-thinning and biomass of trees killed by natural disturbances). Carbon transfer out of the deadwood pool is carbon emitted in decomposition. The rate of decomposition of deadwood under conditions representing conditions of boreal forests was modelled by Tarasov and Birdsey (2001), Kruys et al. (2002) and Mäkinen et al. (2006), and decomposition of deadwood is also included in some soil carbon models, for example in Yasso (Liski et al. 2005).

10.4.3   Litter and Soil Carbon

The changes in litter and soil carbon are difficult to monitor (Chap. 9) because the changes in upland forest soils (forests remaining as forests) are very small in comparison to the size of carbon stocks (Chap. 9). Forest soils are also highly heterogeneous; most of the variance in mean stock is already present over very short distances (Conen et al. 2004). Consequently, large numbers of samples and monitoring sites, or a long-time interval, are required until a significant change in soil carbon can be detected (Conen et al. 2003; Smith et al. 2004)(Chap. 9). For organic soils, such as peatlands drained for forestry, changes in the soil carbon may be fast during the first decades after drainage, but uncertainty of change estimates is still high even after 30 years of drainage (Minkkinen et al. 1999). Even though the expected rate of change in soil carbon stock may be small, the vast areas make them important components of national and global carbon budgets (Chap. 12). Furthermore, change in the soil carbon stock may be in the direction opposite to that of the vegetation and overall change in the ecosystem (Liski et al. 2006).

If no repeated measurements of soil are available or measuring is deemed impractical due to high costs, modelling of soil carbon can be used to close the entire carbon budget of forests. In this approach, biomass models are applied to forest inventory data to obtain biomass estimates and the biomass estimates are further converted to litter (foliage, branches, fine roots, etc.) with biomass turnover rates. Each biomass component has a distinctive lifespan (a) that can be used to estimate the biomass turnover rate (a^{-1}). Various estimates of biomass turnover rates are available from literature and for some ecosystem types and some components, the estimates of biomass turnover are rather comprehensive (Afman and Kaipainen 2005).

The litter input is then decomposed with a dynamic soil carbon model. The inventory-based approach was previously presented by Kurz and Apps (1994) and Liski et al. (2002, 2006). Another modelling option for estimating soil carbon stock changes is to link the soil carbon or its changes directly to NFI variables with statistical models (Johnson and Kern 2002; Smith and Heath 2002; Amichev and Galbraith 2004). A large variety of applicable dynamic soil models is available, with a variety of input demands, parameterisations for different types of soils and climatic ranges (e.g. Powlson et al. 1996; Smith et al. 1997; Peltoniemi et al. 2007). In the nationwide inventories, selection of a soil model is often determined by the availability of representative input data covering the entire country.

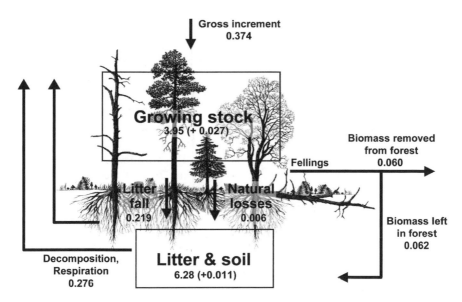

Fig. 10.1 Average carbon budget of the Finnish forests in the 1990s (according to Liski et al. 2006). The average carbon stocks (kg C m^{-2}) and the annual changes in carbon stocks (kg C m^{-2} year^{-1}) are shown in boxes and annual carbon fluxes (kg C m^{-2} year^{-1}) are illustrated by arrows (graph by Essi Puranen).

In general, the dynamic soil models that are applicable for mineral soils and for the litter layer of peatlands are not capable for modelling of decomposition of peatland soils. Furthermore, organic soils may also be sources of non-CO_2 GHG emissions and, for example, the emissions of N_2O may be significant in their relative GHG impact (Alm et al. 2007b). In the GHG inventory of the organic peatland soils, fluxes of CO_2, CH_4 and N_2O need to be estimated (IPCC 2003, 2006). For that, flux measurements must be available for different peatland ecosystems (Alm et al. 2007a) and the flux estimates have to be added to the modelled litter decomposition rate.

The benefit of the dynamic modelling approach is that it takes into account the decomposition of carbon that has accumulated in the soil over time. However, initialisation of soil carbon models with measured data is problematic because the data on soil are inaccurate and imprecise, and measured pools rarely correspond to model pools. The problem in determining the initial state of carbon can be partly avoided by initiating the calculation as far from the past as the data allow (Chen et al. 2000) and letting the model find the initial state with corresponding inputs. When the calculation is initiated far enough from the past, the error in the upward or downward trend due to initialisation is likely to be smaller than other uncertainties governing the modelled sink estimates (Peltoniemi et al. 2006).

10.4.4   Inventory-Based Soil Carbon Budget of Finland

In an application that was run for Finland, input data provided by the NFI was grouped into classes of tree species, region and age-class (Liski et al. 2006). Grouping by age was included, since age-class distribution of forests in Finland has changed notably in the last century and biomass allocation is known to be age-dependent (Satoo and Madgwick 1982; Lehtonen 2005; Metla 2006).

In addition to the above- and belowground litter produced by standing vegetation, natural disturbances (severe storms or pests) and harvestings leave decomposing material into forests. In Finland, natural disturbances have played a small role and most of the trees killed are collected from the disturbed sites. Commercial harvests in Finland are remarkable (currently about 70% of the growth is used) and a large amount of litter is left in the forest in the form of harvest residues (Metla 2006). Because the data related to timber collection are tabulated with high precision and accuracy, the carbon input of harvest residues to the soil was derived based on this information (Liski et al. 2006). Currently, Finland is targeting to higher efficiency in biofuel production with harvesting of logging residues. The effect of the harvesting of branches and root system to the forest carbon balance can be accounted with this modelling approach. Ground vegetation also produces a significant amount of litter, but little is known of its annual variability. In the application for Finland, this flow of litter to the soil was represented with mean estimates of biomass and with published estimates of its growth, which were used for its turnover.

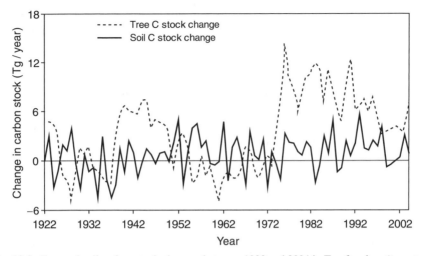

Fig. 10.2 Tree and soil carbon stock changes between 1922 and 2004 in Tg of carbon (tree stock from all soils, while soil stock comprised mineral soils) (adopted from Liski et al. 2006).

In addition to the estimates of litter input, the Yasso decomposition model used in this study required information on temperature sum and drought that was calculated as a difference between potential evaporation and precipitation (Liski et al. 2006). The soil model predicts the carbon content of litter and soil down to 1 m (on mineral soils), which was the depth used in its parameterisation. Yasso is based on the concept that chemical compounds existing in litter determine the rate at which the litter decomposes (Liski et al. 2005). This process was parameterised and tested only for mineral soils with no peat cover (Liski et al. 2005; Palosuo et al. 2005), and other models should be used for peatland soils.

The above-described approach allowed calculation of the carbon budget of forest soils in Finland for the period between 1922 and 2004 (Fig. 10.2). This study shows that all pools must be accounted for in the forest carbon inventory to avoid biased conclusions. At the same time, forest soils may be a source of carbon while trees act as a sink.

10.5   Evaluation of Inventory-Based Methods

10.5.1   Measures for Assessing Accuracy and Precision

The error sources in large-scale carbon budget compilations are numerous. The NFI data and other input data involve errors that are characteristic of measurements and sampling protocol. Models that convert measured variables to target variables are often developed with inadequate sampling of the population or the data quality may be poor (see Sect. 10.4.1). The model structure used may not represent adequately the phenomenon we are interested in (for general discussion on uncertainties, see Morgan and Henrion 1990 and Cullen and Frey 1999). Uncertainty analysis of the modelled results should be a part of any carbon budget calculation. Here we will first examine the uncertainty in the structure and submodels of the inventory-based carbon balance calculation, mostly in the form of subjective assessment. Secondly, we will view the results of a Monte Carlo analysis of uncertainty and discuss the differences in factors that affected the uncertainty of carbon sink and carbon stock estimates of the inventory-based carbon balance. Subjective assessment of the uncertainty of carbon stocks and their changes provides insight on factors that may affect the accuracy (bias) and Monte Carlo analysis the precision (random error) of the carbon sink and stock estimates.

10.5.2   Assessment of Applicability and Accuracy of the Model in Estimating Forest Carbon Budgets

Forest carbon inventories have usually been prepared with aggregated input data, that is by summing or averaging the inputs over large spatial or temporal domains before they are fed to a model (UNFCCC, National Inventory Reports, http:// unfccc.int/). This was also the case in the application for Finland, where the NFI data were grouped into several classes and into the two regions of southern and northern Finland. Depending on the structure of the model, aggregation can bias the results markedly or only to a limited extent, or may not affect the results (Rastetter et al. 1992; Paustian et al. 1997). In the application for forests in Finland, aggregation does not affect the results of biomass sinks, because the summation of the biomass estimates is linear. For soil carbon stocks and sinks, however, summation may have some effect due to the non-linear effect of climate on decomposition.

A justified application of models requires that the models are used in a population similar to that for which they were developed. In practice, this is rarely the case and subjective assessment must be used in model selection. Here, the biomass of trees was estimated with BEFs (Lehtonen et al. 2004), which were developed based on Swedish biomass equations (Marklund 1987; Marklund 1988) and tested for their applicability for conditions in Finland and Sweden (Jalkanen et al. 2005; Lehtonen 2005). The major limitation of both these biomass equations and consequently BEFs was possible underestimation of the root biomass (Petersson and Ståhl 2006).

Biomass turnover models are based on data of varying quality (Afman and Kaipainen 2005). Typically, the best data are available for temperate and boreal zone countries. Similarly the data on the aboveground components of foliage and branch litter are generally better known than belowground components. Belowground turnover estimates are often lacking, which is most problematic for the rapidly regenerating component of fine roots. Because fine roots constitute a substantial percentage of the total net primary production (NPP) of forests (Jackson et al. 1997), quantification of their contribution to the carbon budget is crucial. However, the estimation of fine-root turnover is challenging because the fine-root longevity varies widely, for example, with tree species and the size of fine roots (Matamala et al. 2003). Furthermore, different methods that have been used to assess the turnover produce different results (Gaudinski et al. 2001; Tierney and Fahey 2002).

Component-wise tests of empirical data reveal whether the components give accurate and precise estimates, but they do not assess the system as a whole. Therefore, system-level tests are required but such tests are often limited by data availability and quality. The modelling approach that was used to estimate the carbon budget for forests in Finland was tested on empirical data from mineral soils in the region of southern Finland (Peltoniemi et al. 2004). Despite the difficulties in comparisons of model predictions of soil with measured material, the results showed similar trends and the same magnitude of change, that is the simulated rate of change in soil carbon stock was similar to the rate derived from chronosequence of Scots pine and Norway spruce stands.

10.5.3 Assessment of Precision of the Forest Carbon Budget with Monte Carlo Analysis of Uncertainty

Analysis of random error with the Monte Carlo simulations does not take into account failures of the model structure to explain the phenomenon, but assesses only the random part of the error as related to model parameters and input variables. In this analysis, the calculation is repeated thousands of times, each time drawing random samples from the probability density distributions of the model parameters and input variables (Morgan and Henrion 1990). As a result, probability densities are obtained for the resulting variables.

Sensitivity and uncertainty analysis of the carbon budget for forests in Finland (excluding soils of peatlands) showed that the data on annual growth variation and harvests were the most critical components of the biomass carbon sink (Peltoniemi et al. 2006). The biomass carbon stocks were mostly affected by carbon content, inventory data on growing stock and BEFs. In the Finnish application, the random error in simulated soil carbon stocks was large and mostly the result of uncertain soil model parameters. However, the random error of the soil carbon sink was only somewhat larger than that of the biomass sink. Most of the uncertainty in the soil carbon sink was caused by soil model initialisation, a source of uncertainty that decreased in effect with the time elapsed from initialisation (Fig. 10.3). The decrease was rather rapid after which two inputs, harvest residues and temperature, began to control the soil sink uncertainty. Model parameters of the major litter fractions (biomass and turnover of ground vegetation and fine roots) also contributed notably to the uncertainty in the soil sink.

Many of the variables that contributed largely to the uncertainties in the carbon budget had features in common. The input data played an important role in the sink uncertainties of vegetation and soil, because of uncorrelated error estimates for each year of the calculation period (e.g. temperature, growth variation and drain).

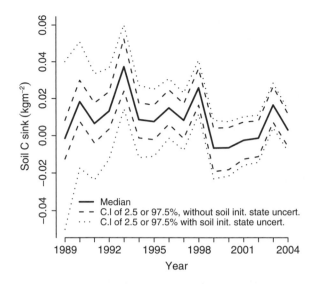

Fig. 10.3 Mean changes in the carbon stock of forest soils in Finland from 1989 to 2004 (mean and S.D. in kg C m^{-2} year^{-1}, calculated with and without soil initial state uncertainty) (adopted from Peltoniemi et al. 2006).

Use of temporal autocorrelations estimated from the input data series did not significantly affect the uncertainty estimates. The central position of the variables was another important feature in common with those variables that were important for the uncertainty in sinks or stocks of carbon. For example, the effect of carbon content was emphasised because it was a shared parameter for all tree species and biomass components.

10.5.4   Land-Use Changes

The rate of change in the soil carbon resulting from land-use shifts is large in comparison to the rate of changes in the forested areas remaining as forests (Schlesinger 1990; Post and Kwon 2000). Therefore, these land-use changes may release large quantities of carbon, which are detected in atmospheric measurements. The role of forest inventories in this assessment is to provide data on areas of different land-use categories and on their mean carbon stocks. However, sampling of forest inventories may not be sufficient to cover small changes in land use and must be appended with other data sources such as survey and satellite data. Furthermore, exact assessment of the effects of land-use change requires that each parcel of land is tracked through history, since the previous land use affects current soil carbon. Following the modelling approach for estimating regional carbon sinks and sources, the simulations should be performed for each of the land parcels, with a tool suitable for modelling the soil carbon stocks in various land-use classes.

The inventory-based approach as applied for Finland used aggregated inventory data, which provided only information on net changes in the forested areas. As such, the method is not optimised to account for the effects of land-use changes on forest carbon sinks. Land-use changes in Finland are relatively small, and originally their effect was assumed to be small. However, uncertainty analysis revealed that even small changes in area are important for soil carbon budgets. This indicates a need for more detailed accounting of the effects of land-use changes. A complementary accounting of land-use changes should be performed by preparing a land-use transition matrix that classifies and aggregates different types of land and applies conversion factors for a period of 20 years, as suggested by the IPCC 2003. For example, Heath et al. (2002) presented the methodology applied in the Forest Carbon model (FORCARB) in the USA, and Tate et al. (2003) used a modified IPCC method to estimate the effect of land-use change for soils in New Zealand.

10.6   Use in Carbon Cycle Research

Forest inventories themselves are able to provide little ecological interpretation of carbon cycle research. Forest inventories measure the realised stocks rather than the process fluxes themselves. Factors affecting the fluxes are implicitly recorded in measured stocks of timber and growth. Comparison of the estimates of large-scale forest carbon sinks obtained with different methods is challenging, and forest inventory-based carbon budgets of forests provide valuable reference data for comparisons, especially in a situation where the carbon stocks of tree biomass are facing trend-like changes. Long-term changes in biomass stocks may be related to silviculture, age-structure of forests or climate change. Recurrent inventories with sound sampling are able to detect these changes in the tree biomass stocks.

10.6.1   Ecological Equivalences can be Estimated Based on the Inventory Data

Data collection in forest inventories is typically slow and may require up to 10 years until the entire country has been surveyed. Some countries are using or have recently begun to use (Finland, Hungary, Sweden and USA) continuous inventory of forest resources, in which the entire country is surveyed each year, but with smaller sampling density. This improves the prospects of monitoring the annual variation in carbon stocks in a representative area, based on forest inventory data. However, even this leads to a minimum time step in inventory data of 1 year with which atmospheric measurements can be compared.

Before annually updated nationwide inventory data can be made available, periodic mean estimates of growing stock and growth need to be processed into annual estimates that are comparable with results obtained with other methods (such as annual flux measurements based on the eddy covariance technique). In the Finnish application, the annual estimates of timber volume were interpolated using data on removal of timber from forests and on annual variation in growth, which were considered the most important factors affecting the annual variation in forest carbon stocks (Liski et al. 2006). The estimates of annual growth were based on growth indices measured from tree rings collected from several hundred NFI sample plots. This variable conveys multiple factors that affect annual variation in growth, such as early start of growing season or otherwise favourable growing conditions during the previous or current season.

In biological science, the main ecosystem processes have been named and defined according to their function. Net primary production (NPP) refers to the net production of carbon by plants and it equals the gross primary production (GPP) minus the carbon respired by plants, where GPP refers to the total amount of carbon assimilated by plants. Net ecosystem production (NEP) refers to NPP minus the carbon losses in heterotrophic respiration, which is the carbon lost by organisms other than plants (e.g. microbes). Net biome production (NBP) equals NEP minus sudden carbon losses from the ecosystem, such as forest fires or harvestings. These terms are commonly used when carbon fluxes of well-measured ecosystems are reported. In forest inventories, the terminology is different although the ecosystems are the same. The fact that the inventory-based approach is able to produce estimates for ecological equivalences, such as NPP, NEP and NBP, makes comparison with other approaches possible (see, however, Roxburgh et al. 2005).

In conjunction with carbon balance of Finnish forests, Liski et al. (2006) estimated NPP, NEP and NBP based on NFI data and modelling for upland soils, by defining:

$$NPP = \Delta C + L + M + F \tag{10.1}$$

where ΔC is the change in carbon stocks, L is annual litter production by vegetation, M represents amount of natural losses and F fellings.

The estimate of NEP was obtained by subtracting from NPP the outflows of carbon, R_h (most of it occurs as heterotrophic respiration), that were simulated using the soil model Yasso (Liski et al. 2005; Palosuo et al. 2005):

$$NEP = NPP - R_h \tag{10.2}$$

The NBP was calculated by subtracting from NEP removals RE, which represented felled roundwood removed from the forests. In the case of Finland, forest fires are negligible and the impact of windthrows is included in the NFI growing stocks:

$$NBP = NEP - RE \tag{10.3}$$

Liski et al. (2006) showed that the NBP of forests in Finland growing in mineral soils has been positive since the 1970s, that is, the ecosystem has been a net sink of

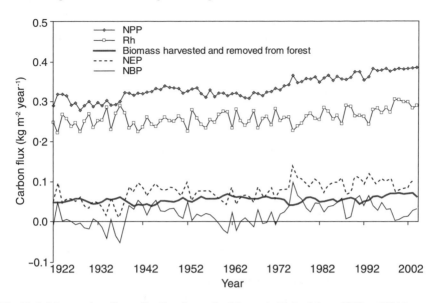

Fig. 10.4 Mean carbon fluxes (kg C m^{-2} year^{-1}) of forests in Finland from 1922 to 2004 in terms of NPP, heterotrophic respiration, NEP, removals and NBP (adopted from Liski et al. 2006).

carbon. These forests were a source of carbon during periods of heavy harvesting in the 1960s and in the 1930s, when NBP was negative (Fig. 10.4). These estimates include other main carbon fluxes occurring at large scale, but exclude estimates for the peatland soils as well as all non-CO$_2$ fluxes.

10.6.2   Future Scenarios of Forest Carbon Balance

Future scenarios for forest carbon sequestration are required by the Climate Convention as well as by national policy processes, especially as a part of climate and energy policy and strategic planning. In the national communications under the Climate Convention, countries report projections of GHG emissions and removals (sinks) for the next 20 years (e.g. Kuusisto and Hämekoski 2001). In general, national and regional scenarios of forest growth and harvests are available and have been updated for the European countries, because these provide basic information for the strategic planning of the forest industry. The scenario of forest carbon balance for the coming decades can be derived with inventory-based methods using information on projected growth and harvests as input data (Karjalainen et al. 2003). The findings of the scenario may vary considerably, depending on the assumptions put forward in both the input data and model simulations. Models that are not validated to conditions of changing climate (and changes in litter quality) are, however, limited to predictions under current climatic variation.

10.7 Conclusions

Forests are considered as an important mechanism for mitigation of climate change, because they can accumulate carbon from the atmosphere to biomass and soil carbon stock. The carbon inventory for the forestry sector should cover the annual changes in the carbon stocks of biomass, deadwood, litter and soil organic matter (IPCC 2003). Methods and datasets needed for the estimation of changes in the carbon stock of tree biomass are somehow available, especially those concerning the aboveground parts. On European scale, however, variation in local conditions and in the tradition of forest inventories between countries makes it challenging to compile a complete GHG inventory for the entire continent with coherent methods. In general, the major challenge in forest carbon inventories is monitoring of the carbon pools of litter and soil organic matter. Because repeated measurements of soil carbon stock on a national scale are generally lacking (Chap. 9), preliminary estimates of soil carbon sink/source can be provided by modelling only. The magnitude of soil carbon changes is such that it should be accounted for in the national forest carbon budget estimates (Liski et al. 2006). Furthermore, the carbon sequestration potential of forests can be affected by forest management practices and management may have opposing influences on biomass and soil carbon stocks. Therefore, forest carbon monitoring must cover all pools to avoid biased conclusions and mitigation strategies.

The inventory guidance (IPCC 2003, 2006) suggested that countries should accommodate their GHG inventories with uncertainty and key category analysis that show the most important sector for the overall GHG inventory as well as the major sources of uncertainties within the sectors. These types of analyses aid in prioritising efforts to develop and improve the quality of the GHG inventories (Peltoniemi et al. 2006; Monni et al. 2007). Furthermore, verification of the inventory by independent methods can provide insights to unknown factors or excluded components that may contribute notably to GHG balance (IPCC 2003). For example, inverse modelling that calculates fluxes from concentration measurements and atmospheric transport models can be used for verification of compiled inventory-based carbon balance on a continental scale (Bousquet et al. 2000). Janssens et al. (2003) compared inverse modelling and inventory-based approaches on a European scale by calculating separate estimates for the forest and agricultural sectors. Verification of the entire inventory may be impossible, but several methods can be used to test the validity of some parts of the inventory; for example, remote-sensing data can be used for verification of sampling-based estimates of the land-use changes on a regional scale. In addition to regional scale, direct measurements of ecosystem fluxes on a local scale can be used for evaluation of the annual variation in the inventory-based carbon balance (IPCC 2003).

The default methods used in the national carbon inventories do not detect annual variations in the carbon balance very accurately, because they interpolate changes over longer time spans (e.g. Tomppo 2000). Based on forest inventory data, it is possible to calculate more detailed carbon balance with the help of growth indices that indicate interannual variations in growth as a result of variation in the climatic conditions measured from sample trees of the NFI (Lehtonen 2005; Liski et al. 2006). In this application for Finland, the effects of the annual variation in growth and harvests on carbon balance were accounted for, but the litter input was calculated as constant proportions of the biomass of each component. It is, however, evident that annual biomass production and litter input from biomass to soil vary from year to year according to climatic conditions, and the accuracy of the annual carbon balance could be improved by taking this variation into account.

Most of the European countries have reported increases in the biomass carbon stock during recent decades (UNECE 2000; Liski et al. 2003) but have not reported estimates of the soil carbon sink/source. Uncertainty estimates for soil are often lacking, or if they are reported, are larger than those of the biomass (Ståhl et al. 2004; Peltoniemi et al. 2006; Monni et al. 2007). The complete carbon balance calculated for forests in Finland showed that the biomass carbon pool increased by a mean of $27\,g\,C\,m^{-2}$ annually in the 1990s, while the carbon pool of upland forest soils increased by $11\,g\,C\,m^{-2}$ (Liski et al. 2006). However, both the tree and soil carbon pools are known to vary widely on an interannual basis. Because the carbon sinks of forest soils have been smaller (and close to zero in some years) and the uncertainty of their estimate is larger than that of vegetation, they may have acted as sources as well as sinks during the last decade.

The importance of soils in the forest GHG balance indicates that all European countries should develop their inventories in a way that soil carbon balance is included before continental-scale inventory-based carbon balance covering the entire forest ecosystem can be compiled. Methods applicable for such an inventory of upland forests are already available (e.g. Lehtonen 2005; Liski et al. 2005, 2006).

Acknowledgements We thank the Academy of Finland (project 108328) and the Forest Focus Programme of the European Commission (through a pilot project 'Monitoring changes in the carbon stocks of forest soils') for the funding of this study. We acknowledge Dr. Jari Liski and Dr. Jukka Alm for discussions and valuable comments during our work on nationwide carbon inventories and Dr. James Thompson for checking the English language.

References

Afman, A., and Kaipainen, T. 2005. Database of biomass turnover rates, http://ghgdata.jrc.it/carboinvent/cidb_bioturn.cfm. EFI, JRC.
Alm, J., Shurpali, N.J., Minkkinen, K., Aro, L., Hytönen, J., Laurila, T., Lohila, A., Maljanen, M., Mäkiranta, P., Penttilä, T., Saarnio, S., Silvan, N., Tuittlia, E.-S., and Laine, J. 2007a. Emission factors and their uncertainty for the exchange of CO_2, CH_4 and N_2O in Finnish managed peatlands. Boreal Environment Research 12: 191–209.

Alm, J., Shurpali, N.J., Tuittila, E.-S., Laurila, T., Maljanen, M., Saarnio, S., and Minkkinen, K. 2007b. Methods for determining emission factors for the use of peat and peatlands—flux measurement and modelling. Boreal Environment Research 12: 85–100.

Amichev, B.Y., and Galbraith, J.M. 2004. A revised methodology for estimation of forest soil carbon from spatial soils and forest inventory data sets. Environmental Management 33: 74–86.

Bousquet, P., Peylin, P., Ciais, P., Le Quere, C., Friedlingstein, P., and Tans, P.P. 2000. Regional changes in carbon dioxide fluxes of land and oceans since 1980. Science 290: 1342–1346.

Cannell, M.G.R., and Dewar, R.C. 1995. The carbon sink provided by plantation forests and their products in Britain. Forestry 68: 35–48.

Chen, W., Chen, J., Liu, J., and Cihlar, J. 2000. Approaches for reducing uncertainties in regional forest carbon balance. Global Biogeochemical Cycles 14: 827–838.

Ciais, P., Reichstein, M., Viovy, N., Granier, A., Ogee, J., Allard, V., Aubinet, M., Buchmann, N., Bernhofer, C., Carrara, A., Chevallier, F., De Noblet, N., Friend, A.D., Friedlingstein, P., Grunwald, T., Heinesch, B., Keronen, P., Knohl, A., Krinner, G., Loustau, D., Manca, G., Matteucci, G., Miglietta, F., Ourcival, J.M., Papale, D., Pilegaard, K., Rambal, S., Seufert, G., Soussana, J.F., Sanz, M.J., Schulze, E.D., Vesala, T., and Valentini, R. 2005. Europe-wide reduction in primary productivity caused by the heat and drought in 2003. Nature 437: 529–533.

Conen, F., Yakutin, M.V., and Sambuu, A.D. 2003. Potential for detecting changes in soil organic carbon concentrations resulting from climate change. Global Change Biology 9: 1515–1520.

Conen, F., Zerva, A., Arrouays, D., Jolivet, C., Jarvis, P.G., Grace, J., and Mencuccini, M. 2004. The carbon balance of forest soils: detectability of changes in soil carbon stocks in temperate and boreal forests. In Griffith, H., Jarvis, P.G. eds., The carbon balance of forest biomes, pp. 233–247. Garland Science/BIOS Scientific Publishers.

Cullen, A.C., and Frey, H.C. 1999. Probabilistic techniques in exposure assessment. Plenum Press, New York.

Fang, J.-Y., Chen, A., Peng, C., Zhao, S., and Ci, L. 2001. Changes in forest biomass carbon storage in China between 1949 and 1998. Science 292: 2320–2322.

Gaudinski, J., Trumbore, S., Davidson, E., Cook, A., Markewitz, D., and Richter, D. 2001. The age of fine-root carbon in three forests of the eastern United States measured by radiocarbon. Oecologia 129: 420–429.

Goodale, C.L., Apps, M.J., Birdsey, R.A., Field, C.B., Heath, L.S., Houghton, R.A., Jenkins, J.C., Kohlmaier, G.H., Kurz, W., Liu, S.R., Nabuurs, G.J., Nilsson, S., and Shvidenko, A.Z. 2002. Forest carbon sinks in the Northern Hemisphere. Ecological Applications 12: 891–899.

Harmon, M.E., Franklin, J.F., Swanson, F.J., Sollins, P., Gregory, S.V., Lattin, J.D., Anderson, N.H., Cline, S.P., Aumen, N.G., Sedell, J.R., Lienkaemper, G.W., Cromack, K., and Cummins, K.W. 1986. Ecology of coarse woody debris in temperate ecosystems. Ecological Research 15: 133–302.

Heath, L.S., Birdsey, R.A., and Williams, D.W. 2002. Methodology for estimating soil carbon for the forest carbon budget model of the United States, 2001. Environmental Pollution 116: 373–380.

Hökkä, H., Kaunisto, S., Korhonen, K.T., Päivänen, J., Reinikainen, A., and Tomppo, E. 2002. Suomen suometsät 1951–1994. Metsätieteen aikakausikirja 2B.

IPCC 2000. A special report of the IPCC. Land use, land-use change, and forestry. Cambridge University Press.

IPCC 2003. Good practice guidance for land use, land-use change and forestry. IPCC, National Greenhouse Gas Inventories Programme, Kanagawa, Japan.

IPCC 2006. Guidelines for national greenhouse gas inventories. IPCC, National Greenhouse Gas Inventories Programme, Kanagawa, Japan.

Jackson, R.B., Mooney, H.A., and Schulze, E.D. 1997. A global budget for fine root biomass, surface area, and nutrient contents. Proceedings of the National Academy of Science of the United States of America 94: 7362–7366.

Jalkanen, A., Mäkipää, R., Ståhl, G., Lehtonen, A., and Petersson, H. 2005. Estimation of biomass stock of trees in Sweden: comparison of biomass equations and age-dependent biomass expansion factors. Annals of Forest Science 62: 845–851.

Color Plate 1

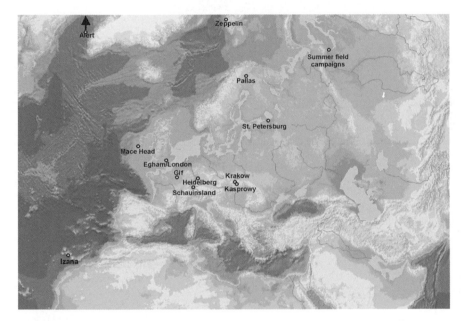

Fig. 6.1 Some existing CH_4 and $\delta^{13}C$-CH_4 measurement sites in Europe, Siberia and the Arctic. Reproduced from Meth-MonitEUr (2005), with kind permission from Meth-MonitEUr (2005) (*See* page 98)

Color Plate 2

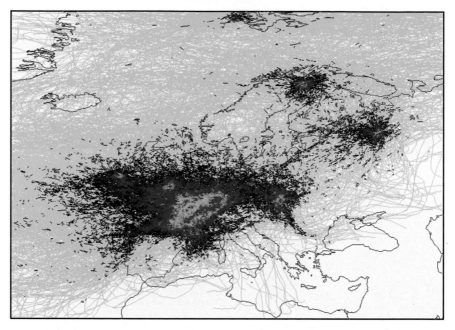

Fig. 6.3 A composite of daily 12.00 GMT 5-day back trajectories for 2002 indicating density of European coverage. Sites in Network are Heidelberg, Kasprowy, Mace Head, Ny-Ålesund, Pallas, Paris, Puy de Dome, RHUL, Schauinsland, and St. Petersburg. Red represents the highest density and yellow the lowest. Reproduced from NUI Galway, report to Meth-MonitEUr, with kind permission from Meth-MonitEUr 2005 (*See* page 100)

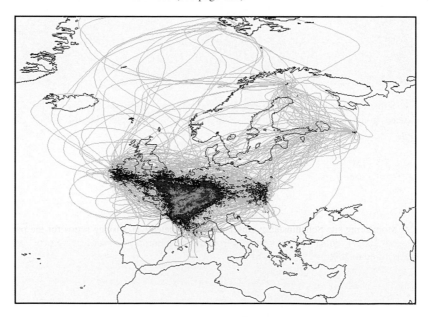

Fig. 6.4 Segments of trajectories as shown in Fig. 6.3 where prior to arrival at the sampling site, the air mass has passed within two degrees of another site in the network. Reproduced from NUI Galway, report to Meth-MonitEUr, with kind permission from Meth-MonitEUr 2005 (*See* page 101)

Color Plate 3

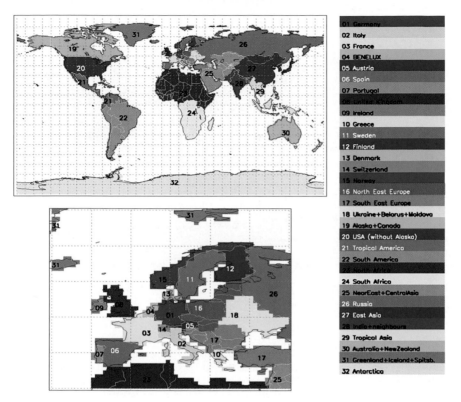

01 Germany	16 North East Europe
02 Italy	17 South East Europe
03 France	18 Ukraine+Belarus+Moldova
04 BENELUX	19 Alaska+Canada
05 Austria	20 USA (without Alaska)
06 Spain	21 Tropical America
07 Portugal	22 South America
08 United Kingdom	23 North Africa
09 Ireland	24 South Africa
10 Greece	25 NearEast+CentralAsia
11 Sweden	26 Russia
12 Finland	27 East Asia
13 Denmark	28 India+neighbours
14 Switzerland	29 Tropical Asia
15 Norway	30 Australia+NewZealand
	31 Greenland+Iceland+Spitsb.
	32 Antarctica

Fig. 7.1 Global and European regions used for the inversion (Bergamaschi et al. 2005) (*See* page 117)

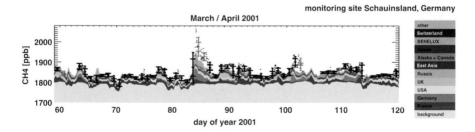

Fig. 7.2 Monitoring site Schauinsland: Observed and modeled CH_4 mixing ratios for the period March–April 2001. Colors highlight influence from different European countries or global regions (Bergamaschi et al. 2005) (*See* page 118)

Color Plate 4

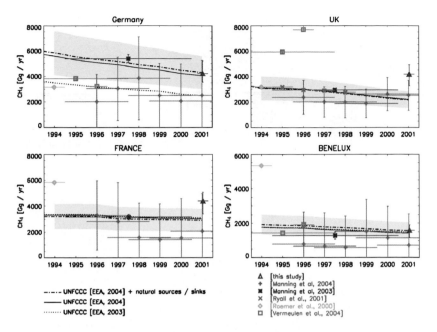

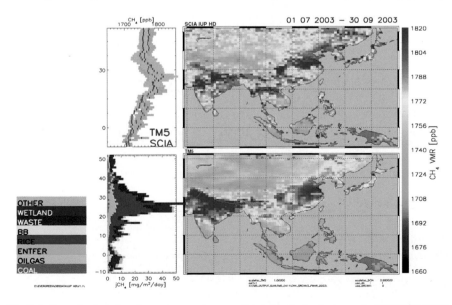

Fig. 7.3 Comparison with other inverse modeling studies and United Nations Framework Convention on Climate Change (UNFCCC) values. UNFCCC estimates (black solid curve) are augmented by our bottom-up estimate of net natural sources (including soil sink), displayed by the black dash-dotted line. The gray-shaded area indicates an assumed 30% uncertainty of the total of UNFCCC and natural emissions. Colored symbols represent different top-down estimates, according to the legend below. For further details, see Bergamaschi et al. (2005) (*See* page 118)

Fig. 7.4 Scanning imaging absorption spectrometer for atmospheric chartography (SCIAMACHY) measurements (right top) and TM5 simulations (right bottom) over Asia (*See* page 120)

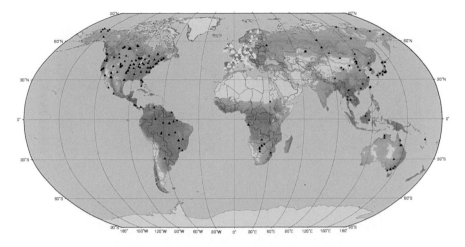

Fig. 11.1 Locations of the 266 FLUXNET measurement sites available on January 1, 2005 (*See* page 216)

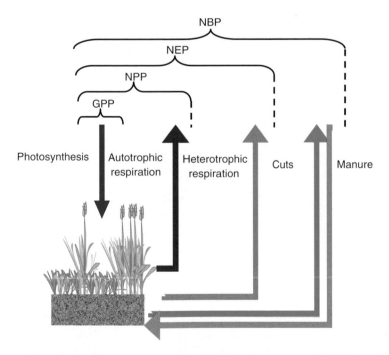

Fig. 13.1 C fluxes in a managed grassland system with imports (manure application) and exports (cuts and harvests) of organic carbon. GPP, gross primary productivity; NPP, net primary productivity; NEP, net ecosystem productivity; NBP, net biome productivity (*See* page 273)

Color Plate 6

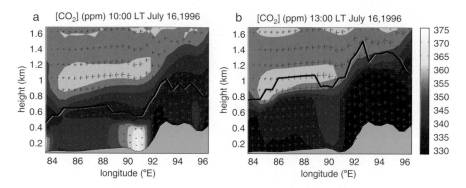

Fig. 14.3 Cross section of the simulated CO_2 concentration field of the coarsest grid at (a) 10:00 h LT and (b) 13:00 h LT on 16 July 1996. The colours and contours indicate concentration differences at 5 ppm intervals; the arrows indicate longitudinal and vertical wind velocity and the solid black line represents the boundary layer height. The surface topography is also shown. Note that the horizontal extent of the domain covers 864 km, so that the horizontal is much compressed relative to the vertical. The cross sections are drawn at 59.2 °N (*See* page 293)

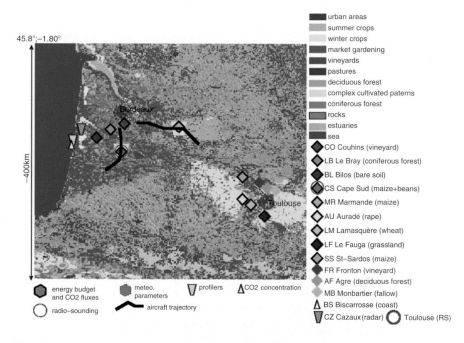

Fig. 14.5 Land cover map at 250-m resolution for the experimental domain in the south-west region of France showing the different location of summer and winter agricultural crops (classification by Champeaux et al. 2005). Also shown are the locations of the ground-based observation sites of surface fluxes and boundary layer. Flight tracks indicate the path flow by the Sky Arrow flux aircraft for agriculture and forested regions (*See* page 297)

Color Plate 7

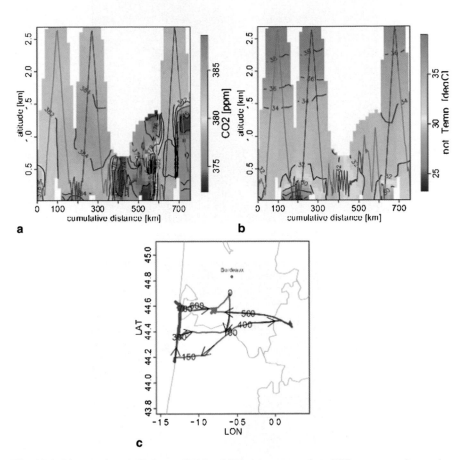

Fig. 14.6 Dimona aircraft flights on 27 May 2003. (**a**) contour plot of CO_2 concentration against distance flown, (**b**) contour plot of potential temperature against distance flown, (**c**) the flight track. Red colour codings indicate agricultural (summer crops) areas, black forested areas and blue the sand dunes of the coastline (*See* page 298)

Color Plate 8

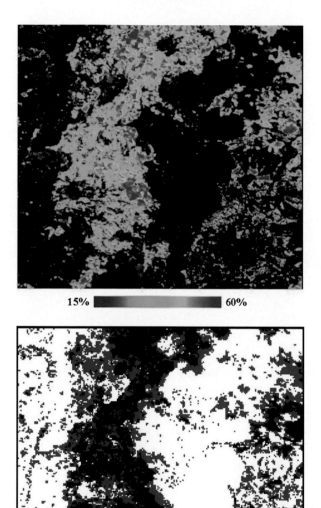

15% 60%

Fig. 15.7 (**a**) Percentage change in near-infrared reflectance on day of burn, measured from moderate-resolution imaging spectroradiometer (MODIS) reflectance data. This will usually be larger for higher intensity burns or where a bigger proportion of the pixel is burned. (**b**) Burnt areas detected by both MODIS and global burnt area (GBA)2000 are marked in black, those by only MODIS in red. The MODIS product detects a significantly larger burnt area than GBA2000, although the latter successfully detects the large fires (*See* page 323)

Color Plate 9

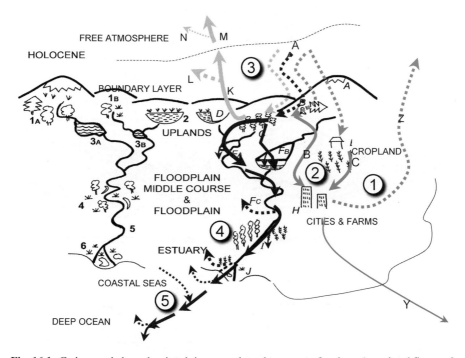

Fig. 16.1 Carbon cycle branches involving some lateral transport of carbon. Associated fluxes of atmospheric CO_2 are in dotted lines, fluxes of transported carbon in solid lines. **Green**. Cycle associated to photosynthesis (A), harvest of wood and crop products, transport by domestic (B, C) and international (Y) trade circuits, and consumption (Z) of carbon in crop products (1) and forest wood products (2). **Brown**. Cycle associated to photosynthesis (A), emissions and atmospheric transport in the boundary layer (K) and transport in free atmosphere (M), and oxidation in the boundary layer (L) and in the free atmosphere (N) of chemically reactive reduced carbon compounds (RCCs). **Blue**. Transport of carbon of atmospheric origin as dissolved inorganic carbon (DIC), dissolved organic carbon (DOC), and particulate organic carbon (POC) by river systems from river uplands to inner estuaries (see text for letters explanation) (4). **Purple**. Fluxes of carbon and CO_2 from coastal seas (5) (*See* page 343)

Color Plate 10

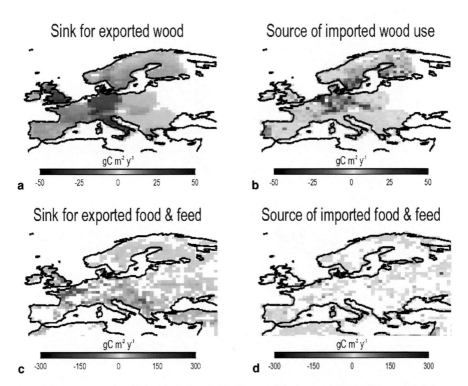

Fig. 16.3 Spatial patterns of trade-induced CO_2 fluxes with the atmosphere. Sources of CO_2 are positive and sinks of CO_2 are negative. **A** and **B** Sinks and sources of CO_2 associated to harvest, trade and consumption of crop products. **C** and **D** Same for forest products CO_2 fluxes (*See* page 345)

Color Plate 11

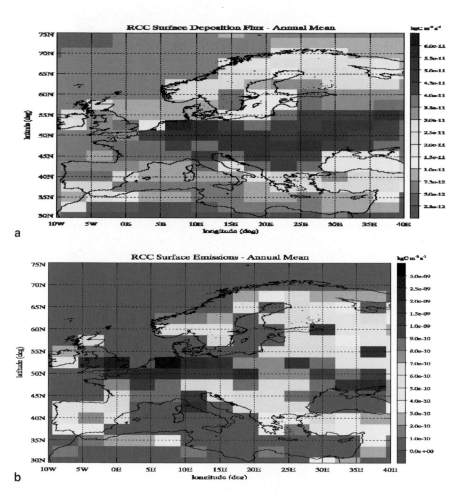

Fig. 16.4 **A** Spatial patterns of the surface deposition (sink) of carbon from reduced carbon compounds. **B** Patterns of reduced carbon compounds emissions to the atmosphere from antropogenic and biospheric sources (*See* page 349)

Color Plate 12

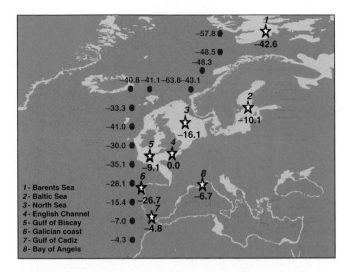

Fig. 16.7 Compilation of annually integrated air–sea CO_2 (gC m^{-2} year^{-1}) fluxes in European coastal seas (stars and black numbers) (adapted from Borges et al. 2006) and adjacent open ocean grid nodes from the Takahashi et al. (2002) climatology (red circles and numbers) (*See* page 356)

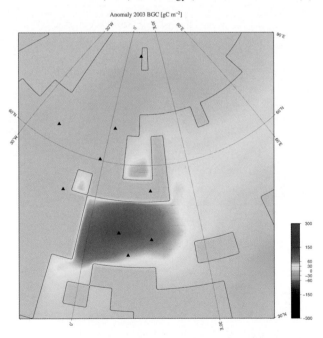

Fig. 17.1 Example of a top-down inversion-based surface–atmosphere CO_2 flux estimate for the European domain. Displayed is the anomalous flux in 2003 (in gC m^{-2}) with respect to the reference period 1998–2002 integrated over the growing season May–September. The black triangles denote the location of the stations whose atmospheric concentration observations were included in the inversion (*See* page 363)

Color Plate 13

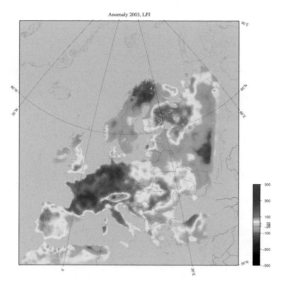

Fig. 17.2 Example of a bottom-up process-based model simulation of the surface–atmosphere CO_2 flux for the European domain. Displayed is the anomalous flux (in gC m^{-2}) in 2003 with respect to the reference period 1998–2002 integrated over the growing season May–September as computed by the LPJ model (Sitch et al. 2003) (*See* page 367)

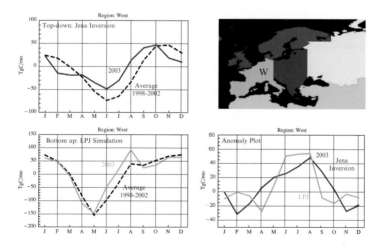

Fig. 17.3 Example of a quantitative comparison between top-down and bottom up flux estimates over Europe. Left hand panels: time series of the seasonal cycle of the net surface–atmosphere CO_2 flux integrated over the "western region" (green region indicated by the letter "W" in the upper right panel). Black dashed lines: 5 year average (1998–2002) monthly fluxes, colored lines: monthly fluxes in 2003. Top-down approach (upper left panel): fluxes determined by the Jena Inversion modeling system. Bottom-up approach (lower left panel): simulation by the LPJ process-based model. Lower right panel: time series of the monthly anomalous fluxes in 2003, blue: top-down inversion, green: bottom-up simulation (*See* page 370)

Color Plate 14

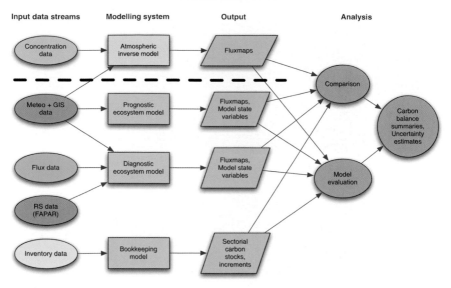

Fig. 17.4 Schematic of the single carbon cycle data stream analysis as currently performed for estimating regional carbon budgets. The dashed black line separates the top-down from the bottom-up approaches (*See* page 371)

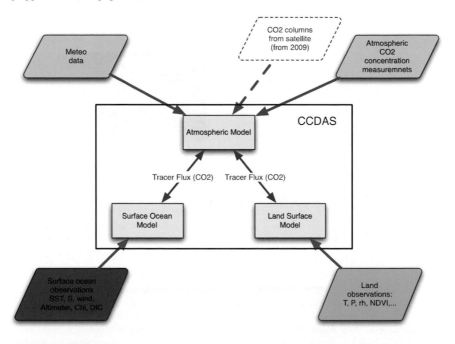

Fig. 17.5 Simplified schematic of a future comprehensive carbon cycle data assimilation system, including atmospheric, oceanic, and land components (*See* page 372)

Color Plate 15

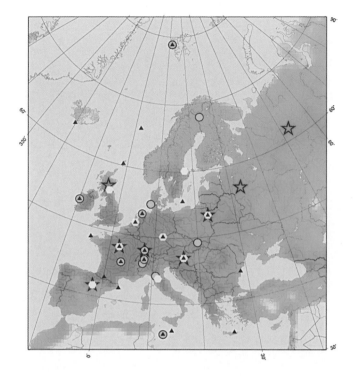

Fig. 18.1 Existing European research network of atmospheric concentration among which the Integrated Carbon Observing System (ICOS) main sites and associated sites will be selected and essential new sites implemented. Stars are regular lower troposphere profiling by aircraft, diamonds the continuous CO_2, CH_4, CO, SF_6, N_2O measurements on tall towers. Diamonds in grey: continuous measurements on intermediate size towers. Circles: continuous surface stations. Triangles: flask sampling sites (*See* page 382)

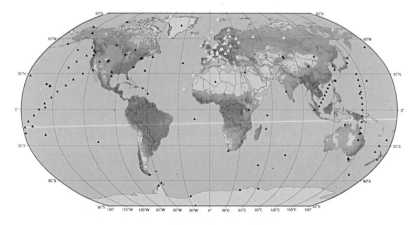

Fig. 18.2 Existing global atmospheric concentration network (ref.) (*See* page 383)

Color Plate 16

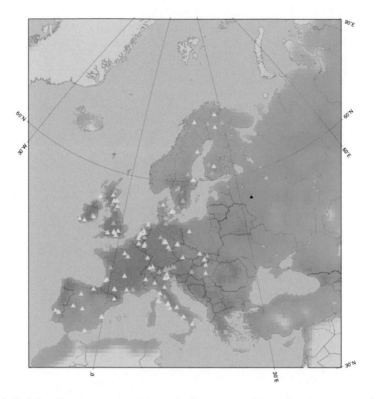

Fig. 18.3 Existing European research network of ecosystem observation sites among which the Integrated Carbon Observing System (ICOS) main sites and associated sites will be selected and essential new sites implemented. White triangles are CARBOEUROPE flux sites and black triangles flux sites from other projects (*See* page 383)

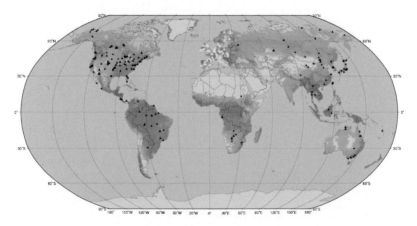

Fig. 18.4 Existing global research network of ecosystem observation sites (ref.) (*See* page 384)

Janssens, I.A., Freibauer, A., Ciais, P., Smith, P., Nabuurs, G.-J., Folberth, G., Schlamadinger, B., Hutjes, R.W.A., Ceulemans, R., Schulze, E.D., Valentini, R., and Dolman, A.J. 2003. Europe's terrestrial biosphere absorbs 7 to 12% of European anthropogenic CO_2 emissions. Science 300: 1538–1542.

Jenkins, J.C., Chojnacky, D.C., Heath, L.S., and Birdsey, R. 2003. National-scale biomass estimators for United States tree species. Forest Science 49: 12–35.

Johnson, M.G., and Kern, J.S. 2002. Quantifying the organic carbon held in forested soils of the United States and Puerto Rico. In Kimble, J.M., Heath, L.S., Birdsey, R.A., Lal, R. eds., The potential of U.S. forest soils to sequester and mitigate the greenhouse effect, pp. 47–72. Lewis, Boca Raton.

Jones, R.J.A., Hiederer, R., Rusco, E., Loveland, P.J., and Montanarella, L. 2004. The map of organic carbon in topsoils in Europe, Version 1.2, September 2003: Explanation of Special Publication Ispra 2004 No. 72 (S.P.I.04.72). European Soil Bureau Research Report 17: 1–26.

Jonsson, B.G., Kruys, N., and Ranius, T. 2005. Ecology of species living on dead wood—lessons for dead wood management. Silva Fennica 39: 1–21.

Karjalainen, T., Pussinen, A., Liski, J., Nabuurs, G.J., Eggers, T., Lapveteläinen, T., and Kaipainen, T. 2003. Scenario analysis of the impacts of forest management and climate change on the European forest sector carbon budget. Forest Policy and Economics 5: 141–155.

Kauppi, P.E., Mielikäinen, K., and Kuusela, K. 1992. Biomass and carbon budget of European forests, 1971 to 1990. Science 256: 70–74.

Krankina, O.N., Treyfeld, R.F., Harmon, M.E., Spycher, G., and Povarov, E.D. 2001. Coarse woody debris in the forests of the St. Petersburg region, Russia. Ecological Bulletins 49: 93–104.

Kruys, N., Jonsson, B.G., and Ståhl, G. 2002. A stage-based matrix model for decay-class dynamics of woody debris. Ecological Applications 12: 773–781.

Kurz, W.A., and Apps, M.J. 1994. The carbon budget of Canadian forests: a sensitivity analysis of changes in disturbance regimes, growth rates, and decomposition rates. Environmental Pollution 83: 55–61.

Kuusela, K. 1979. Forest balance on the national level. Silva Fennica 13: 265–268.

Kuusisto, E., and Hämekoski, K. 2001. Finland's third national communication under the United Nations framework convention on climate change. Ministery of Environment, Hämeenlinna, p. 197.

Laitat, E., Karjalainen, T., Loustau, D., and Lindner, M. 2000. Towards an integrated scientific approach for carbon accounting in forestry. Biotechnology, Agronomy, Society and Environment 4: 241–251.

Lehtonen, A. 2005. Carbon stocks and flows in forest ecosystems based on forest inventory data. Dissertationes Forestales 11: 1–51.

Lehtonen, A., Cienciala, E., Tatarinov, F., and Mäkipää, R. 2007. Uncertainty estimation of biomass expansion factors for Norway spruce in the Czech Republic. Annals of Forest Science 64: 133–140.

Lehtonen, A., Mäkipää, R., Heikkinen, J., Sievänen, R., and Liski, J. 2004. Biomass expansion factors (BEF) for Scots pine, Norway spruce and birch according to stand age for boreal forests. Forest Ecology and Management 188: 211–224.

Levy, P.E., Hale, S.E., and Nicoll, B.C. 2004. Biomass expansion factors and root: shoot ratios for coniferous tree species in Great Britain. Forestry 77: 421–430.

Liski, J., Korotkov, A.V., Prins, C.F.L., Karjalainen, T., Victor, D.G., and Kauppi, P.E. 2003. Increased carbon sink in temperate and boreal forests. Climatic Change 61: 89–99.

Liski, J., Lehtonen, A., Palosuo, T., Peltoniemi, M., Eggers, T., Muukkonen, P., and Mäkipää, R. 2006. Carbon accumulation in Finland's forests 1922–2004—an estimate obtained by combination of forest inventory data with modelling of biomass, litter and soil. Annals of Forest Science 63: 687–697.

Liski, J., Palosuo, T., Peltoniemi, M., and Sievänen, R. 2005. Carbon and decomposition model Yasso for forest soils. Ecological Modelling 189: 168–182.

Liski, J., Perruchoud, D., and Karjalainen, T. 2002. Increasing carbon stocks in the forest soils of Western Europe. Forest Ecology and Management 169: 163–179.

Löwe, H., Seufert, G., and Raes, F. 2000. Comparison of methods used within member states for estimating CO_2 emissions and sinks according to UNFCCC and EU monitoring mechanism: forest and other wooded land. Biotechnology, Agronomy, Society and Environment 4: 315–319.

Mäkinen, H., Hynynen, J., Siitonen, J., and Sievänen, R. 2006. Predicting the decomposition of Scots pine, Norway spruce and birch stems in Finland. Ecological Applications 16: 1865–1879.

Mäkipää, R., and Tomppo, E. 1998. Suomen metsät ovat hiilinielu—vaikka Kioton sopimuksen mukaan muulta näyttää. Folia Forestalia 2: 268–274.

Marklund, L.G. 1987. Biomass functions for Norway spruce (Picea abies (L.) Karst.) in Sweden. Sveriges Lantbruksuniversitet, Rapporter-Skog, p. 123.

Marklund, L.G. 1988. Biomassafunktioner för tall, gran och björk i Sverige. Sveriges Lantbruksuniversitet, Rapporter-Skog 45: 1–73.

Matamala, R., Gonzalez-Meler, M.A., Jastrow, J.D., Norby, R.J., and Schlesinger, W.H. 2003. Impacts of fine root turnover on forest NPP and soil C sequestration potential. Science 302: 1385–1387.

MCPFE 2003. State of Europe's Forests. MCPFE, Vienna.

Metla 2006. Metsätilastollinen vuosikirja 2006, Finnish Statistical Yearbook of Forestry. Metla.

Minkkinen, K., Korhonen, R., Savolainen, I., and Laine, J. 2002. Carbon balance and radiative forcing of Finnish peatlands 1900–2100—the impact of forestry drainage. Global Change Biology 8: 785–799.

Minkkinen, K., Vasander, H., Jauhiainen, S., Karsisto, M., and Laine, J. 1999. Post-drainage changes in vegetation composition and carbon balance in Lakkasuo mire, Central Finland. Plant and Soil 207: 107–120.

Monni, S., Peltoniemi, M., Palosuo, T., Lehtonen, A., Mäkipää, R., and Savolainen, I. 2007. Uncertainty of forest carbon stock changes—implications to the total uncertainty of GHG inventory of Finland. Climatic Change 81: 391–413.

Morgan, M., and Henrion, M. 1990. Uncertainty. A guide to dealing with uncertainty in quantitative risk and policy analysis. Cambridge University Press, Cambridge.

Muukkonen, P. 2007. Generalized allometric volume and biomass equations for some tree species in Europe. European Journal of Forest Research 126: 157–166.

Muukkonen, P., and Mäkipää, R. 2006a. Biomass equations for European trees: addendum. Silva Fennica 40: 763–773.

Muukkonen, P., and Mäkipää, R. 2006b. Empirical biomass models of understorey vegetation in boreal forests according to stand and site attributes. Boreal Environment Research 11: 355–369.

Muukkonen, P., Mäkipää, R., Laiho, R., Minkkinen, K., Vasander, H., and Finér, L. 2006. Relationship between biomass and percentage cover in understorey vegetation of boreal coniferous forests. Silva Fennica 40: 231–245.

Nabuurs, G.J., Päivinen, R., Sikkema, R., and Mohren, G.M.J. 1997. The role of European forests in the global carbon cycle—a review. Biomass and Bioenergy 13: 345–358.

Nabuurs, G.J., Schelhaas, M.J., Mohren, G.M.J., and Field, C.B. 2003. Temporal evolution of the European forest sector carbon sink from 1950 to 1999. Global Change Biology 9: 152–160.

Palosuo, T., Liski, J., Trofymow, J.A., and Titus, B. 2005. Litter decomposition affected by climate and litter quality—testing the Yasso model with litterbag data from the Canadian intersite decomposition experiment. Ecological Modelling 189: 183–198.

Paustian, K., Levine, E., Post, W.M., and Ryzhova, I.M. 1997. The use of models to integrate information and understanding of soil C at the regional level. Geoderma 17: 227–260.

Peltoniemi, M., Mäkipää, R., Liski, J., and Tamminen, P. 2004. Changes in soil carbon with stand age—an evaluation of a modeling method with empirical data. Global Change Biology 10: 2078–2091.

Peltoniemi, M., Palosuo, T., Monni, S., and Mäkipää, R. 2006. Factors affecting the uncertainty of sinks and stocks of carbon in Finnish forests soils and vegetation. Forest Ecology and Management 232: 75–85.

Peltoniemi, M., Thürig, E., Ogle, S., Palosuo, T., Shrumpf, M., Wützler, T., Butterbach-Bahl, K., Chertov, O., Komarov, A., Mikhailov, A., Gärdenäs, A., Perry, C., Liski, J., Smith, P., and Mäkipää, R. 2007. Models in country scale carbon accounting of forest soils. Silva Fennica 41: 575–602.

Petersson, H., and Ståhl, G. 2006. Functions for below-ground biomass of *Pinus sylvestris*, *Picea abies*, *Betula pendula* and *Betula pubescens* in Sweden. Scandinavian Journal of Forest Research 21, Supplement 7: 84–93.

Post, W.M., and Kwon, K.C. 2000. Soil carbon sequestration and land-use change: processes and potential. Global Change Biology 6: 317–327.

Powlson, D.S., Smith, P., and Smith, J.U. 1996. Evaluation of soil organic matter models. Springer-Verlag, Berlin Heidelberg.

Rastetter, E.B., King, A.W., Cosby, B.J., Hornberger, G.M., O'Neill, R.V., and Hobbie, J.E. 1992. Aggregating fine-scale ecological knowledge to model coarser-scale attributes of ecosystems. Ecological Applications 2: 55–70.

Roxburgh, S.H., Berry, S.L., Buckley, T.N., Barnes, B., and Roderick, M.L. 2005. What is NPP? Inconsistent accounting of respiratory fluxes in the definition of net primary production. Functional Ecology 19: 378–382.

Satoo, T., and Madgwick, H.A.I. 1982. Forest biomass. Martinus Nijhoff/Dr W. Junk Publisher, The Hague.

Schlesinger, W.H. 1990. Evidence from chronosequence studies for a low carbon-storage potential of soils. Nature 348: 232–234.

Siitonen, J. 2001. Forest management, coarse woody debris and saproxylic organisms: Fennoscandian boreal forests as an example. Ecological Bulletins 49: 11–41.

Smith, J.E., Heath, L.S., and Woodbury, P.B. 2004. How to estimate forest carbon for large areas from inventory data. Journal of Forestry 102: 25–31.

Smith, J.E., and Heath, L.S. 2002. A model of forest floor carbon mass for United States forest types. Res. Pap. NE-722 USDA Forest Service, Northeastern Research Station, Newtown Square, PA.

Smith, P., Smith, J.U., Powlson, D.S., McGill, W.B., Arah, J.R.M., Chertov, O.G., Coleman, K., Franko, U., Frolking, S., and Jenkinson, D.S. 1997. A comparison of the performance of nine soil organic matter models using datasets from seven long-term experiments. Geoderma 81: 153–225.

Somogyi, Z., Cienciala, E., Mäkipää, R., Lehtonen, A., Muukkonen, P., and Weiss, P. 2006. Indirect methods of large scale forest biomass estimation. European Journal of Forest Research 126(2): 197–207.

Ståhl, G., Boström, B., Lindkvist, H., Lindroth, A., Nilsson, J., and Olsson, M. 2004. Methodological options for quantifying changes in carbon pools in Swedish forests. Studia Forestalia Suecica 214: 1–46.

Ståhl, G., Ringvall, A., and Fridman, J. 2001. Assessment of coarse woody debris—a methodological overview. Ecological Bulletins 49: 57–70.

Stokland, J.N. 2001. The coarse woody debris profile: an archive of recent forest history and an important biodiversity indicator. Ecological Bulletins 49: 71–83.

Tarasov, M.E., and Birdsey, R.A. 2001. Decay rate and potential storage of coarse woody debris in the Leningrad region. Ecological Bulletins 49: 137–147.

Tate, K.R., Scott, N.A., Saggar, S., Giltrap, D.J., Baisden, W.T., and Newsome, P.F. 2003. Land-use changes alters New Zealand's terrestrial carbon budget: uncertainties associated with estimates of soil carbon change between 1990–2000. Tellus 55B: 364–377.

Tierney, G.L., and Fahey, T.J. 2002. Fine root turnover in a northern hardwood forest: a direct comparison of the radiocarbon and minirhizotron methods. Canadian Journal of Forest Research 32: 1692–1697.

Tomppo, E. 2000. National forest inventory in Finland and its role in estimating the carbon balance of forests. Biotechnology, Agronomy, Society and Environment 4: 241–320.

UN-ECE/FAO 1985. The Forest Resources of the ECE region (Europe, the USSR, North America). United Nations, New York and Geneve.

UNECE 2000. Forest Resources of Europe, CIS, North America, Australia, Japan and New
    Zealand. United Nations, Geneva.
UNFCCC 1992. United Nations Framework Convention on Climate Change. http://unfccc.int/
UNFCCC 1997. Kyoto Protocol. http://unfccc.int/
Wirth, C., Schumacher, J., and Schulze, E.D. 2004. Generic biomass functions for Norway spruce
    in Central Europe—a meta-analysis approach towards prediction and uncertainty estimation.
    Tree Physiology 24: 121–139.
Zianis, D., Muukkonen, P., Mäkipää, R., and Mencuccini, M. 2005. Biomass and stem volume
    equations for tree species in Europe. Silva Fennica Monographs 4: 1–63.

Chapter 11
Flux Tower Sites, State of the Art, and Network Design

A. Johannes Dolman, Riccardo Valentini, Margriet Groenendijk, and Dimmie Hendriks

11.1 Introduction

Direct observations of the exchange fluxes of water, energy, carbon dioxide, trace gases, and momentum have been possible since the 1960s. Most of this early work comprised campaigns of short duration, and has contributed substantially to our process understanding of surface atmosphere exchange. The data obtained in these studies were fundamental in assessing the validity of Monin-Obhukov similarity over natural surfaces, in testing resistance analogue formulations (e.g., Monteith 1965), and were also used to calibrate flux models against single site data. The seminal two volume work by J.L. Monteith "Vegetation and the Atmosphere" provides an excellent summary of the micrometeorology of natural vegetation of this period (Monteith 1976). It was also during this period that climate modelers became increasingly aware of the important role of the biosphere in climate. In the 1980s and 1990s, the early observational work was extended to study variability along climatic or other transects. The analysis of Valentini et al. (2000) describing the variation of net ecosystem exchange (NEE) of CO_2 along a North–South gradient in Europe typifies this stage. New developments in technology (e.g., Moncrieff et al. 1998), such as the availability of "off the shelf" fast response instrumentation, gave opportunity to the development of the eddy-covariance technique. The "la Thuile workshop" in 1994 which resulted in a special issue of Global Change Biology (Baldocchi et al. 1996) was fundamental in bringing together micrometeorologists and ecologists in studying exchange fluxes, not only of water and energy but also of carbon. At the political level, the Kyoto process provided extra impetus to these studies, particularly for carbon, although the direct relevance of flux measurements to the Kyoto process remains somewhat controversial (Körner 2003; Dolman et al. 2003). The data coverage (Fig. 11.1) of this FLUXNET network (Baldocchi et al. 2001) is now near global, from Siberian sites (Dolman et al. 2004) to tropical rainforest (Araújo et al. 2002). Since around 2000 the analysis has begun to take advantage of this near global coverage of sites, and studies have emerged of interannual variability, data assimilation, and continent scale estimation of carbon uptake (Knorr and Kattge 2005; Papale and Valentini 2003).

215

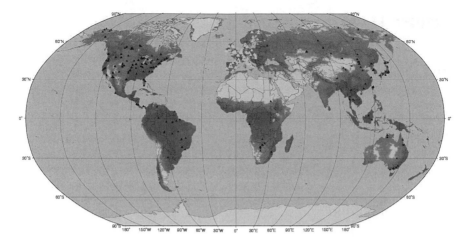

Fig. 11.1 Locations of the 266 FLUXNET measurement sites available on January 1, 2005 (*See Color Plate* 5).

This chapter will briefly describe the technical details of modern eddy-covariance systems, and will describe some of the important developments in our understanding of the magnitude of fluxes of different biomes and their spatial variability. It will then concentrate on the interannual variability of the fluxes and the related questions of vulnerability of the terrestrial carbon cycle. As an example of a vulnerability study, the analysis of the European heat wave and associated drought in 2003 (Ciais et al. 2005) is shown. That analysis clearly shows the value of a dense network in picking up these synoptic scale disturbances in the carbon cycle. The chapter ends with an assessment of the representativity of the current European flux sites and how a future system could be designed to be able to address as many questions as possible that contribute to improved understanding of the continental scale carbon cycle.

11.2   State of the Art

11.2.1   The Eddy-Covariance Technique

Figure 11.2 shows a schematic of the application of the eddy-covariance technique based on Aubinet et al. (2000). The technique is based on the fact that the flux of a quantity like CO_2, F_{co2}, can be expressed as the covariance (the average of the product of the fluctuating, turbulent quantities) of the vertical wind speed, w, with that quantity, for example, the mixing ratio of CO_2, p_c.

$$F_c = \overline{w'\rho_c}'$$

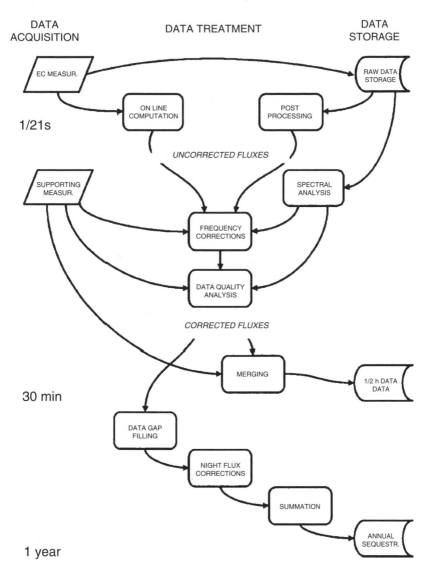

DATA ACQUISITION

DATA TREATMENT

DATA STORAGE

EC MEASUR.

RAW DATA STORAGE

1/21s

ON LINE COMPUTATION

POST PROCESSING

UNCORRECTED FLUXES

SUPPORTING MEASUR.

SPECTRAL ANALYSIS

FREQUENCY CORRECTIONS

DATA QUALITY ANALYSIS

CORRECTED FLUXES

MERGING

1/2 h DATA DATA

30 min

DATA GAP FILLING

NIGHT FLUX CORRECTIONS

SUMMATION

ANNUAL SEQUESTR.

1 year

Fig. 11.2 The long-winding road of eddy covariance in practice, showing the sequence of data aquisition, treatment, and control. Note that to arrive at 30-min data several corrections need to be applied. To progress toward annual sums, gap filling of missing data becomes an issue. Reproduced from Aubinet et al. (2000), with kind permission from Prof. Marc Aubinet.

This requires fast sensors that are able to observe the fluctuation up to 20 Hz. Instruments that achieve this are the sonic anemometer and closed or open path infrared sensors for H_2O and CO_2. This principle is well established and commercial instruments that can perform these measurements with adequate accuracy (Aubinet et al. 2000) are available since the early 1990s. Once these measurements are obtained, a set of data quality checks and assessment are required. These are briefly described below.

For instance, it appears notoriously difficult to distinguish the real vertical velocity from pseudovertical wind velocities that occur due to misalignment of the anemometer or distortion in the flow due to topography. These need to be corrected by rotating the calculated wind speeds in such a way that the mean vertical wind speed is zero. In practice, this implies that the mean vertical wind speed is set perpendicular to the mean flowline. The real vertical wind speed fluctuations can be then estimated (Kaimal and Finnigan 1994; Finnigan et al. 2003).

One of the fundamental hypotheses that underlies the equivalence between the eddy-covariance flux and the NEE of carbon is the existence and stationarity of the turbulent process. In practice, environmental conditions change continuously during the day, for instance, due to the diurnally varying solar input and associated temperature changes, and this assumption is thus easily violated. The nonstationarity test is a means to detect whether the nonstationarity is such that basic assumptions are violated (Wichura and Foken 1995).

In the past, largely because analogue processing was used, running means were applied to calculate the fluctuating parts. Defining the average of such an operation is not trivial and may strongly affect the calculated final covariance. The simplest and most widely used and recommended technique today is the block averaging (Finnigan et al. 2003). One has to take care in defining the length of the block period, as low frequency contributions to the flux may become more important at tall vegetation (forests) than at low vegetation (e.g., grasslands) (Finnigan et al. 2003). Once the averaging period is determined, raw fluxes or covariances can be computed.

Eddies advected past a sonic anemometer do so at various angles. Because the sonic anemometer has a fixed shape, at some angles shading occurs and the measurements are affected. Gash and Dolman (2003) postulated that this angle would depend on the roughness length of the vegetation and that the neglect of this error could be responsible for the observed bad energy balance closure over tall vegetation. They showed furthermore that, although large angles of attack do not occur frequently, a large portion of the turbulent fluxes is carried by eddies under those large angles. This flux angle distribution can be shown to depend on surface (vegetation) roughness, measuring height, and stability regime. With such large parts of the fluxes carried by eddies with angles of attack outside the manufacturer's specifications, the potential for errors to affect the measurements becomes large. It can be shown that outside the recommended angles, the response of a widely used anemometer is bad. This, in combination with the fact that most of the flux is carried at large angles, suggests that the turbulent fluxes measured with these anemometers can be seriously underestimated. Wind tunnel experiments can determine the errors and a so-called angle of attack dependent calibration was developed by Van der Molen et al. (2004).

The all-important step in Fig. 11.2 is the step from uncorrected to corrected fluxes, and complete textbooks have been devoted to this subject (e.g., Lee et al. 2004). Key eddy-covariance corrections (e.g., Lee et al. 2004; Aubinet et al. 2000) include the Webb correction for the existence of an apparent vertical velocity induced by the sensible heat flux from the surface and frequency loss corrections. Measuring the turbulent fluctuations requires fast sensors that are able to sense all the fluctuations in the frequency range (spectrum) in which turbulent transport takes place. Optimum measurement conditions are hard to achieve in the real world due to limitations of the sensor speed, the delays that occur during data acquisition, averaging over a path rather than a point, and physical separation between the sensors. It is possible for each of these situations to derive analytical expressions (transfer functions) about the magnitude of the individual effects for an ideal system. As mentioned before, the spectrum shows which frequencies of transport really contribute to the flux. By deriving the ratio of the frequency correction for a particular measurement setup and the universal spectra for a particular (co)variance, one can calculate the frequency loss of that setup and afterwards add it to the observed flux (Aubinet et al. 2000). Several other smaller corrections also need to be applied before the final flux is recovered from the raw turbulent signals.

The various correction factors result in substantial changes to the raw flux. In particular, the Webb correction can be substantial in dry hot environments. Angle of attack and frequency response corrections normally change the flux by up to 20%. Since the sign of the correction is not always the same, the net effect is of the order 5–10% depending on the flux of interest.

Once the flux is calculated for typically half-hourly values, periods for which no data is available need to be gap filled. Various methodologies exist for this purpose and range from simple linear interpolation, using functional relationships between fluxes and environmental-data (radiation, C-uptake, temperature, C-respiration, etc.) neural networks, looking for flux data under similar environmental conditions, to the use of look up tables and detailed process modeling (e.g., Falge et al. 2001; Reichstein et al. 2005). It appears that those gap-filling techniques that use neural networks generally perform well, surprisingly sometimes better than those based on process modeling.

11.2.2   Key Issues

11.2.2.1   Separation of Respiration and Assimilation

Because the eddy-covariance technique measures only net exchange fluxes, some methodology is required to separate the NEE into its two main components, Gross Ecosystem Production (GEP) and Ecosystem Respiration (R_{eco}). This is nontrivial. Separation is required not only for process understanding but also for a good gap-filling methodology, which obviously depends on adequate calculation of the component fluxes. Existing flux partitioning algorithms can be classified into those that use only nighttime data for the estimation of R_{eco} and those that use daytime data or both daytime and nighttime data using light response curves (see Reichstein et al. 2005 for a review). The two approaches have been compared by Falge et al. (2001), resulting in generally good agreement between the two methods, except in ecosystems where large soil carbon pools exist. Under those conditions, the light curves derived from daytime data may not well represent the nighttime respiratory processes.

The standard light-response curve method only allows the estimation of the daily average respiration without estimation of a temperature-dependent diurnal course. A regression model, where GEP and R_{eco} are described in one equation, is often used where R_{eco} is explicitly dependent on air or soil temperature. Once regression parameters are fitted, GEP and R_{eco} can be computed separately on a half-hourly time step. While this is an elegant approach, it suffers from a number of problems (Reichstein et al. 2005). First, similar model predictions of NEE can be obtained by several parameter settings, as long as R_{eco} and GEP are calibrated in opposite directions. In particular, the estimation of the temperature response of R_{eco} is perturbed by the response of GEP to water vapor pressure deficit (VPD), which can lead to a bias in the estimation of R_{eco}. An afternoon drop of NEE can be caused by a VPD-related drop of GEP or by a temperature-related increase in R_{eco}. If a regression model only ascribes the effect of R_{eco} to the drop of NEE, this will lead directly to an overestimation of R_{eco} (e.g., Fig. 11.3). Second, with this method, both R_{eco} and GEP are modeled and based on certain assumptions (e.g., hyperbolic light response of GEP). If the flux-partitioned data is then used to evaluate models, one is essentially comparing two models, which can lead to circular arguments (e.g., a model

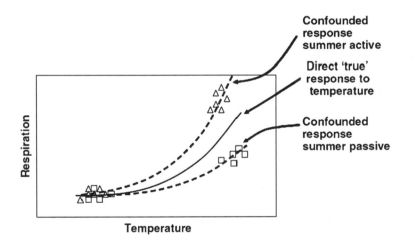

Fig. 11.3 Schematic representation of the various confounding effects introduced into the temperature dependence of ecosystem respiration (R_{eco}) derived from annual data. Triangles and squares are hypothetical data for summer active and summer passive ecosystems, respectively. The true middle curve is the sensitivity of respiration to short-term variations; the other two are based on long-term data input as confounded by other seasonally varying factors, resulting in an overestimation of the temperature sensitivity in summer active and underestimation in summer passive ecosystems. Reproduced from Reichstein et al. (2005), with kind permission from Blackwell Publishing.

with the same hyperbolic assumption will be more likely validated than other models). This is a rather fundamental problem, and unfortunately virtually excludes using this method when the aim of the flux partitioning is model evaluation (Reichstein et al. 2005). Using separate estimates of soil and plant respiration is unfortunately the only objective, but costly and complicated, method to circumvent this issue. This involves taking measurements with dark flux chambers at the eddy-correlation site. Combining these flux data, which only comprise of R_{eco} fluxes, with soil temperatures, an Arrhenius relation with a functional dependence on temperature (Q_{10}) can easily be determined (Dore et al. 2003; Hendriks et al. 2007).

Finally, CO_2 fluxes near sunrise and sunset are rapidly changing; involve nonstationarity problems and often storage changes occur. If these are not corrected for in the data, incorrect attributions to either respiration or GEP may occur. A final problem arises in the poor availability of soil moisture observations at flux sites, which often makes it virtually impossible to derive meaningful relationships between drought and respiration reduction.

Reichstein et al (2005) suggest that the VPD-specific problem can be tackled by including a VPD response of GEP to the regression model. It is interesting to note that the response at canopy scale to VPD was not fully observed until 1988 when Stewart (1988) used micrometeorological techniques above a forest to determine canopy scale characteristics. Schulze et al. (1972) had already earlier observed such a response at leaf level, but the question how strong that response would be at canopy level remained controversial (Jarvis and McNaughton 1986; Schulze et al. 2005). Reichstein et al. (2005) developed a flux-partitioning algorithm that first estimates the temperature sensitivity from short-term periods, and then applies this short-term temperature sensitivity to extrapolate the R_{eco} from nighttime to daytime. In this case, one can introduce seasonally varying temperature sensitivity or apply site-specific constant temperature sensitivity.

Another elegant fundamental analysis of temperature sensitivity of respiration can be found in Davidson and Janssens, (2006). They show the complexity of unraveling the "true" temperature sensitivity of respiration from apparent sensitivities that are confounded by various climatic and edaphic factors. In general, there appears not yet to be a "best" separation algorithm, but when comparing sites, a general applicable methodology may be found in Papale et al. (2006), but also see Van der Molen et al. (2007a, b).

11.2.2.2 Nighttime Fluxes and Advection

A particular complex problem is caused by the nighttime build up of CO_2 under conditions of low wind speed and low turbulence. This leads potentially to a build up of CO_2 near the ground and the possibility of subsequent lateral drainage in terrain with small (1–5% slope) topography. This "nighttime flux loss" is neither seen by the eddy-covariance system at the top of the tower nor by any concentration gradient measurements system on the tower. Correction techniques are hard to define and generally great care has to be taken when using eddy-covariance techniques to estimate nighttime fluxes. Indications of the potential size of errors vary from 20% to 100% (e.g., Kruijt et al. 2004; Aubinet et al. 2005; Araújo et al. 2006). The nighttime flux loss is a systematic error, and a neglect of say $1 \mu mol$ C m^{-2} s^{-1} over a 12-h night every night per year leads to an yearly offset of $189 g$ C m^{-2} that is comparable to the net average annual uptake of ecosystems! Standard practice so far has been to eliminate the fluxes observed under low turbulence conditions (established by a friction velocity threshold) and replace these with estimates of nighttime respiration based on observations under high turbulence (but see also Van Gorsel et al., 2007).

More insight into this problem can be gained by writing the conservation equation for a scalar quantity-like CO_2 in a two-dimensional flow situation (x, y horizontal direction, $v = 0$) as:

$$\underbrace{\int_0^h S dz}_{I} = \underbrace{\overline{w' \varsigma'}}_{II} + \underbrace{\int_0^h \frac{\delta \overline{\varsigma}}{\delta t} dz}_{III} + \underbrace{\int_0^h \overline{u} \frac{\delta \overline{\varsigma}}{\delta x} dz}_{IV} + \underbrace{\int_0^h \overline{w} \frac{\delta \overline{\varsigma}}{\delta z} dz}_{V}$$

where the term denoted I represents the scalar source/sink term that corresponds to the NEE of CO_2. Term II represents the turbulent (eddy) flux at height h (the flux which is measured by eddy-covariance systems) with w the vertical velocity and the prime denoting the fluctuating part, term III represents the storage of the scalar below the measurement height, and IV and V represent the fluxes by horizontal and vertical advection. Under conditions of atmospheric stationarity (no major changes, no trends) and horizontal homogeneity, the last three terms of the right hand side can be ignored, and the eddy flux equals the source/sink term. In practice, these conditions are often not met, and the two advection terms become important. For vertical advection, term V, corrections have been suggested by Lee (1998), but these were later criticized by for lumping all errors, including the horizontal advection, into a single term.

Aubinet et al. (2005) analyzed data from several sites with significant topography. They found that the relative importance of advection and vertical storage was not the same for all sites. At steeply sloping sites, advection appeared to be dominant over storage, but at sites with intermediate topography, both processes were equally important. They introduced a new criterion, the advective velocity that combines, among others, net radiation (as a proxy for radiative, nighttime cooling) and elevation angle, to determine the relative importance of storage versus horizontal advection. Still, the nighttime flux loss problem is not fully resolved until more precise measurements of horizontal and vertical gradients and wind speed and CO_2 concentration become available for more sites. This means that sites prone to advection need additional measurements, and it seems difficult to know a priori which site is prone to which complicating circumstances. Replacement of poor turbulence values with estimated respiration rates still remains the best option for flat sites with low-estimated advective velocities, and for sites with high advective velocities additional observations on drainage may be required.

11.3 Magnitude of Fluxes

This section gives a brief overview of the magnitude of the fluxes of three main land use types in Europe: forest, grasslands, and Northern wetlands (peat lands). Only data obtained by eddy-covariance techniques is used. Data based on other methods and further comparison can be found in the respective chapters of this book (Chaps. 12 and 13 this book).

11.3.1 European Forests

Van Dijk et al. (2005) used the results of eddy-covariance measurements of 15 sites in Europe to study the spatial variability of forest NEE. In contrast to an earlier study (Valentini et al. 2000), they used more data and a consistent methodology to separate the observed NEE into gross primary production (GPP) and respiration. The methodology to split NEE into respiration and assimilation used nighttime flux data only (Van Dijk and Dolman 2004). We used the current data and reevaluated that work using the Reichstein et al (2005) splitting algorithm for GEP and R_{eco}. On average, European forests appear to be a sink of $212\,g\,C\,m^{-2}\,yr^{-1}$. The range around this average value is, however, considerable with a south Italian beech forest taking up $570\,g\,C\,m^{-2}\,yr^{-1}$ with northern forest being small sources of $50\,g\,C\,m^{-2}\,yr^{-1}$. The average GEP of European forests is $736\,g\,C\,m^{-2}\,yr^{-1}$ and the average respiration $530\,g\,C\,m^{-2}\,yr^{-1}$. The average rates given here are not weighted averages taking into account the area covered with a forest species and age structure, but a simple average of the 15 measurement locations. Direct comparison with the forest inventory measurements is therefore problematic, but an average value of $212\,g\,C\,m^{-2}\,yr^{-1}$ for the Net Ecosystem Production of European forest derived from eddy-covariance data is reasonable. This value is about twice as high as that obtained from forest inventory data. Indeed Janssens et al. (2003) used a factor of 0.47 to scale eddy-covariance data to forest inventory data. This factor takes into account that eddy-covariance measurements only measure net ecosystem fluxes without disturbance effects such as harvest, forest fires, and windthrows.

The original data published by Valentini et al. (2000) suggested an increasing NEE with decreasing latitude. Subsequent analysis (Valentini 2003; Janssens et al. 2001; Van Dijk and Dolman 2004) consistently showed the same latitudinal trend. In Fig. 11.4a, the mean annual totals of NEE, GEP, and R are plotted against latitude as in Van Dijk and Dolman (2004). While net uptake tends to decrease toward the pole as found by Valentini et al. (2000), particularly above 60°N, Van Dijk and Dolman (2004) concluded that this is not because respiration increases, but because GPP decreases faster than respiration does.

A more detailed analysis of the same data set led Van Dijk et al. (2005) to conclude that radiation, temperature, and leaf area were the main drivers of the observed variation between sites in GEP, R, and NEE. They also indicated that these

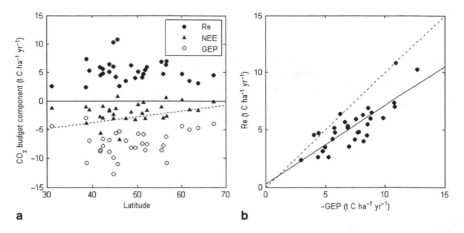

Fig. 11.4 Net Ecosystem Exchange (NEE) and respiration and assimilation versus latitude for 15 European forests (a) and respiration (b) as a function of gross ecosystem production (GEP) of the same set of 15 forests.

drivers were responsible for much of the large-scale interannual variability of the fluxes. Supporting evidence comes from Reichstein et al. (2003) who concluded that respiration could be adequately simulated by a model that contained information on temperature, precipitation, and leaf area index.

11.3.2   European Grasslands

Soussana et al. (2004, Chapter 13) present first results of the Greengrass project concerning the balance of CO_2, N_2O, and CH_4 exchanges between grasslands and the atmosphere at nine contrasted European sites. These cover a major climatic gradient over Europe and include four grassland types (sown grass, sown grass-clover, intensive permanent grassland, and seminatural grassland), three types of herbage use (rotational cattle grazing, continuous cattle grazing, and mowing), and contrasted N fertilizer supplies. At each site, CO_2 fluxes were monitored close to the ground by open-path and closed-path eddy-covariance systems. The results show that for CO_2 all the grasslands of Europe are a sink in terms of NEE, excluding C losses by harvest and grazing. However, when the emission of other GHGs such as N_2O and CH_4 is taken into account some of the sites become a smaller sink (a negative GHG balance, i.e., a net uptake of carbon), but all are able to maintain an overall negative GHG balance. Using data from the Europe, the US and Asia, a close relation between GPP and total R_{eco} with respect to the total annual precipitation has been found, while no significant relationship with temperature was obtained.

For managed grasslands in Europe, the annual C exports are approximately of the same magnitude as the NEE observed by eddy covariance, and the uncertainties on the amounts of harvested C and of organic C supplied by manure thus play a significant role in the total uncertainty of the net C storage (i.e., NBP). This implies that the annual NEE cannot be compared directly among sites with contrasted managements (e.g., grazed vs. cut sites) and that in most cases modeling is required to derive a full carbon balance. Chapter 16 gives more details on the amounts of carbon involved in lateral transport.

11.3.3   European Peat Lands and Northern Wetlands

There is little data available from eddy-covariance measurements of peat land. Most data appear to be related to (sub) arctic ecosystems, and are summarized by Aurela et al. (2002). In northern areas, the length of the growing season puts constraints on the capacity to take up carbon through photosynthesis. Peak uptake is during a very short growing. Table 11.1 shows the partitioning of NEE into its component fluxes as a function of season for a subarctic fen in Northern Finland (Aurela et al. 2002). Clearly, most of the uptake takes place in the summer. Noticeable are the very substantial losses of carbon in the winter period of 105 g C m^{-2}. Winter CO_2 losses are a significant major gap in the understanding of C losses from these and other arctic ecosystems. Although winter CO_2 losses have long been recognized, more recent research indicates that these losses are greater than was previously thought and may be the product of significant respiratory activity during the winter (Aurela et al. 2002), when recently fixed C is respired. The snow cover acts as a very good thermal insulator and provides an environment for respiratory bacteria to continue decomposing organic material, well after the air temperatures have dropped substantially below zero.

In contrast to arctic fens, temperate fens show a more prolonged growing season, resulting in a larger uptake. For a Dutch fen meadow site in the center of the Netherlands, Hendriks et al. (2007) found that the growing season in 2004, defined as the period when NEE is negative, lasted from April 7 to October 14 and has a net uptake of 543 g C m^{-2}. The NEE for 2004 was cumulated, resulting in a net annual CO_2 uptake of 446 ± 83 g C m^{-2}. The growing season in 2005 lasted from March 25 to September 8 and has a net uptake of 327 g C m^{-2}, while the annual NEE amounts 311 ± 58 g C m^{-2}. The growing season in 2006 lasted from April 25 to September 2 and has a net uptake of 283 g C m^{-2}, while the annual NEE amounts to 232 ± 58 g C m^{-2}. Table 11.2 shows the partitioning of NEE into its component

Table 11.1 Carbon uptake (g C m^{-2}) for several periods of the year for a subarctic fen (after Aurela et al. 2002)

	Thaw	Preleaf	Summer	Autumn	Winter	Year
Length of period	23	45	71	40	186	365
DOY of period	112–134	135–179	180–250	251–290	291–111	112–111
NEE	17.1	31.5	−273.8	51.8	105.2	−68.1
GEP	−7.9	−84.6	−786.1	−31.4	0	−910
R	26.7	118.9	515.8	81.9	109.5	852.8

Table 11.2 CO_2 fluxes (g C m^{-2}) for growing and nongrowing periods of the 2004, 2005, and 2006 for a temperate fen meadow (after Hendriks et al. 2007)

	Growing season 2004	Nongrowing season 2004	Year 2004	Growing season 2005	Nongrowing season 2005	Year 2005	Growing season 2006	Nongrowing season 2006	Year 2006
Length of period	192	174	366	168	197	365	129	236	365
DOY of period	96–288	288–96	1–366	83–251	251–83	1–365	115–244	244–115	1–365
NEE (gC-CO_2 m^2 yr)	−543	97	−446	−327	16	−311	−283	511	−232
GEP (gC-CO_2 m^2 yr)	−1213	−101	−1314	−877	−300	−1177	−804	−352	−1156
R_{eco} (gC-CO_2 m^2 yr)	670	199	869	550	316	866	521	403	924

fluxes as a function of season for a temperate fen meadow in the Netherlands (Hendriks et al. 2007).

These numbers are of similar size as those of the forest in the same climate region, but cannot be directly compared as C uptake. A third of the NEE in the peat land is lost as CH_4, another fraction lost belowground by water movement. As in the case of grasslands, care must be taken to interpret the results as suggesting that peat lands act to sequester carbon. In particular, methane emissions may act as GHG sources Chap. 12, and turn the ecosystem into a overall GHG source rather than sink (see also Van der Molen et al. 2007a, b).

11.3.4 Croplands

Croplands play an important part in the carbon cycle, but few direct measurements are available of cropland NEE. Croplands in Europe occupy 126.5 Mha (EU25+ Norway and Switzerland). Modeling studies suggest that European croplands are sources of carbon (Vleeshouwers and Verhagen 2002, Janssens et al. 2003). In the past, some efforts at measurements of fluxes by eddy covariance were made, but these served largely to estimate water balances at short periods or when they were executed also for carbon, important losses through soil and residue respiration were not taken into account (Baldocchi, 1994). Particularly for cropland, it is important to cover several full rotation cycles. Using a combination of modeling, yield data, and direct eddy-covariance estimates, Anthoni et al. (2004) were able to determine the NEE and net biome production of a winter wheat crop in East Germany at 185–245 and −45 ± 11 g C m^{-2} yr^{-1}, respectively. Three quarters of the source of carbon were due to soil carbon losses. This latter finding agrees with other estimates that suggest the European agricultural soils to be a source of carbon (e.g., Janssens et al. 2003). Given the potential to use agricultural practices to mitigate climate change under the Kyoto protocol, it is still surprising how little effort is put into obtaining reliable eddy-covariance data in combination with yield and soil respiration measurements within Europe.

11.4. Variability of Fluxes

Climate is a main forcing agent for the exchange fluxes and consequently the fluxes show considerable variation from year to year. For example, NEE in a 60- to 80-year-old deciduous forest in the northeastern USA ranged from 1.2 to 2.5 Mg C ha^{-1} yr^{-1} during the period from 1993 to 2000 (Barford et al. 2001). The interannual variation in NEE is an omnipresent phenomenon observed at almost all of the flux sites across the world (Baldocchi et al. 2001). Understanding the causes of interannual variation may improve our understanding and future predictions of global carbon cycling. Observed interannual variation in NEE at flux sites has been related to several sources (e.g., Hui et al. 2003):

(1) changes in climatic variables per se such as temperature, cloud cover, summer drought, winter snow depth, and the time of snowmelt;

(2) changes in physiological and ecological processes such as growing season length, natural or artificial changes in stand structure, the timing of leaf emergence, and coupling between carbon and nutrient cycling with time delay; and

(3) altered balance between canopy photosynthesis and R_{eco}. Empirical evidence suggests that climatic variability among years is among the factors causing interannual variation in NEE, the latter being also strongly influenced by climate-induced changes in ecosystem physiological parameters.

As an example, we show in Fig. 11.5 the annual courses of NEE for evergreen coniferous and deciduous forests in Europe. The data show the average annual trends in NEE together with the standard deviation. The data is based on multiyear values of 14 coniferous forest sites and 9 deciduous forest sites, and includes for northern evergreen coniferous forest 26 site years, for temperate coniferous forest 35 site years, and for southern coniferous forest 19 site years. For northern deciduous

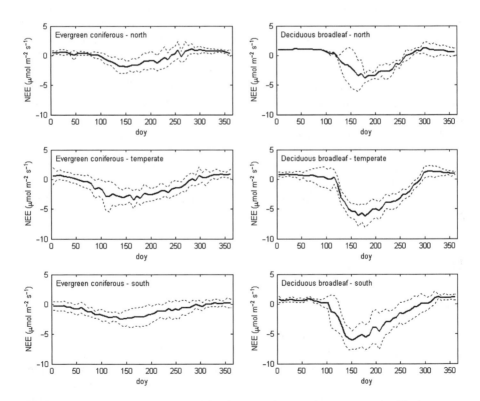

Fig. 11.5 Mean annual trends in NEE for European forests, shown are geographical averages for deciduous and coniferous forests (solid line) and standard deviation around the average (broken lines).

broadleaf forest: the data set contains 7 site years, for temperate deciduous forest 18 site years, and for southern deciduous broadleaf forest: the data set contains. The standard deviation around the mean suggests periods of the year when the variability is high, and that will drive the changes in NEE between years. A few issues stand out. The variation in NEE in conifers as indicated by the broken lines is less than deciduous forest. Particularly for deciduous forest, there is a shift in variation from early spring in northern forests to late summer in southern European forest. This suggests that in Southern Mediterranean forests, the summer period with erratic rainfall is the most variable (vulnerable) in NEE. A similar trend is seen in the coniferous forest, albeit that the northern forests do not really show a marked variation around the spring period. In central European temperate forests, we observe a larger variation at the beginning of the growing season than during summer and autumn. Overall, Fig. 11.5 suggests that care must be taken in generalizing interannual variation controls in forests. Depending on the climate and forest type, different responses of NEE to interannual variation in climate can be observed.

The NEE response is confounded by the two large fluxes in and out of the ecosystem, the GEP and respiration. We used the Reichstein et al. (2005) algorithm to separate the fluxes of the same data as used in Fig. 11.5 to plot the GEP and respiration (Re) separately in Figs. 11.6 and 11.7. It appears that for GEP of coniferous forests, the largest variation is found in the southern forest in the summer. We suggest that drought responses to variable rainfall are the main cause for this. Surprising is the low variation of GEP in spring for northern forests. Northern deciduous forests show very large variation during the spring leaf budding period. The variable timing of the start of the growing season with associated large changes in phenology finds expression in the large GEP variation around spring. Both deciduous and coniferous temperate forests show little variation throughout the year. There is some additional variability around the late summer, autumns for southern deciduous forest, that, as for coniferous forest, might be associated with periodic drought conditions. A further noteworthy phenomenon is that the maximum average fluxes for deciduous forest are always larger than those for coniferous forest. This holds in all climate zones and only for GEP and NEE, suggesting that variation in NEE of European forest is dominated by changes in GEP.

The variation in respiration fluxes is generally large and of similar order as the average flux. The largest variation is found in summer for the coniferous forest, suggesting temperature-related effects in the north and drought effects in the south. Of the deciduous forests, the variation in respiration in the northern forests is very large, with apparently surprisingly little variation in the southern deciduous forests. For the latter forest, the total respiration flux is, however, relatively small compared to the temperate and northern forests. In fact, the variation in the summer and late autumn is still up to 50% of the mean value and thus in relative terms quite large. For all southern forests, drought responses to respiration would seem a likely explanation. Differences in seasonal respiration between deciduous and coniferous forest might be related to changes in the balance between heterotrophic and autotrophic soil respiration, but this remains a speculation in the absence of full carbon balance observations.

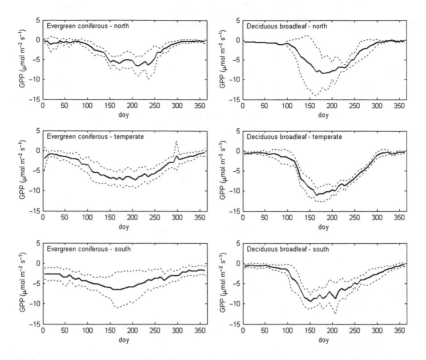

Fig. 11.6 Mean annual trends in GEP for European forests, shown are geographical averages for deciduous and coniferous forests.

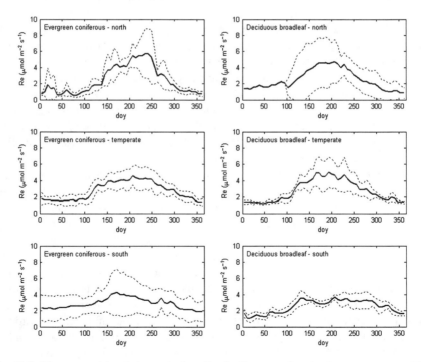

Fig. 11.7 Mean annual trends in respiration, Re, for European forests; shown are geographical averages for deciduous and coniferous forests.

11.5   Vulnerability

11.5.1   Vulnerability from Eddy-Covariance Flux Variability

Modeling results (Cramer et al. 2001; Cox et al. 2000; Friedlingstein et al. 2001, 2006) suggest that the sink strength of the biosphere may deteriorate and change from a sink to a source. This may have potentially dramatic impacts on the CO_2 content of the future atmosphere. This potential for positive feedback is based on the different temperature sensitivity of assimilation and respiration. This hypothesis is contested by both short-term and long-term experiments that show acclimation to higher temperatures. Knorr et al. (2005) present an analysis of the temperature sensitivity. They conclude that the hypothesis of a positive feedback of increased C-losses from soils due to rising temperature is fully consistent with the current experimental evidence.

Eddy-covariance data can be helpful in two aspects: showing where typical vulnerabilities of NEE lie, and to check and validate models that are used to predict future vulnerabilities. Table 11.3 shows the sensitivity of the NEE of carbon of European ecosystems to changes in fire frequency, drought, length of growing season, temperature, vapor pressure deficit, and snow coverage. This table is based on eddy-covariance data of the FLUXNET network (e.g., Baldocchi and Valentini 2004 and this chapter).

The main sensitivity of temperate forest appears to be in the response to the length of the growing season. An early spring adds quickly a number of extra days of carbon uptake to the season, amplifying the sink strength of these forests. The interannual variability at this period of time also suggest that this is important (Fig. 11.5). Luyssaert et al (2007) suggest, however, that these extra events hardly ever dominate the interannual variability. On the negative side increasing drought stress in summer season may decrease the uptake capacity (southern forest in Fig. 11.6.). The Scandinavian boreal forests are very sensitive to small changes in temperature and fire frequency, whereas Mediterranean forest and woodland are very sensitive to drought and fire stress. Fires in the boreal regions, however, have a larger effect on the carbon emissions as the boreal forests contain much more biomass than the Mediterranean woodlands. Wetlands appear to be mostly sensitive to temperature change, although changes in the hydrological cycle may enhance such sensitivities. In general, these vulnerabilities coincide with the areas where the fluxes show their largest variation.

11.5.2   The Value of a Network, the Summer Heat Wave of 2003 in Central Europe

Europe experienced a particularly extreme climate anomaly during 2003, with July temperatures up to 6 °C above long-term means, and annual precipitation deficits up to 300 mm yr[-1], 50% below the average. The presence of an extensive network

Table 11.3 Sensitivity of NEE to environmental change

Vegetation type	Annual NEE (g C m⁻² yr⁻¹)	Max. daily uptake (g C m⁻² day⁻¹)	Sensitivity to environment					
			Fire	Drought	Length of growing season	Temperature	Vapor pressure deficit	Snow
Temperate broadleaf	100–600	6	+	+	++	+	+	+
Temperate needle leaf	300–400	4	+	++	++	+	++	−
Boreal needle leaf	100–200	3	+++	+	++	+++	+++	++
Mediterranean woodland	300	0.7	+++	+++	+−	++	−	
Northern wetlands	100	0.8	±	±	++	+++	−	++
Agriculture	−100	4	NA	++	±	−	−	
Peat lands	250	4	NA*	++	±	−	±	
Grassland	100	4	+	++	+	±	−	±

The plusses indicate weak or strong sensitivity; the minus sign indicates no sensitivity (note that sensitivity to fires indicates primarily a sensitivity of the net biome exchange, and less so of NEE); and the asterisk indicates that sensitivity of natural peat land to fires is small (only drained peat lands burn, the others are too wet, or burn only in the top few centimeters) of eddy-covariance sites with continuous records of CO_2, water, and energy fluxes allowed Ciais et al. (2005) to assess the impact of such extreme events on the carbon balance. They analyzed CO_2 fluxes from 14 forest sites and 1 grassland site for 2002–2003. Hourly fluxes of photosynthesis (gross primary productivity, GPP) and total R_{eco} were separated from the observed NEE fluxes using the same method for each site. In general, a spatially uniform response of GPP to the abnormal conditions of 2003 would not be expected, given the intersite differences in drought duration and intensity, soil characteristics, vegetation state, and species-specific responses to climate variation. Yet, Fig. 11.8 clearly shows that nearly all sites in the affected area experienced a significant GPP reduction in 2003 compared to 2002. The GPP drop coincides with reduced evapotranspiration and soil drying due to the rainfall deficit. Generally, below a threshold of 0.4 in relative extractable water, water stress occurs causing both GPP and transpiration to decrease in response to stomatal closure. Particularly large reductions in GPP were found in temperate deciduous beech and northern Mediterranean forests (Hesse, Roccarespampani, San Rossore), together with reductions in canopy conductance (in 2003 conductance reached only 15% of its 2002 value at Hesse). These productive temperate and Mediterranean forests were greatly affected by extreme drought and/or heat. Moreover, the GPP did not entirely recover from the summer stress during the remainder of the growing season. Southern evergreen forest sites El Saler (pine) and Puéchabon (oak) also experienced a reduction in GPP during the heat wave. Southern forest sites with an herbaceous layer that normally dries in summer were less affected by the heat wave and drought (Pianosa). It is noteworthy that forest sites in the Netherlands and Denmark showed higher NEE and GPP, than in 2002, because the temperature stimulated GPP more than R_{eco} in these sites, where drought appeared less important. These results reinforce our conclusions above from the analysis of interannual variability, which shows high variability in summer/autumn for southern forests. They also suggest that the 2003 response is a mixed reaction to drought that altered the partitioning between TER and GPP.

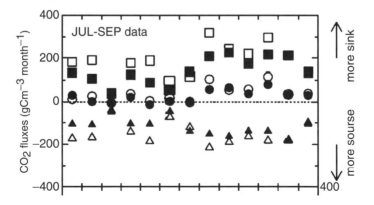

Fig. 11.8 Eddy-covariance forest site measurements of GPP, R_{eco}, and NEP from July to September in 2002 and in 2003 (from Ciais et al. 2005). Circles, NEP, squares GPP, triangles R_{eco}, open values refer to 2002.

11.6. Representativity and Network Design

11.6.1 Representativity

The key issue that stands out is the assessment of the representativeness of the network. How good is the current FLUXNET network distributed along biomes and does it capture the climatic (interannual) variability within a biome, and perhaps more importantly, does it capture the intrabiome variability? There is little known about the optimum structure of a (the FLUXNET) network and it is likely that what defines an optimum depends on the type of questions asked. The vulnerability study of Ciais et al. (2005) showed the value of a dense network to pick up synoptic scale disturbances. An assessment of the Ameriflux network suggested that coverage of the important climate zones and ecosystems was achieved. Globally, this is unlikely to be the case, given the different densities of the network in parts of the world other than Europe and the USA (see Fig. 11.1). The current network is not well distributed, there is severe overlapping in Western Europe, USA, and there is a dominance of forest sites that are in the high growth rate part of their life cycle.

The CarboEurope-IP plan gives an indication of the European distribution, both in number of sites and when weighted with NBP estimates. For this example, the distribution is good, although for a larger EU, the grassland and cropland sites would be underrepresented. Note also that this comparison only indicates that fractional coverage is correct, that is, the percentage of forest sites equals the percentage of forest area in Europe. It tells us nothing of how dense such a network should be to pick up synoptic scale disturbances, changes in age of forest, management, etc., or about the number of sites and their spatial distribution required to sample a single land use type. To resolve synoptic disturbances with typical size of 500 km, a network with spacing of 250 km (half the size of the disturbance) would be required. This, however, is independent of soil and biome distribution.

Globally, most measurement sites are located in the deciduous forest type, and this type seems to be overrepresented with more than 40% of all sites, compared to a global coverage of the biome of only 16%. These numbers can also depend considerably on the land use map used to derive the statistics. When the productivity of the different vegetation types is taken into account, forests (with most sites) have the highest net primary production (NPP). Viewed in this context, the high percentage of forest sites in the FLUXNET is less of a problem. Also, the variation of the fluxes between the forests sites is high. The distribution furthermore covers quite reasonably the global climate space, as defined by annual rainfall and temperature. Most of the sites are located in the temperate climate region with latitudes of 15–50° (Central Europe and Northern America). There are almost no sites located in the more extreme cold, warm, or wet regions.

An important question is whether there is more variability between biomes, or biomes in different climates than within a single biome. If there is more or similar variation within a single biome or climate zone, than we may need to deploy more resources in obtaining insight into this variability than trying to cover the full climate space, as generally is noted in discussion on the design of the flux network. We may investigate this issue by looking at the variability of annual estimates of flux sites. One way to calculate the random error between sites, that is, the variance within a biome is given by Richardson et al. (2006). Such analysis performed for all FLUXNET sites should give a fair indication of the random variation of the measurements. This estimate can then be used in a priori estimates using flux data in inverse models. Importantly, such an analysis could provide guidelines on where to put sites in an effort to reduce uncertainty, and where site could be removed. If the variability between sites in a given climate band is larger than that between climate band, we need to conclude that issues like land use history and management play a more important role than climate and vegetation type per se. The implication for network design follows immediately.

For 63-FLUXNET sites, there is enough data available to calculate the average annual NEE and the annual evaporation flux. We plotted these against latitude as a proxy for the combined effect of the annual average temperature and the annual precipitation (Fig. 11.9). From this figure, it is evident that there is as much variation between sites in a single latitude band as between latitude bands. When separating NEE into GPP and respiration a single picture emerges. Factors, such as land use, history and management, and soil type, are thus likely to play a more important role in determining annual NEE, GPP, and R than climate. Analysis of two towers in Brazil 30 km from each other apart, suggest that flux sites will never be able to capture this local variation (Harris et al. 2004).

The interannual variation at a single site is also often much larger or of similar size as the variation between sites. This suggests that within biome variability is not yet adequately captured by the network and that techniques should be further developed to capture this. High spatial resolution satellite observation would be required to correctly interpret the causes of this variation. It also suggests that long-term observations are key to understanding the variability of the global carbon cycle.

11.6.2   Network Design

We can conclude that the coverage of FLUXNET sites seems appropriate in terms of climate space (precipitation and temperature), and appears generally appropriate in terms of fractional coverage of land use types, be it with an overrepresentation of midlatitude forest sites. The current data, however, also suggest that there is more variability between sites at climate bands than between adjacent climates. Although various studies attribute key controlling factors to the NEE such as radiation and temperature and moisture (e.g., Van Dijk et al. 2005), these appear to be true only in a general sense. When such data or parameters are used to derive fluxes at similar sites, differences will occur that are related to small-scale differences in land use history, management, and climate. Particularly in Europe, land use history and small-scale climate differences are important. Information on management

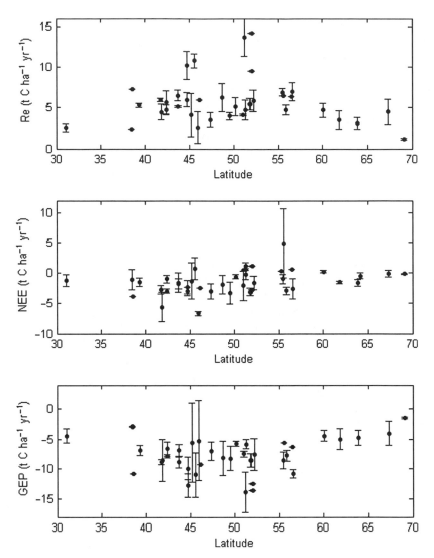

Fig. 11.9 Distribution of GEP. NEE and LE fluxes (annual average values) with latitude for FLUXNET sites (dots, error bars indicate standard deviations of single sites for multiple years.

practice and on previous land use history is essential to interpret carbon fluxes, but hardly measured or adequately described. This may imply that the current setup needs to be redesigned in such a way that the network still captures the climatic variability, but that it also samples within supersites, different forms of land management, and history, age, and class of the forests. Rather than expanding the network into new areas, an intensification of the observation in the vicinity of a few single sites may be appropriate.

To make any further progress in designing the network further, synthetic type data studies need to be performed. For instance, studies could be performed in future on the effect of loosing information in a data assimilation network (e.g., Chap. 3, this book). This can be done using more sites within a single biome in a data assimilation study, and showing whether there is redundancy or underrepresentation in the network. Even with relatively simple data-assimilation techniques such as the neural networks used by Papale and Valentini (2003), these studies could be performed.

11.7 Conclusions and Recommendations

In the USA and Europe, the flux measurements have long started to form an integral part of the experimental programs that are based around the multiple constraint approach to close the carbon balance. The key rationale behind these efforts is that, ultimately, fluxes from the land, both the slow and the fast ones, affect atmospheric concentrations of CO_2 at a regional scale and that the slow and fast fluxes ideally should match. Currently, the results of eddy-correlation measurements are considered reliable, when they meet the standards of CarboEurope or FLUXNET minimum requirements. These requirements that have been developed in the past decades comprise of necessary corrections on the measurements and gap-filling strategies.

Flux towers yield a consistent picture of the controls on NEE at local scale, and have improved our process understanding tremendously. They do not in isolation contribute to the Kyoto goals of monitoring, as is often erroneously claimed. On the contrary, as correctly pointed out by Körner (2003), their strength is in the combined use as a network or in a multiple constrains approach. Eddy-covariance networks have substantially improved our understanding of biosphere–atmosphere exchange of carbon, water, and energy. Additionally, flux towers play an extra role of importance in measurement programs because they provide data for model validation and comparison (Van Dijk et al. 2005), as well as vulnerability analysis (Ciais et al. 2005). Various methods have been developed for the separation of main flux components as well as for the calculating nighttime fluxes, although this last topic subject is still not fully solved. Especially in mountainous areas and areas with strong advection more effort should be put in interpreting the nighttime data. Magnitude and variation of fluxes is well described for a large part of Europe. The existence and quality of the flux network plus the databases reflect the evolvement of the networks and maturity of the field. Nonetheless, improvements can be made on the representativity of the network design (land use types and geographical spread) as well as descriptions of land use history of the sites. The main paradox of the current network is that it will never be feasible to sample adequately the full variation in the climate, land history, and management matrix, but that the design of a minimum size of the network is also not immediately obvious (Chap. 3, this book).

Despite the complexity of the issue, it is possible to set a number of criteria for future new site level studies:

- They should be operated as multitower sites to sample more than one land use (history) type, say peat lands and different types of crops within a single climate area.
- They should adhere to the full CarboEurope, FLUXNET minimum requirements for measurements, calculations, and corrections. This pertains particular to the execution of soil moisture and biomass measurements.
- They should collect adequate auxiliary data (soil, leaf area index, N-profiles, canopy age, etc.), so that the flux measurements can be meaningfully interpreted. Without this data, a site is meaningless.
- More sites could be established in the former East European countries as well as the Balkans and Greece.
- They should run at least for a minimum of 5 years to capture interannual variability in climate.
- An effort should be put into generating a realistic (carbon) land use history for all existing sites as well as future sites.
- Data submission to a central database should be timely and consistently.
- Measurements in adjacent ecosystems to cover the full development cycle or management range of ecosystems (e.g., forest age classes, major crop types).

The eddy-covariance network can provide uniform data sets for sites now and in the future. But possible problems here are the predictability of the future status of sites, and thus the continuity of the current network. For most sites, there is no long-term plan for measurements. Most sites are used for several years within a project framework and are either kept operational with new funding in another project, are kept partly operational, or are abandoned. International organizations Global Climate Observing System, Integrated Global Carbon Observation, Global Carbon Project (GCOS, IGCO, and GCP) are actively seeking support to keep an optimum number of sites operational, well after the first Kyoto commitment period.

References

Anthoni, P.M., Freibauer, A., Kolle, O. and Schulze, E.-D., 2004. Winter wheat carbon exchange in Thuringia. Agricultural and Forest Meteorology 121, 55–67.

Araújo, A.C., Nobre, A.D., Kruijt, B., Elbers, J.A., Dallarosa, R., Stefani, P., von Randow, C., Manzi, A.O., Culf, A.D., Gash, J.H.C., Valentini, R. and Kabat, P., 2002. Comparative measurements of carbon dioxide fluxes from two nearby towers in a central Amazonian rainforest: The Manaus LBA site. Journal of Geophysical Research 107 (D20), 8090, DOI: 10.1029/2001JD000676.

Araújo, A.C., Kruijt, B., Nobre, A.D., Dolman, A.J., Waterloo, M.J., Moors, E.J., de Souza, J.S., 2006. CO2 underneath a tropical forest canopy and nocturnal accumulation in a topographical gradient. Ecological Applications (in press).

Aubinet, M., Grelle, A., Ibrom, A., Rannik, Ü., Moncrieff, J., Foken, T., Kowalski, A.S., Martin, P.H., Berbigier, P., Bernhofer, C., Clement, R., Elbers, J., Granier, A., Grünwald, T., Morgenstern, K., Pilegaard, K., Rebmann, C., Snijders, W., Valentini, R., and Vesala, T.,

2000. Estimates of the annual net carbon and water exchange of forests: The EuroFlux methodology. Advances in Ecological Research 30, 113–176.

Aubinet, M., Berbigier, P., Bernhofer, C., Cescatti, A., Feigenwinter, C., Granier, A., Grünwald, T., Havrankova, K., Heinesch, B., Longdoz, B., Marcolla, B., Montagnani, L., Sedlak, P., 2005. Comparing CO_2 storage and advection conditions at night at different carboeuroflux sites. Boundary-Layer Meteorology 116 (31), 63–93.

Aurela, M., Laurila, T. and Tuovinen, J.-P., 2002. Annual CO_2 balance of a subarctic fen in northern Europe: Importance of the wintertime efflux. Journal of Geophysics Research 107 (D21), 4607, DOI: 10.1029/2002JD002055.

Baldocchi, D. 1994. A comparative-study of mass and energy-exchange rates over a closed C-3 (wheat) and an open C-4 (corn) crop. 2. CO_2 exchange and water-use efficiency. Agricultural and Forest Meteorology 67, 291–321.

Baldocchi, D. and Valentini, R., 2004. Geographic and Temporal Variation of Carbon Exchange by Ecosystems and Their Sensitivity to Environmental Perturbations. In: C. B. Field and M. Raupach (Eds.) The Global Carbon Cycle. Island Press, Washington.

Baldocchi, D.D., Valentini, R., Running, S.R., Oechel, W. and Dahlman, R. 1996. Strategies for measuring and modelling CO_2 and water vapor fluxes over terrestrial ecosystems. Global Change Biology 2, 159–168.

Baldocchi, D.D., Falge, E., Gu, L., Olson, R., Hollinger, D., Running, S., Anthoni, P., Bernhofer, C., Davis, K., Fuentes, J., Goldstein, A., Katul, G., Law, B., Lee, X., Malhi, Y., Meyers, T., Munger, J.W., Oechel, W., Pilegaard, K., Schmid, H.P., Valentini, R., Verma, S., Vesala, T., Wilson, K. and Wofsy, S., 2001. FLUXNET: A New Tool to Study the Temporal and Spatial Variability of Ecosystem-Scale Carbon Dioxide, Water Vapor and Energy Flux Densities. Bulletin of the American Meteorological Society 82, 2415–2435.

Barford, C.C., Wofsy, S.C., Goulden, M.L., Munger, J.W., Pyle, E.H., Urbanski, S.P., Hutyra, L., Saleska, S.R., Fitzjarrald, D., and Moore, K., 2001. Factors controlling long- and short-term sequestration of atmospheric CO_2 in a midlatitude forest. Science 294, 1688–1691.

Ciais, P., Reichstein, M., Viovy, N., Granier, A., Ogée, J., Allard, V., Aubinet, M., Buchmann, N., Bernhofer, C., Carrara, A., Chevallier, F., De Noblet, N., Friend, A.D., Friedlingstein, P., Grünwald, T., Heinesch, B., Keronen, P., Knohl, A., Krinner, G., Loustau, D., Manca, G., Matteucci, G., Miglietta, F., Ourcival, J.M., Papale, D., Pilegaard, K., Rambal, S., Seufert, G., Soussana, J.F., Sanz, M.J., Schulze, E.D., Vesala, T. and Valentini, R., 2005. Europe-wide reduction in primary productivity caused by the heat and drought in 2003. Nature 437, 529–533.

Cox, P.M., Betts, R.A., Jones, C.D., Spall, S.A. and Totterdell, I.J., 2000. Acceleration of global warming due to carbon-cycle feedbacks in a coupled climate model. Nature 408, 184–187.

Cramer, W., Bondeau, A., Woodward, F.I., Prentice, I.C., Betts, R.A., Brovkin, V., Cox, P.M., Fisher, V., Foley, J.A., Friend, A.D., Kucharik, C., Lomas, M.R., Ramankutty, N., Sitch, S., Smith, B., White, A. and Young-Molling, C., 2001. Global response of terrestrial ecosystem structure and function to CO_2 and climate change: Results from six dynamic global vegetation models. Global Change Biology 7 (4), 357–373. DOI: 10.1046/j.1365-2486.2001.00383.x.

Davidson, E.A. and Janssens, I.A. 2006. Temperature sensitivity of soil carbon decomposition and feedbacks to climate change. Nature 440 (7081), 165–173.

Dolman, A.J., Schulze, E.-D. and Valentini, R., 2003. Analyzing Carbon Flux Measurements. Science 301, 916–917.

Dolman, A.J., Maximov, T.C., Moors, E.J., Maximov, A.P., Elbers, J.A., Kononov, A.V., Waterloo, M.J. and van der Molen, M.K., 2004. Net ecosystem exchange of carbon dioxide and water of far eastern Siberian Larch (Larix cajanderii) on permafrost. Biogeosciences 1, 133–146.

Dore, S., Hymus, G.J., David, J.P., Hinkle, C.R., Valentini, R., Drake, B.G. (2003). Cross validation of open-top chamber and eddy covariance measurements of ecosystem CO_2 exchange in a Florida scrub-oak ecosystem. Global Change Biology 9 (1), 84–95.

Falge, E., Baldocchi, D., Olson, R.J., Anthoni, P., Aubinet, M., Bernhofer, C., Burba, G., Ceulemans, R., Clement, R., Dolman, A.J., Granier A., Gross, P., Grünwald, T., Hollinger, D., Jensen, N.-O.,

Katul, G., Keronen, P., Kowalski, A., Ta Lai, C., Law, B.E., Meyers, T., Moncrieff, J., Moors, E., Munger, J.W., Pilegaard, K., Rannik, Ü., Rebmann, C., Suyker, A., Tenhunen, J., Tu, K., Verma, S., Vesala, T., Wilson, K., Wofsy, S., 2001. Gap filling strategies for defensible annual sums of net ecosystem exchange. Agricultural Forest Meteorology 107, 43–69.

Finnigan, J.J., Clement, R., Mahli, Y., Leuning, R. and Cleugh, H., 2003. A re-evaluation of long-term ux measurement techniques: Part 1. Averaging and coordinate rotation. Boundary-Layer Meteorology 107, 1–48.

Friedlingstein, P., Bopp, L., Ciais, P., Dufresne, J.-L., Fairhead, L., LeTreut, H., Monfray, P., Orr, J., 2001. Positive feedback between future climate change and the carbon cycle. Geophysics Research Letters 28 (8), 1543–1546. DOI: 10.1029/2000GL012015.

Friedlingstein, P., Cox, P., Betts, R., Bopp, L., von Bloh, W., Brovkin, V., Cadule, P., Doney, S., Eby, M., Fung, I., Govindasamy, B., John, J., Jones, C., Joos, F., Kato, T., Kawamiya, M., Knorr, W., Lindsay, K., Matthews, H.D., Raddatz, T., Rayner, P., Reick, C., Roeckner, E., Schnitzler, K.-G., Schnur, R., Strassmann, K., Weaver, A.J., Yoshikawa, C., and Zeng, N., 2006. Climate-carbon cycle feedback analysis: Results from the C4MIP model intercomparison. Journal of Climate 19, 3337–3353.

Hendriks, D.M.D., van Huissteden, J., Dolman, A.J., van der Molen, M.K., 2007. The full greenhouse gas balance of an abandoned peat meadow. Biogeosciences 4, 411–424.

Harris, P.P., Huntingford, C., Gash, J.H.C., Hodnett, M.G., Cox, P.M., Malhi, Y. and Araújo, A. C., 2004. Calibration of a land-surface model using data from primary forest sites in Amazonia. Theoretical and Applied Climatology 78, 27–45, http://dx.doi.org/10.1007/s00704-004-0042-y, DOI: 10.1007/s00704-004-0042-y.

Gash, J.H.C. and Dolman, A.J., 2003. Sonic anemometer (co)sine response and flux measurement I. The potential for (co)sine error to affect sonic anemometer-based flux measurements. Agricultural and Forest.

Hui, D., Luo, Y. and Katul, G., 2003. Partitioning interannual variability in net ecosystem exchange between climatic variability and functional change. Tree Physiology 23, 433–442.

Janssens, I.A., Lankreijer, H., Mateucci, G., Kowalski, A.S., Buchmann, N., Epron, D., Pilegaard, K., Kutsch, W., Longdoz, B., Grünwald, T., Montagnani, L., Dore, S., Rebmann, C., Moors, E. J., Grelle, A., Rannik, Ü., Morgenstern, K., Oltchev, S., Clement, R., Guðmundsson, J., Minerbi, S., Berbigier, P., Ibrom, A., Moncrieff, J., Aubinet, M., Bernhofer, C., Jensen, N.O., Vesala, T., Granier, A., Schulze, E.-D., Lindroth, A., Dolman, A.J., Jarvis, P.G., Ceulemans, R. and Valentini, R., 2001. Productivity and disturbance overshadow temperature in determining soil and ecosystem respiration across European forests. Global Change Biology 7, 269–278.

Janssens, I.A., Freibauer, A., Ciais, P., Smith, P., Nabuurs, G.-J., Folberth, G., Schlamadinger, B., Hutjes, R.W.A., Ceulemans, R., Schulze, E.-D., Valentini, R., Dolman, H., 2003. Europe's biosphere absorbs 7–12% of anthrogogenic carbon emissions. Science 300, 1538–1542.

Jarvis, P.G. and McNaughton, K.G., 1986. Stomatal control of transpiration: Scaling up from leaf to region. Advance Ecology Research 15, 1–49.

Kaimal, J.C. and Finnigan, J.J., 1994. Atmospheric Boundary Layer Flows: Their Structure and Measurement. Oxford University Press, New York, 289 pp.

Knorr, W. and Kattge, J., 2005. Inversion of terrestrial ecosystem model parameter values against eddy covariance measurements by Monte Carlo sampling. Global Change Biology 11 (8), 1333.

Knorr, W., Prentice, I.C., House, J.I., and Holland, E.A., 2005. Long term sensitivity of soil carbon turnover to global warming. Nature 433, 298–301.

Körner, C., 2003. Slow in, rapid out—carbon flux studies and Kyoto targets. Science 300, 1242–1243.

Kruijt, B., Elbers, J.A., von Randow, C., Araújo, A.C., Oliveira, P.J., Culf, A., Manzi, A.O., Nobre, A.D., Kabat, P., and Moors, E.J. 2004. The robustness of eddy covariance fluxes for Amazon rain forest conditions. Ecological Applications 14, S101–S113.

Lee, X., 1998. On micrometeorological observations of surface-air exchange over tall vegetation. Agricultural and Forest Meteorology 91 (1–2), 39–49.

Lee, X., Massmann, W.B. and Law, B., 2004. Handbook of Micrometeorology: A Guide for Surface Flux Measurement and Analysis. Kluwer Academic Press, Rotterdam, 250 pp.

Luyssaert, S., Inglima, I., Jung, M., Richardson, A.D., Reichstein, M., Papale, D., Piao, S.L., Schulze, E.-D, Wingate, L., Matteucci, G., Aragao, L., Aubinet, M., Beer, C., Bernhofer, C., Black, K.G., Bonal, D., Bonnefond, J.-M., Chambers, J., Ciais, P., Cook, B., Davis, K.J., Dolman. A.J., Gielen, B., Goulden, M., Grace, J., Granier, A., Grelle, A., Griffis, T., Grünwald, T., Guidolotti, G., Hanson, P.J., Harding, R., Hollinger, D.Y., Hutyra, L.R., Kolari, P., Kruijt, B., Kutsch, W., Lagergren, F., Laurila, T. (2007). CO_2 balance of boreal, temperate, and tropical forests derived from a global database Global Change Biology 13 (12), 2509–2537. doi:10.1111/j.1365–2486. 2007.01439.x

Moncrieff, J.B., Massheder, J.M., de Bruin, H., Elbers, J., Friborg, T., Heusinkveld, B., Kabat, P., Scott, S., Soegaard, H. and Verhoef, A., 1998. A system to measure surface fluxes of momentum, sensible heat, water vapour and carbon dioxide. Journal of Hydrology 188/189, 589–611.

Monteith, J.L., 1965. Evaporation and environment. In G.E. Fogg (Ed). The state and movement of water in living organisms. Cambridge University Press, Cambridge, UK. pp. 205–234.

Monteith, J.L., 1976 (Ed). Vegetation and the Atmosphere—Volume 1: Principles. Academic Press. London, NY.

Papale, D. and Valentini, R., 2003. A new assessment of European forests carbon exchanges by eddy fluxes and artificial neural network spatialization. Global Change Biology 9, 525–535.

Papale, D., Reichstein, M., Aubinet, M., Canfora, E., Bernhofer, C., Kutsch, W., Longdoz, B., Rambal, S., Valentini, R., Vesala, T. and Yakir, D., 2006. Towards a standardized processing of Net Ecosystem Exchange measured with eddy covariance technique: Algorithms and uncertainty estimation. Biogeosciences 3, 571–583.

Reichstein, M., Rey, A., Freibauer, A., Tenhunen, J., Valentini, R., Banza, J., Casals, P., Cheng, Y., Grünzweig, J., Irvine, J., Joffre, R., Law, B., Loustau, D., Miglietta, F., Oechel, W., Ourcival, J.-M., Pereira, J., Peressotti, A., Ponti, F., Qi, Y., Rambal, S., Rayment, M., Romanya, J., Rossi, F., Tedeschi, V., Tirone, G., Xu, M., Yakir, D., 2003. Modelling temporal and large-scale spatial variability of soil respiration from soil water availability, temperature and vegetation productivity indices. Global Biogeochemical Cycles 17, 1104, DOI: 10.1029/2003GB002035.

Reichstein, M., Falge, E., Baldocchi, D., Papale, D., Aubinet, M., Berbigier, P., Bernhofer, C., Buchmann, N., Gilmanov, T., Granier, A., Grünwald, T., Havrankova, K., Ilvesniemi, H., Janous, D., Knohl, A., Laurila, T., Lohila, A., Loustau, D., Matteucci, G., Meyers, T., Miglietta, F., Ourcival, J.-M., Pumpanen, J., Rambal, S., Rotenberg, E., Sanz, M., Tenhunen, J., Seufert, G., Vaccari, F., Vesala, T., Yakir, D., and Valentini, R., 2005. On the separation of net ecosystem exchange into assimilation and cosystem respiration: Review and improved algorithm. Global Change Biology 11, 1424–1439.

Richardson, A.D., Hollinger, D.Y., Burba, G.G., Davis, K.J., Flanagan, L.B., Katul, G.G., Munger, J.W., Ricciuto, D.M., Stoy, P.C., Suyker, A.E., Verma, S.B., and Wofsy, S.C., 2006. A multi-site analysis of random error in tower-based measurements of carbon and energy fluxes. Agricultural and Forest Meteorology 136, 1–18.

Schulze, E.-D., Lange, O.L., Buschbom, U., Kappen, L., Evenari, M., 1972. Stomatal response to humidity in plants growing in the desert. Planta 108, 259–270.

Schulze, E.-D., Beck, E., and Müeller-Hohenstein, K., 2005. Plant Ecology. Springer Verlag.702 pp.

Soussana, J.-F., Saletes, S., Smith, P., Ambus, P., Amezquita, M.C., Andren, O., Arrouays, D., Ball, B., Boeckx, P., Brüning, C., Buchmann, N., Buendia, L., Cellier, P., Cernusca, A., Clifton-Brown, J., Dämmgen, U., Ewert, F., Fiorelli, J.L., Flechard, C., Freibauer, A., Fuhrer, J., Harrison, R., Hensen, A., Hiederer, R., Janssens, I.A., Jayet, P.A., Jones, M., Jouany, J.P., Jungkunst, H.F., Kuikman, P., Lagreid, M., Leffelaar, P.A., Leip, A., Loiseau, P., Martin, C., Milford, C., Neftel, A., Oenema, O., Ogle, S., Olesen, J.E., Pesmajoglou, S., Peterson, S.O., Pilegaard, K., Raschi, A., Rees, B., Schils, R.L.M., Sezzi, E., Stefanie, P., Sutton, M., Van Amstel, A., Van Cleemput, O., Van Putten, B., Van Wesemael, B., Verhagen, A., Viovy, N., Vuichard, N., Weigel, H.J., Weiske, A., Willers, H.C., Tuba, Z., 2004. Greenhouse gas emissions from European grasslands. Report 4/2004 CarboEurope-GHG Concerted action. University of Tuscia, Viterbo, Italy, 93 pp.

Stewart, J.B., 1988, Modelling surface conductance of pine forest. Agricultural and Forest Meteorology 43 (1), 19–35.

Valentini, R. (Ed.), 2003. Fluxes of Carbon, Water and Energy of European Forests. Ecological Studies Series 163, Springer-Verlag, Heidelberg.

Valentini, R., Matteucci, G., Dolman, A.J., et al., 2000. Respiration as the main determinant of carbon balance in European forests. Nature 404, 861–865.

Van der Molen, M.K., Gash, J.H.C., Elbers, J.A., 2004. Sonic Anemometer cosine response and flux measurement. II. The effect of introducing an angle of attack dependent calibration. Agricultural and Forest Meteorology 122, 95–109.

Van der Molen, M. K., et al., 2007a. The carbon balance of the Boreal Eurasia consolidated with eddy covariance observations. Global Change Biology (in review).

Van der Molen, M.K., van Huissteden, J.C., Parmentier, F.J., Petrescu, A.M.R., Dolman, A.J., Maximov, T.C., Kononov, A.V., Karsanaev, S.V., Suzdalov, D.A., 2007b. The seasonal cycle of the greenhouse gas balance of a continental tundra site in the Indigirka lowlands, NE Siberia. Biopgeosciences Discussions 4, 2329–2384.

Van Dijk, A.I.J.M. and Dolman, A.J., 2004. Estimates of CO_2 uptake and release among European forests based on eddy covariance data. Global Change Biology 10 (9), 1445–1459. DOI: 10.1111/j.1365-2486.2004.00831.x.

Van Dijk, A.I.J.M., Dolman, A.J., Schulze, E.-D., 2005. Radiation, temperature, and leaf area explain ecosystem carbon fluxes in boreal and temperate European forests. Global Biogeochemical Cycles 19, GB2029, DOI: 10.1029/2004GB002417.

Van Gorsel, E., Leuning, R., Cleugh, H.A., Keith, H., Suni, T., 2007. Nocturnal carbon efflux: reconciliation of eddy covariance and chamber measurements using an alternative to the u*-threshold filtering technique Tellus B 59 (3), 397–403. doi:10.1111/j.1600-0889.2007.00252.x

Vleeshouwers, L.M. and Verhagen, A., 2002. Carbon emission and sequestration by agricultural land use: A model study for Europe. Global Change Biology 8 (6), 519.

Wichura, B. and Foken, T., 1995. Anwendung integraler Turbulenzcharakteristiken zur Bestimmung von Beimengungen in der Bodenschicht der Atmosphäre. DWD, Abteilung Forschung, Arbeitsergebnisse no 29, 52 pp.

Chapter 12
Observations and Status of Peatland Greenhouse Gas Emissions in Europe

Matthias Drösler, Annette Freibauer, Torben R. Christensen, and Thomas Friborg

12.1   Introduction

A peatland is a type of ecosystem where carbon (C) along with nitrogen and several other elements has been accumulated as peat originating from the plant litter deposited on the site. A logical consequence of the above definition of peatlands is that they are ecosystems, which by way of nature are a sink for atmospheric carbon dioxide (CO_2). This is the case because more C is accumulated through photosynthesis than is released through respiration. As a consequence of this, organic matter accumulates as peat. The C accumulated in peatlands is equivalent to almost half the total atmospheric content, and a hypothetical sudden release would result in an instantaneous 50% increase in atmospheric CO_2. While this scenario is unrealistic, it nevertheless highlights the central role of peatlands where huge amounts of CO_2 have almost entirely been "consumed" since the last glacial maximum, but could respond differently as a result of future changes in climatic conditions. Peatlands have, hence, over the last 10,000 years helped to remove significant amounts of CO_2 from the atmosphere.

A complicating factor in this respect is that in terms of the major greenhouse gases (GHGs), peatlands are not just acting as a sink for CO_2. The wet conditions that lead to the slow decomposition of organic material and enable peat accumulation to occur, also cause significant amounts of the powerful GHG methane (CH_4) to be formed. Indeed global wetlands (predominantly peatlands) are considered to be the largest single source of atmospheric CH_4 also when considering all anthropogenic emissions. Peatlands are, therefore, also a key player in the atmospheric CH_4 budget and as a result also influence the global climate.

12.2   GHG Fluxes in the European Peatlands

12.2.1   Peatland Areas in Europe

Peatlands cover 5–6% of the European continental land surface. They are concentrated in northern and temperate lowlands where water drainage is impeded, precipitation is high, and/or low temperatures limit evapotranspiration. Joosten and Clarke

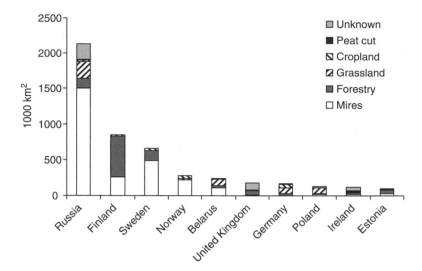

Fig. 12.1 Peatland area and use in the top ten peat countries of Europe. Mires are undrained natural peatlands. The categories forestry, grassland, cropland, and peat cut are peatland use with drainage. The unknown area results from inconsistencies in the data sources and incomplete information and is considered to be undrained. Data from Joosten and Clarke (2002), Lappalainen (1996), World Energy Council (2001), and national sources.

(2002) give a good overview of the peatland distribution in Europe. The numbers used in this chapter are amended with national data where available. Russia contains half of the European peatlands. In EU-27, most peatlands are found in Scandinavia, UK, Ireland, and the young pleistocene landscapes of Germany and Poland. Half of the peatlands on the European continent are subject to various sorts of land use, which is often associated with drainage. Twenty percent are drained for forestry, mainly in Scandinavia and UK; 16% are drained for agriculture, mainly in Germany, Poland, and the Netherlands; and 0.5% is used for peat extraction. Restoration of abandoned peatlands is an emerging issue but still negligible in terms of areal coverage. The fractions of the various peatland uses greatly vary among countries and regions (Fig. 12.1).

12.2.2   Peatland C Stocks

Despite the small surface cover, peatlands hold large C stocks because of the high organic matter content, which often reaches down to several meters of peat accumulation. Estimates of the European peatlands C stocks are limited by unknown peat depth, peat bulk density, and C fractions in the profiles. Byrne et al. (2004) estimated the C stocks in European peatlands to 42 Pg C but with a high degree of uncertainty. The estimate relied on areal estimates by Joosten and Clarke (2002). The peat resources in volume or mass reported in Lappalainen (1996) were converted into C storage estimates based on the average dry bulk density of 91 g dm^{-3} (Mäkilä 1994) and C concentration of 51.7% (Gorham 1991). The depth estimates were based on Lappalainen (1996) where available. However, for most countries, the average peat depths were not available and a conservative estimate of 1.75 m was used as the base of volume calculations.

Despite these uncertainties, peatlands are those European ecosystems with the highest C storage per area, which for example is on average two to three times higher than the ecosystem C stock in forests.

12.2.3   Typical GHG Flux Rates

In general, natural mires take up CO_2, as part of photosynthesis, and release a part of the C as CO_2 and CH_4 to the atmosphere during microbial decomposition. Natural mires are neutral with respect to N_2O. In contrast, drainage of peatlands results in the release of CO_2 and sometimes N_2O while CH_4 emissions cease or are even changed to small CH_4 uptake. GHG fluxes from peatlands are much more variable in space and time than GHG fluxes in upland ecosystems. Measured emission data are limited in number and rarely cover one or more full years. Nevertheless, a first attempt was made to classify peatlands and calculate typical annual GHG flux rates for the European peatlands. At the first level of classification, bogs (ombrogenic peatlands) were separated from fens (minerogenic peatlands). Bogs and fens were further distinguished according to the following categories:

- Forestry (drained),
- Fresh drainage,
- Grassland (drained),
- Cropland (drained),
- Peat cut (drained),
- Abandoned (drained),
- Restored (rewetted), and
- Mires (natural).

Drained and abandoned peat cut areas are only relevant for bogs.

A literature review was performed to gather all available information on measured GHG fluxes and on calculated regional average and typical GHG fluxes (Minkkinen and Laine 1998; Sundh et al. 2000; Kasimir Klemedtsson et al. 1997; Nilsson et al. 2001). The full set of references is given in Byrne et al. (2004). The available data are strongly biased toward Finland and unequally distributed through the categories listed above (Fig. 12.2). Figure 12.2 demonstrates the shortcomings in the presently available datasets. CO_2 and CH_4 have been studied in some breadth, while N_2O studies focused on managed fens. Few studies have measured the full GHG balance. Data about the C balance in arable, abandoned, and restored peatlands

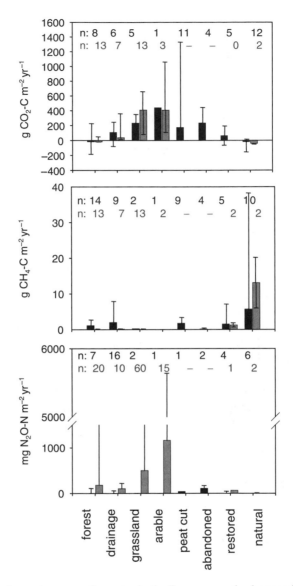

Fig. 12.2 Typical greenhouse gas flux rates in the European peatland categories. Black: bogs, grey: fens. *n* indicates the number of datasets available for bogs and fens. The bars represent the median greenhouse gas flux rates, error bars indicate the maximum and minimum values found in the literature.

and fen mires are particularly scarce. Therefore, the values shown in Fig. 12.2 can only be seen as indicative orders of magnitude and may not necessarily be representative of the European peatlands. In order to avoid bias produced by few exceptionally large measured fluxes within a category, the median of all available data was taken to represent typical GHG flux rates in Fig. 12.2.

In Fig. 12.2, the GHG flux rates for forestry were mainly derived from Finnish datasets, where drainage for forestry is moderate and water tables are still close to natural conditions, in particular in young stands. Therefore, the calculated emission factors for peatlands drained for forestry may not be representative for more intensively managed forests. Grasslands dominate on more fertile peatlands and are sometimes deeply drained and fertilized. The wide range in literature values reflects a huge diversity in management practice in grasslands, from virtually no drainage in extensively grazed grasslands in Ireland and the UK to deep drainage in Germany. Croplands are the deepest drained, most intensively used, fertilized, and ploughed, most disturbed, most fertile sites with strong peat degradation. Peat cut areas on bogs are small, but very variable, depending on treatment prior to and after peat extraction. The peat C loss from removal is not considered here, but only the flux rates on-site is considered. Peatland restoration represents a new development due to peatland conservation programs. Still marginal areas are covered and very little data are available. While peatland restoration is an emerging area of GHG research, GHG fluxes on natural mires rely on many, however mostly seasonal, data from Northern bogs, oligotrophic and mesotrophic mires. There are no data for natural temperate fens.

Taking the three GHGs together, peatlands in the categories of natural mires, restoration, and forestry are close to climate neutral while all other managed peatlands are significant sources of GHGs. As shown in Fig. 12.3, the net climate effect calculated by the global warming potentials (GWP) of the three gases used under the United Nations Framework Convention on Climate Change (UNFCCC) (GWP100, IPCC 1996) is closely related to the net C balance of peatlands. We

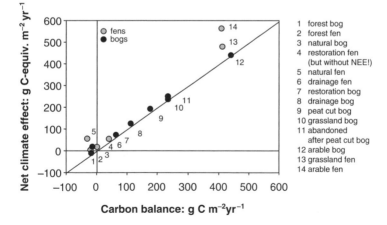

Fig. 12.3 Carbon balance versus net climate effect of peatlands. Positive values indicate C losses and net climate warming. The net climate effect was calculated from typical flux rates in Fig. 12.2 using the global warming potentials (GWP) used under the United Nations Framework Convention on Climate Change (UNFCCC) (GWP100, IPCC 1996): $1\,kg\ CH_4 = 21\,kg\ CO_2$; $1\,kg\ N_2O = 310\,kg\ CO_2$. Carbon balance = Net ecosystem exchange (NEE) $-$ C removal $+ CH_4$-C.

followed this classical GWP approach, as for policy-related expression of the net-climate effect it is still the state-of-the-art (Frolking et al. 2006). The peatland classes roughly follow the 1:1 line of C balance versus GWP. Deviations from this line originate from significant CH_4 emissions in natural peatlands, which account for 30% of the net ecosystem exchange (NEE), and/or N_2O emissions from drained agricultural peatlands. The GWP is typically stronger for fens than for bogs. All peatlands with a strong C source are also strong GWP sources, and all of them are managed (Fig. 12.3, numbers 6–14). In contrast, natural peatlands and those only marginally drained for forestry are relatively balanced with regard to C and GWP (Fig. 12.3, numbers 1–5), with a small exception of natural fens (Fig. 12.3, number 5) where CH_4 emissions dominate the GWP. Peatlands are hot spots not only of C storage but also of GHG emissions, in particular when drained.

Figure 12.4 shows a conservative estimate for the peatland GHG budget of the top ten peat countries in Europe based on typical GHG flux rates per land use class in Fig. 12.2. Because of lack of adequate area statistics, recent drainage for forestry or peat cut, abandoned peat cut, and restoration could not be calculated. While Finland, Sweden, and UK turn out as CO_2 sinks, all countries are net GHG emitters from peatlands.

The results of Fig. 12.4 indicate some interesting features. While mires and forestry dominate the peatland areas (Fig. 12.1), grasslands and croplands dominate the emissions. Russia contributes most to the European peatland GHG balance, which is equivalent to its share in peatland area. Germany turns out as the second largest emitter (12% of European total) although it contains only 3.2% of the

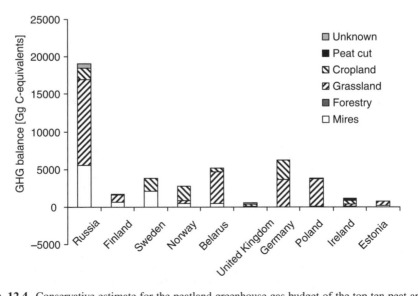

Fig. 12.4 Conservative estimate for the peatland greenhouse gas budget of the top ten peat area countries in Europe.

Table 12.1 Sensitivity of the European peatland greenhouse gas balance to assumptions about land use and drainage intensity (Gg C-equ)

		Continental Europe	EU-15	EU-25
Peatland	Default	51,660	15,291	21,468
Peatland	Maximum	67,512	24,377	32,019
Peatland	Minimum	49,057	14,325	20,324
Forest	Default	−362	−212	−225
Forest	Deeper drainage	9,666	7,133	7,726
Agriculture	Default	40,036	11,925	17,471
Agriculture	100% crop	45,860	13,666	20,071
Agriculture	100% grass	37,433	10,959	16,327

European peatlands. The reason is the intensive cropland and grassland use of most of the peatland area. Clearly, land use intensity disconnects the national GHG balance (Fig. 12.4) from the peat area (Fig. 12.1).

The results in Fig. 12.4 are associated with large uncertainties, which cannot be adequately quantified yet. One of these is the distribution of land use types, particularly in Russia, the largest European peat nation. Sensitivity tests of some crucial assumptions give a rough order of magnitude of possibly higher or lower typical GHG emissions. The reality is likely to lie in between the extremes. The average GHG flux rates for forestry indicate that the peat forests have been drained only slightly, if at all. However, the area data indicate the areas of drained forest where drainage is likely to be repeated periodically. A first sensitivity test (Table 12.1) uses an alternative GHG flux rate for "drainage" (Fig. 12.2). This alternative changes forestry from a small GHG sink to a significant GHG source and increases the overall peatland GHG balance in Europe by 25–50%. The fractional cover of cropland and grassland within agricultural use is highly uncertain. The default represents the best possible estimate by expert judgment. In two extreme scenarios, we assume (1) 100% cropland and (2) 100% grassland. Consequently, GHG emissions from agricultural peatland change by 15–20%. This alters the European peatland GHG estimate by 5% only (Table 12.1). The uncertainty in the European peatland GHG balance resulting from the combined uncertainty in land use fractions and typical GHG flux rates is estimated between −10% and +50% of the default estimate of 52 Gg Cequ.

12.3 Processes of GHG Sources and Sinks in Peatlands

The principal functioning of peatlands in their natural state is the accumulation of C, with the consequence of a long-term growing peat profile. C accumulation results from the gap between gross primary production (GPP), ecosystem respiration (Reco), emissions of CH_4 and losses via dissolved organic carbon (DOC) and dissolved inorganic carbon (DIC) and particulate organic carbon (POC). From the functional aspect of peat growth, the C balance is the key factor. Classically the long-term average C balance is assessed via ^{14}C dating of the profile together with the determination of the corresponding bulk density and the C content. Long-term average C uptake rates (LORCA) in northern peatlands are around 25 g C m^{-2} a^{-1}. However, for the assessment of the climatic relevance of the peatlands, the short-term GHG exchange is more important than the long-term C balance. Specifically in peatlands, not only the fluxes of CO_2 but also the fluxes of CH_4 and N_2O have to be assessed. All three gases together, multiplied with their individual GWP, determine the climatic forcing of the sites.

The dominating factor for the sensitive balance between uptake and release of C and as well for the exchange of the three different gases within the system is the water table. Undisturbed peatlands hold water tables close to the surface as the specific vegetation types together with the peat characteristics and a net surplus of water keep the water level high. Therefore, aerobic conditions in an undisturbed peat profile are limited to a shallow upper layer (acrotelm). Below the mean water table, in the catotelm, the consumption of oxygen creates anaerobic conditions. The thickness of the upper aerobic zone is of major importance for the gas fluxes:

1. The speed of respiration processes is an order of magnitude faster in aerobic than anaerobic conditions. The product of aerobic respiration in the acrotelm is CO_2, produced by soil respiration and autotrophic respiration of the below-ground plant parts. Anaerobic respiration by methanogenic bacteria, in contrast, takes place in the catotelm and leads to CH_4 formation. In an undisturbed peat-land, the decomposition is slower than the supply with fresh material through vegetation growth and decay. Therefore, undisturbed peatlands sequester C in the peat profile. However, if the water table drops down, two effects provoke rising respiration emissions: the zone in the peat profile for aerobic respiration is thicker with the consequence of rising soil respiration and the vegetation at the surface, which requires normally high water tables, starts to decay. Both effects together provoke rising Reco values, with the potential to shift the C sink to a C source.

2. CH_4 is formed in the anaerobic zone of the peat profile, specifically in the water-saturated rooting zone of the wetland plants. Here, fresh carbohydrates and H_2/CO_2 are available in the rhizosphere, and the lack of oxygen creates the conditions for the methanogenic bacteria to produce CH_4. However, the quantity of CH_4 which leaves the ecosystem to the atmosphere depends on several factors, among which the aerated zone seems to be crucial: CH_4 entering the acrotelm is oxidized by methanotrophic bacteria to CO_2. The efficiency of the oxidation of CH_4 is, therefore, mainly dependant on the thickness of the upper aerated layer (acrotelm). A mean water table of -10 cm seems to be a critical level for switching on or off the CH_4 emissions. This shallow aerated layer allows the methanotrophic bacteria to efficiently oxidize CH_4 to CO_2 (Nykänen et al. 1998; Christensen et al. 2003; Freibauer et al. in prep.). Together with diffusion, two other processes determine the amount of CH_4 released from the peat profile, that is, plant-mediated transport and ebullition. Plant-mediated transport takes place,

once helophytes with aerenchyma are present and the water table is higher than the rooting zone. Ebullition, that is, the formation and emission of bubbles, is predominantly a phenomenon of water-logged peatland sites and appears more stochastically than the other two emission pathways. It is difficult, however, to determine the relation of the different emission processes. But water table plays for all of them a vital role as controlling variable.

3. The water table close to the surface is, together with nutrients and pH values, the key factor for the development and the maintenance of the typical peat-forming vegetation types. Vegetation composition reacts sensitively to water level changes both by the shift of dominance structures of the different species as by the vitality of the vegetation layer.

12.3.1 Typical Situations with CO_2 Sources and Sinks

Natural peatlands are the only terrestrial ecosystems which act as continuous and long-term CO_2 sinks. Peat growth models (Clymo 1984) predict a decrease of uptake rates till an equilibrium state between decomposition and sequestration because with rising peat depth, decomposition takes place in rising volumes of the peat profile. However, recent studies of C exchange (Lafleur et al. 2003; Drösler 2005) showed that even old growth mires with peat deposits of up to 10 m depth and development times since the last glaciations demonstrate active uptake rates, which are higher than the LORCA. The reason for this discrepancy may be that the peat growth models underestimate the decrease of the decomposition rates with depth. Hence, gas exchange studies hint to the crucial role of the activity of the top layer (acrotelm). Here again, the water table is the dominating factor, both for the activity and for the vitality of the vegetation, as for the volume of peat where aerobic decomposition takes place. A dropping water table can be nonpermanent because of dry weather situations or permanent of human interference (drainage). Alm et al. (1999) showed that a peatland can lose in a single dry summer the amount of C stored in 5 years of average accumulation. So the sensitive balance between uptake and release was disturbed with the consequence of a part-time shift to a CO_2 source (see Fig. 12.5). However, if the vegetation is not damaged, the system can switch back to a sink in the following year with favorable climate conditions. Severely drained peatlands are typically permanent CO_2 sources because of the deep-aerated zone together with the shift or removal of the vegetation cover on top (see Fig. 12.5). Even if very active grasses or crops are grown after drainage, they cannot (normally) hold against the rising soil respiration, thus converting the system to a CO_2 source. Restoration via elevating the water table has short-term effects on the reduction of the soil respiration and mid- to long-term effects in creating the appropriate site conditions for the development of typical peatland vegetation types. Hence, almost immediately after the elevation of the water table, the CO_2 balance changes because of a decreasing heterotrophic respiration. However, the direction and magnitude of the CO_2 exchange depends principally on the GPP part of the fluxes (see Fig. 12.5). The time

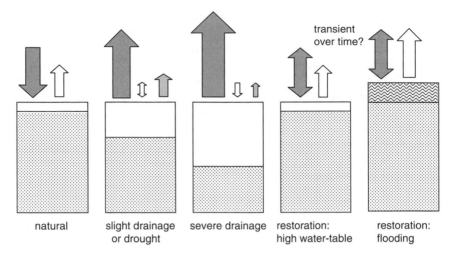

Fig. 12.5 Schematic comparison of flux directions and relative magnitudes of Net-CO_2 (dark grey), CH_4 (white) and N_2O (light grey) fluxes in peatlands. Dotted areas symbolize the water saturated part of the profile. Waves in last panel: flooded.

till a new equilibrium in CO_2 exchange depends on the development speed of the vegetation and can take several years to decades. This development is triggered by the water table (high or flooded), the type of start-vegetation before restoration and potential engineering practices, like the introduction of matrix or target species. Principally, restoration leads instantly to reduced CO_2 emissions (lower source strength) in comparison to drained sites. The time till the sink-function will be restored is up to now not generally predictable, as only very few studies are available on that topic. First results from fen peatlands in the Netherlands (Hendriks et al. 2007) confirmed a strong CO_2 sink (uptake of 270 g CO_2-C m^{-2} a^{-1}) 10 years after restoration of a peatland meadow. Similarly, restoration in the Donauried fen (southern Germany; Bergmann et al. in prep.) leads after 25 years to CO_2 sinks (uptake of 109–142 g CO_2-C m^{-2} a^{-1}). In a southern German bog, however, 12 years after restoration of former peat cut areas, the sites were still acting as CO_2 sources (release of 62–126 g CO_2-C m^{-2} a^{-1}; Drösler 2005).

12.3.2   Typical Situations when High CH₄ Emissions Occur

On the basis of the outlined processes for CH_4 formation and emission, Fig. 12.5 visualizes typical situations when high CH_4 emissions occur, in relation to the water table position. Natural sites as well as restored ones normally act as significant CH_4 sources. Most problematic, according to recent studies, seem to be flooded sites with emission rates up to 205 g CH_4-C in a restored northeastern German fen peatland (Augustin et al. in preparation.). Both the lack of an aerobic layer for CH_4 oxidation as the decay of the former grassland vegetation under anaerobic conditions lead to these huge CH_4 emissions. So restoration without flooding, oriented at the critical mean water level of around -10 cm provoke much lesser CH_4 emissions (see Fig. 12.5). In the southern German Donauried, for example, the difference of -1.5 versus -6.5 cm of water table in sites with a comparable cover of sedges (>90%) was reflected in significantly different CH_4 emissions of 77 versus 13 g CH_4-C m^{-2} a^{-1} (Freibauer et al. in prep.). Very important for the CH_4 emission seems to be as well the period of flooding, when temperature is rising in the year course and the plants are entering in the growing season. Here, rising soil temperatures and the supply of fresh substrates in the rhizosphere create favorable conditions for methanogenesis. NEE seems, therefore, to be a good predictor for CH_4 emission as exemplarily shown for NE Greenland (Joabsson and Christensen 2001) and southern Germany (Drösler 2005). But as the wetland plants with high NEE rates are normally sedges with aerenchyma, the combination of the by-pass effect with the fresh substrate supply may be responsible for the high CH_4 emission rates.

CH$_4$ uptake, however, is a phenomenon limited to drained peatland soils (see Fig. 12.5). In peatlands, the importance for the C balance of this small uptake pathway (limited to ~0.3 g CH_4-C; Jungkunst, unpub. data) is limited, because these sites are big emitters for CO_2, with rates several orders of magnitude higher than the uptake of CH_4.

12.3.3   Typical Situations when High N$_2$O Emissions Occur

N$_2$O is formed via nitrification and denitrification. Principally, bog peatlands show very small emissions because of their low-nutrient contents. Fens, however, can have significant N$_2$O emissions under conditions which are not too wet (natural or restored) and not too dry (severely drained sites). In case of N$_2$O, Fig. 12.5 shows therefore the behavior for a fen peatland.

12.4   Emission Patterns

12.4.1   Spatial Emission Patterns

Peatlands are hot spots for GHG emissions in the landscape. The European continental average GHG flux rate is 1 g C-equ m^{-2} year^{-1}. Russia has average GHG flux rates of 0.23 g C-equ m^{-2} year^{-1}, but Germany has 3.1 g C-equ m^{-2} year^{-1}. The estimated net CO_2 emissions (3.2–5.1 Tg C a^{-1}) from German peatlands under current land use compensate for the recent C sink (5.2 Tg C a^{-1}) in German forests (Benndorf pers. comm. 2005), while German Peatlands hold with 16.520 km^2 just 15% of the forest area.

At a smaller scale, hot spots can be identified within the peatlands, according to the spatial representation of the different driving factors. As shown before, the parameters water table, peat characteristics, and vegetation type and cover dominate the control over the GHG emissions in peatlands. These parameters are partly intercorrelated (like water table and vegetation type). However, the overlay of these parameters creates small-scale patterns of changing site conditions. This small-scale variability of site conditions is not only a phenomenon of disturbed peatlands. Even in natural peatlands, depending on the peatland type, microsite variability is high. Natural bogs, for example, show normally complex patterns of hummocks, lawns, and hollows, all of them colonized by a specific plant composition, sensitively reacting to the water table level and fluctuation. Therefore, Bubier et al. (1995) was successful to extrapolate CH_4 emission rates from the representation of distinct Sphagnum species as indicators for the mean water table.

In peatlands with a former land use history, the variability of the site factors can even be more pronounced. Drainage and especially peat cutting activities lead to severe changes in the site conditions, which create mostly sharp gradients and small-scale patterns of site factors. Site management like restoration via water table manipulation creates even more pronounced small-scale site differentiation, depending on the former microtopography and the magnitude of restoration works.

So the small-scale variability in peatlands, mostly driven by very small differences in the mean water table and the vegetation composition, creates an individual small-scale pattern of GHG exchange behavior at scales of square meter. Hot spots can be direct neighbors of balanced sites. The sites of the schematic Fig. 12.5 can all be represented within a small area of less than 1 ha. This implies the necessity to be able to distinctively measure the individual gas exchange of the different sites, applying techniques with sufficient spatial resolution.

12.4.2   Temporal Emission Patterns

In terms of temporal emission patterns, different sites show individual reactions to the driving variables. For the temporal variability over whole year cycles, Drösler (2005) showed, for example, in southern German bogs that the daily NEE of natural bogs is negative (uptake to the system) over almost the whole growing period, whereas degraded former peat cut sites where sinks just on single days over the same period and demonstrated much higher winter fluxes. Restored sites behave in between, with uptake to the system in spring but a shift back to sources during mid- and late-summer. These differences could be attributed to the dynamic of the water table and to the vegetation cover.

Temporal emission pattern in CH_4 is mostly reflecting the temperature dynamics, as shown, for example, in Bellisario et al. (1999). Therefore, these emissions follow generally a seasonal curve. However, the functional link between NEE and CH_4 emission hints to additional underlying processes, which at the natural sites go parallel with the temperature curve through the year. Here, the detection of the dominating factor is not possible in field studies.

Emission dynamics in N_2O, however, do not follow a simple seasonal temperature or radiation driven curve, for example, in CH_4 emissions or in Reco or GPP. Notable amounts of N_2O are leaving the system in occasions of frost and thaw or after fertilizer applications. Therefore, the modeling of these fluxes is not a simple topic. Reth et al. (2005) found in forest and grassland ecosystems the time since the last rain event as a critical parameter for the prediction of the emissions. The applicability in peatlands is still to be tested, in view of the importance of water table fluctuations for the N_2O emission rates.

As weather is changing from year to year, the interannual dynamics can be significant. This is especially true for CO_2 because NEE is just the small difference between the big terms of GPP and Reco. Lafleur et al. (2003) showed the span of yearly NEE sums for consecutive measurement years in a Canadian bog ecosystem. Single year studies should, therefore, be interpreted with caution in their absolute numbers, but can definitely allow for the comparison of the behavior of different sites, measured within the same year.

12.5   State-of-the-Art and Current Capabilities of GHG Measurements in Peatlands

C stocks in peatlands are very big in relation to annual changes of the C content. Therefore, for the assessment of the climatic relevance, the measurement of the GHG gas exchange is the only viable approach. Especially in peatlands, all three climatic relevant trace gases have to be monitored to get full GHG balances of the sites. Together with the above outlined topics of temporal and spatial variability, the requirements for the applied measurement techniques are

- high-spatial resolution to cope with small-scale mosaics of site conditions and hence spatial emission patterns (specifically, needed for the detection of hot spots and the influence of management upon the fluxes!),
- potential to be applied under all climate conditions for the development of full year balances,
- high temporal resolution to cover the sensitive times for the different gases, and
- in remote conditions (like peatlands normally are) independency from power line connection.

A single technique is not available, which can support all of the above mentioned requirements. Therefore, the combination of different techniques seems to be appropriate to fully cover the specific needs for GHG exchange measurements in peatlands. A general comparison of the characteristics of the different established methods can help to choose the best combination according to the specific needs in a given environment (see Table 12.2).

The combination of eddy covariance for CO_2 and manual chambers for CH_4 exchange monitoring is currently successfully applied, for example, in a restored peatland meadow in the Netherlands (Hendriks et al. 2007). Eddy covariance

Table 12.2 General comparative characteristics of greenhouse gas-measurement techniques

Characteristics	Methods		
	Eddy covariance	Automated chambers	Manual chambers
Undisturbed gas exchange	++	+/− (cross-checks!)	+/− (cross-checks!)
Integration over spatial variability	++	− (quantity of chambers vs. mosaic)	− (quantity of chambers vs. mosaic)
Direct measurement of small-scale spatial variability and management	−−	+	++
Tracking temporal variability	++	++	− (campaigns as basis for modeling)
Costs	−−	−−	++
Workload	++	+	−−
Performance under all climate conditions	+/−	+/−	++

provides the overall integrated signal of the different underlying small-scale ecosystems (dry land, saturated land, and ditches). With the chambers, the individual contribution of these subsystems for CH_4 emissions is assessed. Ebullition of CH_4 from water bodies, however, is difficult to measure with chambers, as these fluxes appear stochastically and may need a continuous measurement system to be detected realistically.

In typical peatland small-scale mosaics of sites, it is of interest, to differentiate the CO_2 exchange for the individual sites (combination of vegetation and management, water table, and peat characteristics) to better understand the underlying processes of NEE. Climate-controlled chambers can help out and allow, for example, to directly link CH_4 emissions to NEE, once the fluxes are measured at the same plot. Those climate-controlled transparent NEE chambers, for example, as used in Finland by Tuittila et al. (1999) or in Germany by Drösler (2005), can successfully be applied to support NEE models for entire year courses, once the parameterization of the models is done frequently (e.g., biweekly) under all climate conditions over the year. In peatlands, according to current topographical settings, very often stable boundary layers develop during nighttime. CO_2 concentrations can rise to 1000 ppm directly above the vegetation. Eddy covariance may have difficulties to realistically track the CO_2 exchange under these nonturbulent conditions. Automated chambers could be used for the direct high frequent measurement of NEE even under those conditions. However, up to now, no climate-controlled automated chambers are available, which would be the best match between climate-controlled manual chambers and eddy covariance systems.

12.6 Strategy Toward a Pan-European GHG (Continental Scale) Peatlands Monitoring

12.6.1 Observation Network

A continental-scale GHG monitoring must take the large, highly variable GHG fluxes on peatlands into account. Observation sites should include LORCA, the present ecosystem C balance and fluxes of all GHG species. In the following, minimum requirements for peatland observation sites are recommended.

For LORCA, at minimum peat depth, peat dry bulk density profile, C content profiles, and 14C at the peat bottom are needed. All of these measurements are scarce in Europe. Finland may be an exception with a long history in peat profiling by the Finish forestry service. The first three variables are particularly relevant as they establish the C stocks in the peatland.

The ecosystem C balance can only be established if all important (>1% of C flux) components of the C exchange in the peatland are observed, that is, CO_2, CH_4, C losses to the ground- and surfacewater via DOC, DIC, and POC, and C import and export by land use and management.

For measuring the GHG budget comprehensive, year-round, multiyear GHG flux measurements (CO_2, CH_4, and N_2O) are required. Frequent measurements in periods of high-flux variability and likely extreme bursts of CH_4 and N_2O are as critical as long-term measurements to cover the interannual variability and sensitivity of GHG fluxes to climate variability and trends. The thawing period is particularly important for CH_4 and N_2O while for CO_2, the entire nongrowing season can significantly contribute to the annual C balance. Each study site should comprise measurements in all relevant subunits (e.g., combination of individual vegetation types, water table situations, peat characteristics). A minimum core set of ancillary variables is needed for interpretation of the measurements and for scaling the site-level measurements to the European peatlands. Scaling usually relies on models based on relationships between climatic, edaphic, and biotic drivers and observed GHG fluxes. The most critical core variables are

- meteorological data, in particular, temperature (air and soil temperature profiles) and radiation,
- water table,
- vegetation type, fractional cover, and biomass of plant functional types (peat-forming mosses, other mosses, vascular plants: aerenchymal plants, others to be specified if appropriate),
- C/N ratio of peat,
- pH, electrical conductivity in soil water, and
- detailed documentation of land use history and present management.

12.6.2 Upscaling

First results of the combination of emission factors with land cover types underline the relevance of peatlands as hot spots for the European GHG budget. However, the assessment and monitoring of this GHG exchange needs major efforts for upscaling methodologies. Apart from site-level GHG flux data, ancillary scaling variables and suitable models, a major constraint to the European peatland GHG budget is set by the poor information about the geographic location and extent of peatlands, their properties (bog vs. fen, nutrient status, water table) and, in particular, their use. Peat resource data are often outdated as peatlands have been degraded so much that a significant fraction of the former peat area is turning to shallow peaty soils, which are, by definition, no peatlands any more. Detailed spatial information about dominant vegetation types, water levels, and management is a prerequisite for reducing the uncertainty in the European peatland budget to levels below 50%.

At present, data gaps about C stocks, GHG fluxes, spatial coverage, and management are most problematic for regions outside EU-15, where vast peatland areas have been subject to significant changes in land use after 1990. There is particular need for more data about GHG budgets of peatlands under particular land uses: (a) bog: grassland, cropland, land abandoned after peat cut, restoration, forest chronosequences, N_2O fluxes in general. (b) fen: abandoned after harvest, restoration, CO_2 fluxes in general.

Hot spots from small scales to regions and typical management regimes appear to be very important for overall European GHG emissions from peatlands. Large emissions isolated in time and space have, for example, been observed associated with organic agricultural soils for N_2O and with permafrost melting and subsequent changes in northern natural mires for CH_4. These hot spots will be crucial to capture in a pan-European GHG monitoring effort.

To support upscaling activities, there is a specific need to

- represent much better the GHG exchange of the span of different peatland ecosystems (measurements),
- update the land cover and land use information of the peatlands (remote sensing), and
- update the actual extent and depth of the peatlands (combination of remote sensing, ground radar, and inventories).

12.6.3 Future Directions for Enhanced Measurements

The very nature of peatlands with their intrinsic heterogeneity of vegetation and site factors, the sharp differences in the GHG exchange of natural versus disturbed sites and, due to topography and very high C stocks, the pronounced CO_2 accumulation in the boundary layer during the night need to be tackled in an enhanced measurement program. Existing techniques should be wisely combined to overcome individual deficits (see Table 12.2). However, automation of chamber measurements with climate control systems and remote control to minimize labor requirements and disturbance during measurements seems a prerequisite for a better operationalization of GHG measurements in peatlands. Measurements of the net C balance including C losses to the groundwater need to be included in the core set of observations. The methodology is established, but still laborious. For integration of site-level observations into a landscape-scale C budget the fate of DOC, DIC, and POC in the adjacent waters and streams needs to be traced.

The recommendations made above can be implemented stepwise toward a systematic, operational modeling/monitoring strategy as follows:

1. Define peat regions with similar sensitivity of GHG fluxes to climate and management, with similar site conditions [e.g., grouping at several levels (i) latitude and altitude, (ii) existing peatland classifications, (iii) site factors and vegetation, and (iv) management impact and land use history].
2. Identify regional hot spots with particular vulnerability to climate and land use changes, which may merit additional attention (e.g., permafrost regions, lowland areas with saltwater intrusion from sea level rise, areas with drastically changing water tables, for example, through restoration or drainage).
3. Identify most important land use types in terms of areas and hot spots (e.g., peat cutting, agriculture, forestry).
4. Set up core set of long-term observation sites in each peat region, each covering a range of subsites to represent the typical small-scale mosaic [here a screening is necessary if measurement sites, which are already included in international networks (Carboeurope-IP), can cover the range of necessary peatland core areas, and where regional (e.g., southern part of Central Europe; Eastern Europe) and thematic gaps (e.g., sea level rising, large-scale restoration) should be tackled].
5. Characterize spatially explicit peatland area, peat types, water tables, and land use (the major goal is to establish a homogeneous European-wide database, based on compatible classification schemes as basis for upscaling procedures; here even within individual countries, like Germany, the database differs between the different regions; this effort should be seen, both as a national and international task, to overcome inconsistencies in the estimates for whole Europe; see Byrne et al. 2004).
6. Improve models to represent the GHG exchange of all major peatland uses (models are especially needed to predict the exchange of all three gases under drastically changing conditions, like drought, sea level rise, restoration, etc., to assess the vulnerability of the peatland GHG exchange against external drivers).

12.7 Conclusion

Peatlands are hot spots for GHG emissions. Even in an industrialized country, like Germany, the estimated peatland GHG exchange accounts for 2.3–4.5% of anthropogenic emissions. Peatlands can, in contrast, be managed more wisely with the potential to contribute significantly to GHG mitigation. Raising the water table to more natural levels will reduce CO_2 losses and hence improve the net C balance of peatlands. Improper water management may, however, increase CH_4 emissions and reduce the overall GHG mitigation efficiency in some cases. The optimal effect can be determined by careful monitoring of CO_2 and CH_4.

To fully cover all major peatland regions, -types, and -ecosystems as well as the ongoing or future management practices, a specific monitoring program is necessary. The key requirements of such a program are outlined above. This program should be geographically differentiated according to the peatland regions to fully cover the ongoing driving forces (e.g., permafrost melting, rising sea level, peat cutting, afforestation, restoration, conservation) in EU peatlands, with impact on the GHG exchange. In-depth assessments of the vulnerability, the hot spot activity, and future management strategies of EU peatlands need a pan-European monitoring program of GHG exchange in peatlands.

References

Alm, J., Schulman, L., Walden, J., Nykanen, H., Martikainen P.J., Silvola, J., 1999. Carbon balance of a boreal bog during a year within an exceptionally dry summer. *Ecology*, 80(1), 161–174.

Augustin et al., in preparation. (pers. com.).

Bellisario, L.M., Bubier, J.L., Moore, T.R., Chanton, J.P., 1999. Controls on CH_4 emissions from a northern peatland. *Global Biogeochemical Cycles*, 13(1), 81–91.

Bergmann, L., Drösler, M., Freibauer, A., Schultz, R., Höll, B., Jungunst, H., Fiedler, S., in prep. Net ecosystem exchange of restored and agriculturally used fen-peatlands, Donauried, southern Germany.

Bubier, J.L., Moore, T., Juggins, S., 1995. Predicting methane emission from bryophyte distribution in northern Canadian peatlands. *Ecology*, 76(3), 677–693.

Byrne, K.A., Chojnicki, B., Christensen, T.R., Drösler, M., Freibauer, A., Friborg, T., et al., 2004. EU Peatlands: Current Carbon Stocks and Trace Gas Fluxes, ISSN 1723-2236. http://gaia.agraria.unitus.it/ceuroghg/ReportSS4.pdf, Viterbo, Italy.

Christensen, T.R., Ekberg, A., Strom, L., Mastepanov, M., Panikov, N., Oquist, M., Svensson, B. H., Nykanen, H., Martikainen, P.J., Oskarsson, H., 2003. Factors controlling large scale variations in methane emissions from wetlands. *Geophysical Research Letter*, 30, 1414.

Clymo, R.S., 1984. The limits to peat bog growth. *Philosophical Transactions of the Royal Society of London, Series B*, 303, 605–654.

Drösler, M., 2005. Trace gas exchange and climatic relevance of bog ecosystems, Southern Germany, Dissertation TUM, 179 pp. Online publication: urn:nbn:de:bvb:91-diss20050901-1249431017.

Freibauer, A., Höll, B.S., Drösler, M., Fiedler, S., Jungkunst, H., in prep. Methane and nitrous oxide budgets in drained and restored calcaric fens.

Frolking, S., Roulet, N., Fuglestvedt, J., 2006. How northern peatlands influence the Earth's radiative budget: Sustained methane emission versus sustained carbon sequestration. *Journal of Geophysical Research*, 111, G01008, doi:10.1029/2005JG000091.

Gorham, E., 1991. Northern peatlands: Role in the carbon cycle and probable responses to climatic warming. *Ecological Applications*, 1, 182–195.

Hendriks, D.M.D., van Huissteden, J., Dolman, A.J., van der Molen, M.K., 2007. Biogeosciences, 277–316. SRef-ID: 1810-6285/bgd/2007-4-277.

IPCC, 1996. Global Change 1995—The Science of Climate Change. Contribution of Working Group I to the Second Assessment Report of the Intergovernmental Panel on Climate Change.

Joabsson, A., Christensen, T.R., 2001. Methane emissions from wetlands and their relationship with vascular plants: An Arctic example. *Global Change Biology*, 7, 919–932.

Joosten, H., Clarke, D. (eds.), 2002. Wise use of mires and peatlands, background and principles including a framework for decision-making, pp. 1–304.

Kasimir Klemedtsson, Å., Klemedtsson, L., Berglund, K., Martikainen, P.-J., Silvola, J., Oenema, O., 1997. Greenhouse gas emissions from farmed organic soils: A review. *Soil Use and Management*, 13, 245–250.

Lappalainen, E. (ed.), 1996. Global Peat Resources, International Peat Society and Geological Survey of Finland, pp. 57–162.

Lafleur, P.M., Roulet, N.T., Bubier, J.L., Frolking, S., Moore, T.R., 2003. Interannual variability in the peatland-atmosphere carbon dioxide exchange at an ombrotrophic bog. *Global Biogeochemical Cycles*, 17(2), 5.1–5.13, 1036, doi:10.1029/2002GB001983.

Minkkinen, K., Laine, J., 1998. Long-term effect of forest drainage on peat carbon stores of pine mires in Finland. *Canadian Journal of Forest Research*, 28, 1267–1275.

Mäkilä, M., 1994. Suon energiasisällön laskeminen turpeen ominaisuuksien avulla. Summary: Calculation of the energy content of mires on the basis of peat properties. Report of Investigation 121, Geological Survey of Finland: 1–73.

Nilsson, M., Mikkela, C., Sundh, I., Granberg, G., Svensson, B. H., Ranneby, B., 2001. Methane emission from Swedish mires: National and regional budgets and dependence on mire vegetation. *Journal of Geophysical Research-Atmospheres*, 106, 20847–20860.

Nykänen, H., Alm, J., Silvola, J., Tolonen, K., Martikainen, P.J., 1998. Methane flux on boreal peatlands of different of different fertility and the effect of long-term experimental lowering of the water table on flux rates. *Global Biogeochemical Cycles*, 12, 53–69.

Reth, S., Hentschel, K., Drösler, M., Falge, E., 2005. DenNit—Experimental analysis and modelling of soil N_2O efflux in response on changes of soil water content, soil temperature, soil pH, nutrient availability and the time after rain event. *Plant-and-Soil*, 272(1–2), 349–363.

Sundh, I., Nilsson, M., Mikkelä, C., Granberg, G., Svensson, B.H., 2000. Fluxes of methane and carbon dioxide on peat-mining areas in Sweden. *Ambio*, 29(8), 499–503.

Tuittila, E.S., Komulainen, V.M., Vasander, H., Laine, J., 1999. Restored cut-away peatland as a sink for atmospheric CO_2. *Oecologia*, 120, 563–574.

World Energy Council, 2001. 2001 Survey of Energy Resources. http:// www.worldenergy.org/ wec-geis/publications/ reports/ ser/ peat/ peat.

Chapter 13
Towards a Full Accounting of the Greenhouse Gas Balance of European Grasslands

Jean-François Soussana

13.1 Introduction

Pastures and livestock production systems are extremely diverse. They occur in a large range of climate and soil conditions and range from very extensive pastoral systems, where domestic herbivores graze and browse rangelands, to intensive systems based on forage and grain crops, where animals are mostly kept indoors. On a global scale, livestock use 3.4 billion ha of grazing land, in addition to animal feed produced on about a quarter of the land under crops (Delgado 2005). Grasslands and pastures contribute to the livelihoods of over 800 million people including many poor smallholders (Reynolds et al. 2005). By 2020, this agricultural sub-sector will produce about 30% of the value of global agricultural output (Delgado 2005).

Pastures include both grasslands and rangelands. Grasslands are the natural climax vegetation in areas (e.g. the Steppes of central Asia and the prairies of North America) where the rainfall is low enough to prevent the growth of forests. In other areas, where rainfall is normally higher, grasslands do not form the climax vegetation (e.g. north-western and central Europe) and are more productive. Rangelands are characterised by low stature vegetation, due to temperature and moisture restrictions, and found on every continent. Worldwide the soil organic C sequestration potential is estimated to be 0.01–0.3 GtC year^{-1} on 3.7 billion ha of permanent pasture (Lal 2004). Thus, soil organic C sequestration by the world's permanent pastures could potentially offset up to 4% of the global greenhouse gas (GHG) emissions.

According to remote sensing data (CORINE LandCover and PELCOM; CORINE 1995, 2000), grasslands cover 20% of the area of the European continent and distribute about equally between western Europe (80 Mha) and eastern Europe (60 Mha). Within the area managed by agricultural practices on the European continent, 37% is devoted to grasslands (EEA 2005). Following the new membership of twelve countries, the EU 27 has an enlarged area of grassland of about 20 million ha (Carlier et al. 2004).

The GHG budget of European grasslands is still highly uncertain. There are only few continental scale modelling estimates of the GHG budget of grasslands, primarily focused on the CO_2 component of the GHG budget. Vleeshouwers and Verhagen (2002), further quoted by Janssens et al. (2003), applied a semi-empirical model of land use induced soil carbon disturbances to the European continent (as far east as the Urals) and inferred a carbon sink of 101 GtC year^{-1} over grasslands (0.52 tC ha^{-1} year^{-1}) with uncertainties above the mean.

Managed European grasslands are often fertilised to sustain productivity and thus emit N_2O to the atmosphere above the background level that is found in natural systems (Flechard et al. 2007). Typical N_2O emissions from grassland soils, converted into CO_2 equivalent sources on a 100 years time horizon (Bouwman 1996), range between 0.1 and 1 teqC ha^{-1} year^{-1} (Machefert et al. 2002; Sozanska et al. 2002). One recent estimate of N_2O fluxes from grasslands indicates a mean emission of 2.0 kg N_2O–N ha^{-1} year^{-1} in 2000, which translates into 0.25 tCO$_2$–C equivalent ha^{-1} year^{-1} (Freibauer et al. 2004).

European grasslands sustain an important number of domestic herbivores, 150 millions of cows and 150 millions of sheep, roughly 15% of the global animal population (FAO 2004). Grazers impact the cycling of C and N within pastures via defoliation, excretal returns and mechanical disturbance and emit CO_2 via their metabolic activity. Ruminants also emit CH_4 with range for annual emission from 0.05 to 0.25 tCH$_4$ (0.3–1.5 teqC, Vermorel 1995).

Currently, the net global warming potential (GWP, in terms of CO_2 equivalent) from the GHG exchanges with European grasslands is not known. It is clear that an integrated approach, that would allow to quantify the fluxes from all three radiatively active trace gases (CO_2, CH_4, N_2O), would be desirable. Management choices to reduce emissions involve important trade-offs: for example, preserving grasslands and adapting their management to improve carbon sequestration in the soil may actually increase N_2O and CH_4 emissions.

The Marrakech Accords, resulting from the 7th Conference of Parties (COP7) to the 1992 United Nations Framework Convention on Climate Change (UNFCCC), allow biospheric carbon sinks (and sources) to be included in attempts to meet Quantified Emission Limitation or Reduction Commitments (QELRCs) for the first commitment period (2008–2012) outlined in the Kyoto Protocol (available at: www.unfccc.de). Under Article 3.4, the following activities are included: forest management, cropland management, grazing land management and re-vegetation. Soil carbon sinks (and sources) can therefore be included under these activities. Further, direct emission reductions of the GHGs N_2O and CH_4 will help parties to meet QELRCs.

Since agricultural management is one of the key drivers of the sequestration and emission processes, there is a potential within the EU to reduce the net GHG flux for grasslands, expressed in CO_2 equivalents. However, it is essential that effects of land management of all three GHGs are evaluated concomitantly.

Until recently, there were very few direct measurements of the GHG fluxes exchanged with grasslands with a sufficiently long-term continuity. A network of nine sites was established as part of the GreenGrass project starting in 2002 (European Commission DG Research Vth Framework Programme—Contract no.

EVK2-CT2001-00105). Other flux sites were established within the 'CarboMont' (Cernusca 2004) and 'CarboEurope IP' (CarboEurope 2003) projects.

Here we review the characteristics which would allow monitoring the GHG balance of European grasslands with sufficient accuracy. We first summarise the main features of the GHG balance in a grassland ecosystem. We then review recent results concerning the GHG balance at three complementary spatial scales: the field, the farm and the region. We finally discuss how to scale up by bottom-up modelling and/or inventories.

13.2 The C Cycle and Non-CO$_2$ Greenhouse Gas Emissions in Grassland Ecosystems

13.2.1 Carbon Cycling in Grasslands

For grasslands, the nature, frequency and intensity of disturbance play a key role in the C balance. In a cutting regime, a large part of the primary production is exported from the plot as hay or silage, but part of these C exports may be compensated for by farm manure and slurry application. Under intensive grazing, up to 60% of the above-ground dry matter production is ingested by domestic herbivores (Lemaire and Chapman 1996). However, this percentage can be much lower during extensive grazing. The largest part of the ingested carbon is digestible (up to 75% for highly digestible forages) and, hence, is respired shortly after intake. Only a small fraction of the ingested carbon is accumulated in the body of domestic herbivores or is exported as milk (Vermorel 1995).

Additional carbon losses (ca 5% of the digestible carbon) occur through methane emissions from the enteric fermentation. The non-digestible carbon (25–40% of the intake according to the digestibility of the grazed herbage) is returned to the pasture in excreta (mainly as faeces). In most European husbandry systems, the herbage digestibility tends to be maximised by agricultural practices such as frequent grazing and use of highly digestible forage cultivars. Consequently, the primary factor which modifies the carbon flux returned to the soil by excreta is the grazing pressure which varies with the annual stocking rate (mean number of livestock units per unit area). Secondary effects of grazing on the carbon cycle of a pasture include: (1) the role of excretal returns, concentrated in patches, for nutrients cycling, which could increase primary production up to a moderate rate of grazing intensity especially in nutrient-poor grasslands (De Mazancourt et al. 1998) and (2) the role of defoliation by animals and of treading both of which reduce the leaf area.

Organic matter is partly incorporated in grassland soils through rhizodeposition (Jones and Donnelly 2004). This process favours carbon storage (Balesdent and Balabane 1996), because direct incorporation into the soil matrix allows a high degree of physical stabilisation of the soil organic matter. Soil carbon stocks display a high spatial variability (coefficient of variation of 50%, Cannell et al. 1999) in grassland as compared to arable land and ca 15% of this variability comes from sampling to different depths (Robles and Burke 1998; Chevallier et al. 2000; Bird et al. 2002).

The Intergovernmental Panel on Climate Change (IPCC) report dealing with this issue (IPCC 2004) provides estimates for the grassland management on SOC storage based on a literature review and meta-analysis of grassland studies (Conant et al. 2001). According to those findings, land use change from grassland to crop-land systems causes losses of SOC in temperate regions ranging from 18% (±4) in dry climates to 29% (±4) in moist climates. Converting cropland back to grassland uses for 20 years was found to restore 18% (±7) of the native carbon stocks in moist climates (relative to the 29% loss due to long-term cultivation) and 7% (±5) of native stocks in temperate dry climates.

Based on the IPCC method for classifying management systems (IPCC 2004), grassland practices are categorised as improved (e.g. sowing legumes, irrigation, fertilisation and planting more productive forage species) or degraded (e.g. overgrazing and planting less productive species relative to native vegetation). Grasslands that are degraded for 20 years typically have 5% (±6) less carbon than native systems in tropical regions and 3% (±5) less carbon in temperate regions. Improving grasslands with a single practice caused a relatively large gain in SOC over 20 years, estimated as 14% (±6) in temperate regions and 17% (±5) in tropical regions, while having an additional improvement led to another 11% (±5) increase in SOC (Ogle et al. 2004).

In Europe, most soils are out of equilibrium as they have been affected by a number of changes in land use/land management practices during the past 100 years. Management practices affecting soil organic C stocks in grassland areas include changes between arable and grassland, grassland and forest, grassland man-agement such as tillage (sown grasslands), grazing and cutting management, inor-ganic and organic fertiliser use, legumes, the type of fertiliser applied and water management.

As a result of periodic tillage and resowing, short duration grasslands tend to have a potential for soil carbon storage intermediate between crops and permanent grassland. Part of the additional carbon stored in the soil during the grassland phase is released when the grassland is ploughed up. The mean carbon storage increases in line with prolonging the lifespan of covers, that is less frequent ploughing (Soussana et al. 2004a).

In France, average soil (0–30 cm) organic C stocks are lower than 45 tC ha^{-1} for arable land and reach nearly 70 tC ha^{-1} for land under permanent grassland and forests (excluding litter). Based on this inventory, the average difference between cropland and pastures is ca 25 tC ha^{-1}. This value represents the range of carbon change involved in land use conversions. Changes in soil carbon through time are non-linear after a change in land use or in grassland management. A simple two parameters model has been used to estimate the magnitude of the soil carbon stock change after a change in grassland management (INRA 2002; Soussana et al. 2004b).

13.2.2 N_2O Emissions from Soils

Biogenic emissions of N_2O from soils result primarily from the microbial processes nitrification and denitrification. N_2O is a by-product of nitrification and an intermediate during denitrification. Nitrification is the aerobic microbial oxidation of ammonium to nitrate and denitrification is the anaerobic microbial reduction of nitrate through nitrite, nitric oxide (NO) and N_2O to N_2. Nitrous oxide is a gaseous product that may be released from both processes to the soil atmosphere.

Major environmental regulators of these processes are temperature, pH, soil moisture (i.e. oxygen availability) and carbon availability (Velthof and Oenema 1997). In most agricultural soils, biogenic formation of N_2O is enhanced by an increase in available mineral nitrogen, which in turn increases nitrification and denitrification rates. Hence, in general, addition of fertiliser N or manures and wastes containing inorganic or readily mineralisable N will stimulate N_2O emission as modified by soil conditions at the time of application. N_2O losses under anaerobic conditions are usually considered more important than nitrification-N_2O losses under aerobic conditions. At a global scale, soils account for 65% of N_2O emissions (IPCC 1996b). For given soil and climate conditions, N_2O emissions are likely to scale with the nitrogen fertiliser inputs. Therefore, the current IPCC (1996a) methodology assumes a default emission factor (EF_1) of 1.25% (range 0.25–2.25%) for non-tropical soils emitted as N_2O per unit nitrogen input N (0.0025–0.0225 kg N_2O–N kg^{-1} N input).

N_2O emissions in soils usually occur in 'hot spots' associated with urine spots and particles of residues and fertiliser, despite the diffuse spreading of fertilisers and manure (Flechard et al. 2007). Nitrous oxide emissions from grasslands tend also to occur in short-lived bursts following the application of fertilisers (Leahy et al. 2004; Clayton et al. 1997). Temporal and spatial variations contribute large sources of uncertainty in N_2O fluxes at the field and annual scales (Flechard et al. 2005). The overall uncertainty in annual flux estimates derived from chamber measurements may be as high as 50% due to the temporal and spatial variability in fluxes, which warrants the future use of continuous measurements, if possible at the field scale (Flechard et al. 2007). In the same study, annual emission factors for fertilized systems were highly variable, but the mean emission factor (0.75%) was substantially lower than the IPCC default value of 1.25% (Flechard et al. 2007).

13.2.3 CH_4 Fluxes Exchanged with Soils and Vegetation

In soils, methane is formed under anaerobic conditions at the end of the reduction chain when all other electron acceptors, such as, for example nitrate and sulphate, have been used. Methane emissions from freely drained grassland soils are, therefore, negligible. In fact, aerobic grassland soils tend to oxidise methane at a larger rate than cropland soil (6 and 3 kg CH_4 ha^{-1} year^{-1}, respectively), but less so than uncultivated soils (Boeckx and Van Cleemput 2001). For drained grasslands, methane oxidation was estimated between 0.1 and 1.1 kg CH_4 ha^{-1} year^{-1} (Van Den Pol-Van Dasselaar 1998). In contrast, in wet grasslands as in wetlands, the development of anaerobic conditions in soils may lead to methane emissions.

Keppler et al. (2006) have shown the emissions of low amounts of CH_4 by terrestrial plants under aerobic conditions. However, this claim has not been confirmed and could be due to an experimental artefact (Dueck et al. 2007). In an abandoned peat meadow, methane emissions were lower in water unsaturated compared to water saturated soil conditions (Hendriks et al. 2007).

13.2.4 CH_4 Fluxes from Enteric Fermentation

Enteric emissions in the world and the EU-15 have been estimated to 60–80 and 7–10 Gt year^{-1} respectively, and contribute around 18% and 29% of total methane emissions, respectively. The enteric methane plays a greater part to total methane emissions in EU than in the world (29% vs. 18%). The direct contribution of enteric methane to the total greenhouse effect has been estimated to 2–3%. Enteric methane emissions tended to stabilise during the last decade at the world level, or even decreased in EU-15. Such a decline within the EU is explained by a reduction of the number of animals and intensification of animal production following the reform of the European CAP. Ruminants are the major methane producers since they account for 95% of the total animal enteric methane emissions. The emissions of methane by ruminants contribute between 16% and 23% of the global emissions of this gas (IPCC 1996b).

The emissions of methane by ruminants are due to the fermentative reactions in the digestive tract, called enteric fermentation. The rumen is a large anaerobic fermentative chamber located at the beginning of the digestive tract of ruminants, which contributes 70% of total organic matter digestion. Enzymes involved in ruminal digestion are solely of microbial origin. Every day a cow produces 300–700 l of methane. The annual emissions of CH_4 originating from enteric fermentation are typically between 80 and 100 kg per animal per year for dairy cattle in Europe (IPCC 1996a), leading to annual emissions equivalent to 0.67–0.84 tC per animal as CO_2 equivalent.

The CH_4 emissions by cattle depend upon the type, age and weight of the animal and the quantity and quality of the feed consumed. Methane emission is positively correlated to the amount of fermented organic matter in the rumen and the intake of digestible energy (Blaxter and Clapperton 1965). It increases with the amount of feed intake but, for the same diet, the proportion of gross energy lost in methane decreases with intake level. Large variations exist in methanogenic power of ingredients and diets. Methane production is closely related to the amount of dietary digestible cellulose content (Pinares-Patino et al. 2003), whereas it decreases with addition of concentrate at a level higher than 30% in the diet (Giger-Reverdin et al. 2000).

Under grazing conditions, most of the variability in the enteric methane production of grassland plots lies in the number of animals, and, therefore, the emissions per unit land area will primarily vary with the stocking rate (Pinares-Patino et al. 2007, Soussana et al. 2007). The methane emission rate per unit liveweight (LW) varies also markedly at grazing between different animal types. This rate, measured with the SF6 dual tracer technique (Johnson et al. 1994), was comprised between 0.33 and 0.45 g CH_4 kg^{-1} LW day^{-1} for heifers and bulls and reached 0.68–0.97 g CH_4 kg^{-1} LW day^{-1} for lactating cows (LE) (Soussana et al. 2007).

The impact of belched CO_2 (digestive + metabolic CO_2) on warming effect has not been considered until now as significant. Using the SF6 method and taking into account the GWP of each gas and their quantitative emissions, the warming CO_2/CH_4 ratio for various animal types has been calculated at grazing. It varies from 1.1 for low-producing dairy cows (Pinares-Patino et al. 2007) to 1.8 for dry cattle. CO_2 emission from cattle has not been taken into account by the IPCC because it is 'short-cycling' carbon, which has been fixed by the plants earlier and has thus no effect on atmospheric concentrations. However, it should be noted that importing animal feed leads to trans-boundary fluxes of carbon and to additional CO_2 emissions in the importing country.

13.2.5 Emissions During Housing: Manures and Livestock Wastes

On-farm emissions from animals and manure must be taken into account when the GHG mitigation potential of grassland management strategies involving grazing is evaluated. GHG emissions from manure management include direct emissions of CH_4 and N_2O, as well as indirect emissions of N_2O derived from NH_3/NO_x. Quantification of GHG emissions from manure are typically based on national statistics for manure production and housing systems combined with emission factors which have been defined by the IPCC or nationally. The quality of GHG inventories for manure management is critically dependent on the applicability of these emission factors.

Animal manure is collected as solid manure and urine, as liquid manure (slurry) or as deep litter, or it is deposited outside in drylots or on pastures. These manure categories represent very different potentials for GHG emissions, as also reflected in the methane conversion factors and nitrous oxide emission factors, respectively. However, even within each category, the variations in manure composition and storage conditions can lead to highly variable emissions in practice. This variability is a major source of error in the quantification of the GHG balance for a system. To the extent that such variability is influenced by management and/or local climatic conditions, it may be possible to improve the procedures for estimating CH_4 and N_2O emissions from manure (Sommer et al. 2004).

Excreta (dung and urine) deposited during grazing influences fluxes of CH_4, N_2O and NH_3 from the pasture. In particular, urine patches are important point sources of NH_3 and N_2O, whereas the N input may locally reduce CH_4 oxidation activity. Ammonia losses from pastures are not specifically represented in the IPCC methodology, which calculates NH_3 volatilisation as a fixed proportion of total N excreted. However, ammonia losses from excretal returns to the pasture increase with N surplus in the diet since this N is mainly excreted as urea in the urine.

13.2.6 Greenhouse Gas Balance

When assessing the impact of land use and land use change on GHG emissions, it is important to consider the impacts on all GHGs (Robertson et al. 2000; Smith et al. 2001). N_2O and CH_4 emissions are often expressed in terms of CO_2 or CO_2–C equivalents, which is possible because the radiative forcing of nitrous oxide, methane and carbon dioxide, can be integrated over different timescales and compared to that for CO_2.

For example, over the 100-year timescale, 1 unit of nitrous oxide has the same GWP as 310 units of carbon dioxide, whereas, on a kilogram for kilogram basis, 1 unit of methane has the same GWP as 21 units of carbon dioxide (IPCC 2001a). An integrated approach is needed to quantify in CO_2–C equivalents the fluxes of all three trace gases (CO_2, CH_4 and N_2O).

13.3 Carbon and Greenhouse Gas Balance of Grasslands in Europe

13.3.1 Methodological Issues

CO_2 fluxes exchanged by grasslands were initially measured in enclosure systems (e.g. Paustian et al. 1990; Casella and Soussana 1997; Aeschlimann et al. 2005). However, given the large spatial variability of the vegetation and the microclimatic artefacts created by enclosures, the results from such experiments were difficult to extrapolate at the field scale. With the advancement of micrometeorological studies of the ecosystem-scale CO_2 exchange (Baldocchi et al. 1996; Baldocchi and Meyers 1998), eddy flux covariance measurement techniques have been applied to grassland and rangelands. In Europe, the proportion of grassland sites in the flux tower network now roughly matches the proportion of the land surface covered by grasslands (CarboEurope IP 2003).

Eddy covariance measurements integrate fluxes over a large area and are therefore very relevant for estimating net ecosystem exchange (NEE) over heterogeneous covers such as pastures. Moreover, since the measurement uses a free air technique, as opposed to enclosures, there is no disturbance of the measured area which can be freely accessed by herbivores. However, eddy covariance techniques may encounter some specific problems when used in grasslands and rangelands.

In Europe most managed grasslands are divided at the landscape scale in a large number of relatively small (often <1 ha) paddocks, each paddock being managed differently by farmers, through grazing, cutting and fertiliser applications. Footprint models are required to separate the fluxes among the paddocks which adds to the uncertainties involved in the calculation of the annual NEE at this scale. By contrast, the large paddock size usually found in rangelands and extensive grasslands is better suited to eddy covariance measurements.

Large domestic herbivores form mobile point sources of CO_2. As long as the herbivores are within the footprint of the sensors, the belched CO_2 will be rapidly diluted in the background air and transported by turbulence to the sensors. However, when grazers are very close to the sensors, they tend to create CO_2 spikes that may be rejected when filtering the data. Therefore, in order to account for the herbivore's CO_2 emissions at grazing two conditions need to be fulfilled: a random distribution of the herbivores within the measured paddock and a minimum distance (a few metres or more) between the herbivores and the sensors.

13.3.2 Net Ecosystem Exchange and Atmospheric CO_2 Balance of European Grasslands

Gilmanov et al. (2007) have analysed tower CO_2 flux measurements from 20 European grasslands covering a wide range of environmental and management conditions. Annual net ecosystem CO_2 exchange (NEE) varies from significant net uptake (>2400 g CO_2 m^{-2} year^{-1}) to significant release (<−600 g CO_2 m^{-2} year^{-1}), though in 15 of 19 cases grasslands performed as net CO_2 sinks. The carbon source was associated with organic rich soils, grazing and heat stress. Maxima of gross primary productivity (GPP) and ecosystem respiration (RE) were found in intensively managed grasslands of Atlantic climate. Extensively used semi-natural grasslands of southern and central Europe had much lower production, respiration and light-use efficiency, while temperate and mountain grasslands of central Europe ranged between these two extremes. Gross primary production (GPP) of European grasslands ranges from 1,700 g CO_2 m^{-2} year^{-1} in dry semi-natural pastures to 6,900 g CO_2 m^{-2} year^{-1} in intensively managed Atlantic grasslands. Ecosystem respiration was in the range 1,800–6,000 g CO_2 m^{-2} year^{-1} (Gilmanov et al. 2007).

Janssens et al. (2003) concluded that European grasslands may constitute a net C sink (−60 ± 80 g C m^{-2} year^{-1}), although the uncertainty surrounding this estimate was larger than the sink itself. On average of nine grassland sites, Soussana et al. (2007) have shown that grassland NEE displayed a significant atmospheric ($n = 9$, sign test, $p < 0.01$) sink activity for CO_2. The magnitude of the sink was higher than the mean estimate by Janssens et al. (2003). With an extended data set covering 20 European grassland sites, Gilmanov et al. (2007) show that 4 sites became C sources in some years, 2 of them during drought events and 2 of them with a significant peat horizon. These findings for European grasslands confirm earlier estimates for North America (Follett 2001) that these ecosystems predominantly act as a sink for atmospheric CO_2.

Nevertheless, as shown by Ciais et al. (2005), the CO_2 sink activity can turn to a source during summer heat waves, which affect both gross photosynthesis and total ecosystem respiration. Gilmanov et al. (2005) have also shown that a source type of activity is not an exception for the mixed prairie ecosystems in North America, especially during years with lower than normal precipitation.

13.3.3 Net Biome Productivity

In grazed-only systems (not supplied with manures), NEE is a good proxy of net C storage. Plant biomass is digested on site by the herbivore and this process contributes to the total ecosystem respiration that can be analysed from eddy covariance data (Gilmanov et al. 2007). By contrast, in cut grasslands, biomass is exported off site and neither this carbon export nor the import of carbon from organic fertilisers is detected by the atmospheric budget. Therefore, accounting for exports and imports of organic carbon is essential to compare cut and grazed grasslands in terms of their net carbon storage (net biome productivity, NBP) (Yazaki et al. 2004) (Fig. 13.1). Numerous studies should be reanalysed following this precept (Novick et al. 2004; Rogiers et al. 2005).

In order to calculate NBP, Soussana et al. (2007) have adapted the definition of Chapin et al. (2002) to a managed grassland:

$$NBP = NEE - F_{import} + F_{harvest} + F_{CH_4} + F_{LW} + F_{leach} \qquad (13.1)$$

where $F_{harvest}$ is the C lost from the system through plant biomass export (mowing), F_{import} the flux of C entering the system through manure and/or slurry application, F_{CH4} the C lost through CH_4 emissions by grazing cattle, F_{LW} the C lost from the system through animal body mass increase and milk production and F_{leach} the C lost through dissolved organic/inorganic C leaching.

Within the GreenGrass project, the full GHG balance of nine contrasted grassland sites covering a major climatic gradient over Europe was measured during 2 complete years (Soussana et al. 2007). The sites include a wide range of management regimes (rotational grazing, continuous grazing and mowing), the three main types of managed grasslands across Europe (sown, intensive permanent and semi-natural grassland) and contrasted nitrogen fertiliser supplies. At all sites, the NEE of CO_2 was assessed using the eddy covariance technique. N_2O emissions were monitored using various techniques (GC-cuvette systems, automated chambers and tunable diode laser) and CH_4 emissions resulting from enteric fermentation of the grazing cattle were measured in situ at four sites using the SF_6 tracer method.

Averaged over the two measurement years, NEE results show that the nine grassland plots displayed a net sink for atmospheric CO_2 of 240 ± 70 g C m^{-2} year^{-1} (mean $\pm$ confidence interval at $p > 0.95$) (Soussana et al. 2007). The average C storage (NBP) was estimated at -104 ± 73 g C m^{-2} year^{-1}, that is 45% of the atmospheric CO_2 sink. This discrepancy can be explained, since more organic C was harvested

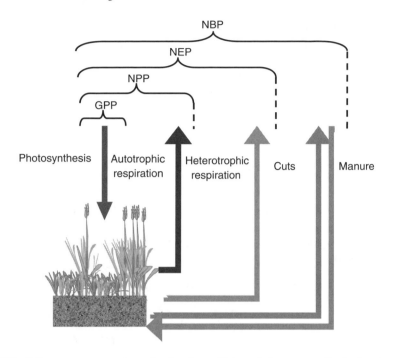

Fig. 13.1 C fluxes in a managed grassland system with imports (manure application) and exports (cuts and harvests) of organic carbon. GPP, gross primary productivity; NPP, net primary productivity; NEP, net ecosystem productivity; NBP, net biome productivity (*See Color Plate* 5).

from, than returned to, the grassland plots (Soussana et al. 2007). Such results show that the balance of horizontal fluxes of organic carbon (exports by mowing and imports by manure spreading) needs to be accurately measured in eddy flux sites in order to be able to estimate the plot scale carbon storage in managed grasslands.

Quite often not all components of the NBP budget are measured. For example, dissolved organic/inorganic C (DOC/DIC) losses as well as C exports in milk and meat products are sometimes neglected. Siemens and Janssens (2003) have estimated at the European scale the average DOC/DIC loss at $11 \pm 8\,g\ m^{-2}\ year^{-1}$. Assuming a value at the upper range of this estimate would reduce the grassland NBP by 20%. In contrast, the role of organic C exports is relatively small with meat production systems (e.g. 1.6% of NBP, Allard et al. 2007) but can be higher with intensive dairy production systems.

The selection of sites is crucial for any network and may induce a bias (Körner 2003). In Europe, most agriculturally managed grasslands are used according to a production potential that is partly determined by local climate conditions. Therefore, N fertiliser supply, manure spreading and forage harvests tend to scale with the length of the herbage growing season (Soussana et al. 2007). This creates a confounding effect between climate and management and reduces our ability to fully understand the determinants of field scale carbon storage in grasslands. Moreover, the long-term history of land use and land management is usually not known in eddy covariance grassland sites. This is also a major limitation as soil respiration is affected by long-term changes in SOC pools as a result of past changes in land use and land management. Mechanistic modelling is needed to overcome these limits and assess the role of both management variables and site history for C fluxes (Soussana et al. 2004b).

13.3.4   Net Greenhouse Gas Exchange

Budgeting equations can be extended to include fluxes (F_{CH4} and F_{N2O}) of non-CO_2 radiatively active trace gases and calculate a net exchange rate in CO_2–C equivalents (13.2), using the global warming potential (GWP) of each gas at the 100 years time horizon (IPCC 2001b):

$$\text{NGHGE} = \text{NEE} + F_{CH_4} \cdot \text{GWP}_{CH_4} + F_{N_2O} \cdot \text{GWP}_{N_2O} \qquad (13.2)$$

where

$$\text{GWP}_{N_2O} = 127, \text{as} \, 1 \, \text{kg} \, N_2O - N = 127 \, \text{kg} \, CO_2 - C$$

$$\text{GWP}_{CH_4} = 8.36, \text{as} \, 1 \, \text{kg} \, CH_4 - C = 8.36 \, \text{kg} \, CO_2 - C.$$

In order to account for (1) the off-site CO_2 and CH_4 emissions resulting directly from the digestion by cattle of the forage harvests and (2) the manure and slurry applications which add organic C to the soil, an attributed net greenhouse gas balance (NGHGB) was calculated by Soussana et al. (2007) as:

$$\text{NGHGB} = \text{NGHGE} + F_{harvest} \cdot (k_{digest} + k_{CH_4} \cdot \text{GWP}_{CH_4}) \qquad (13.3)$$

where k_{digest} is the proportion of the ingested C that is digestible and hence will be respired by ruminants and k_{CH4} the fraction of the ingested C emitted as CH_4 from enteric fermentation.

On average of the nine sites covered by the 'GreenGrass' European project, the grassland plots displayed annual N_2O and CH_4 emissions of 14 ± 5 and 32 ± 7 g CO_2–C equivalents m^{-2} year^{-1}, respectively. Hence, when expressed in CO_2–C equivalents, emissions of N_2O and CH_4 resulted in a 19% reduction of the NEE sink activity. The attributed GHG balance was, however, not significantly different from zero (−85 ± 77 g CO_2–C equivalents m^{-2} year^{-1}). The net exchanges by the grassland ecosystems of CO_2 and of GHG were highly correlated with the difference in carbon used by grazing versus cutting, indicating that cut grasslands have a greater on-site sink activity than grazed grasslands. However, the NBP was significantly correlated to the total C used by grazing and cutting, indicating that, on average, net carbon storage declines with herbage utilisation by herbivores (Soussana et al. 2007).

Taken together, these results show that European grasslands are likely to act as relatively large atmospheric CO_2 sinks. By contrast to forests, approximately half of the sink activity is stored in labile carbon pools (i.e. forage), which will be digested within 1 year by herbivores. When expressed in CO_2–C equivalents, on-site N_2O and CH_4 emissions from grassland plots would apparently not compensate the atmospheric CO_2 sink activity. Nevertheless, the off-site digestion by livestock of the harvested herbage leads to additional emissions of CO_2 and CH_4 which would offset the net GHG sink activity of the grassland plots. These first conclusions are, however, based on a small number of sites and of years. More efforts are clearly needed in order to further reduce the uncertainties in the carbon and GHG budgets in each site, and to reach an optimum number of sites (see below).

13.4 Farm Scale

A grazing livestock farm consists of a productive unit that converts various resources into outputs as milk, meat and sometimes grains too. In Europe, many ruminant farms have mixed farming systems: they produce themselves the roughage and, most often, part of the animal's feeds and even straw that is eventually needed for bedding. Conversely, these farms recycle animal manure by field application. Most farms purchase some inputs as fertilisers and they always use direct energy derived from fossil fuels. The net emissions of GHGs (methane, nitrous oxide and carbon dioxide) are related to carbon and nitrogen flows and to environmental conditions.

Until now, there are only few recent models of the farm GHG balance. Most models have used fixed emission factors for both indoors and outdoors emissions (e.g. FARM GHG Olesen et al., 2006; CNFarm Schils et al. 2005; Lovett et al. 2006). Although, these models have considered the on and off farm CO_2 emissions (e.g. from fossil fuel combustion), they did not include possible changes in soil C resulting from the farm management. Moreover, as static factors are used rather than dynamic simulations, the environmental dependency of the GHG fluxes is not captured by these models.

A dynamic farm scale model called FarmSim has been developed within the framework of the European 'GreenGrass' project. It consists of nine interacting modules concerning the farm structure (area and type of grasslands and of annual crops, herd types), the soils, the meteorology, the grassland management (grazing and cutting dates, stocking rates, organic and inorganic fertiliser applications), the annual crops management, the herd (number of animals per type each fortnight), the feeding system and the waste management system. FarmSim simulates the herbage productivity and greenhouse gas (GHG) balance of each grassland plot using the mechanistic Pasim (Riedo et al. 1998) model. Contrasted scenarios concerning the net organic C storage of both grassland and cropland plots are compared. The IPCC methodology Tier 1 and Tier 2 is used to calculate the CH_4 and N_2O emissions from annual crops and cattle housing. The net GHG balance at the farm gate is calculated in CO_2 equivalents. Emissions induced by the production and transport of major farm inputs (fuel, electricity, N fertilisers and feedstuffs) are calculated using a full accounting scheme based on life cycle analysis. The FarmSim model has been applied to seven contrasted cattle farms in Europe (Salètes et al. 2004). The balance of the farm gate GHG fluxes leads to a sink activity for four of the seven farms. When including pre-chain emissions related to inputs, all farms but one were found to be net sources of GHG. The total farm GHG balance varied between a sink of −70 and a source of +310 kg CO_2 equivalents per unit (GJ) energy in animal farm products. As with other farm scale models, the annual farm N balance was the single best predictor of the farm GHG budget (Schils et al. in press).

The short-term C cycling has sometimes been neglected as the C exchanged with the atmosphere is of biological origin and, hence, is not included in the IPCC methodology. Nevertheless, the atmospheric CO_2 balance at the farm scale is essential for scaling up from the plot to the farm and then to the landscape. Directly scaling the measured NEE of grassland plots up to the regional scale would be misleading in terms of net C storage, since: (1) the NBP (net C storage at the field scale) is much lower than the NEE (see 2.3), and (2) livestock farms may import and export organic C in the form of straw, roughage, concentrate feeds and manure, which means that the farm scale net C storage will not be equal to the atmospheric CO_2 balance.

13.5 Regional Scale

Grasslands and rangelands management in Europe varies with the duration of the plant cover (e.g. sown vs. long-term grasslands), the grazing and mowing patterns, as well as the fertiliser supply, irrigation and drainage. This diversity makes direct comparison among sites or direct upscaling difficult.

A grassland typology was developed for carbon inventories by the IPCC (2004). This typology separates degraded grasslands (e.g. overgrazed, less productive ...), nominally managed (e.g. pasture and rangelands with no grazing problems or inputs, native vegetation) and improved grasslands with medium/high inputs (e.g. sown grasses and legumes, fertiliser supply, liming, irrigation). While these general categories are useful, they need to be adapted in each region, taking into account not only the soil and vegetation types but also the animal stocking density, the organic and inorganic fertiliser application and the number and timing of harvests.

Georeferenced data on small regions need to be obtained every 5–10 years, including the area of grasslands per type, the management data for each grassland type and estimates of the total below and above-ground organic carbon content per grassland type. Therefore, a key message is the need to develop improved systems for the collection of georeferenced statistics on types and timing of agricultural management events for grasslands, information which is currently missing from most national and European statistics (Soussana et al. 2004b).

Grassland productivity should also be known, since regional or sectoral budgets often require knowledge of crop growth or yields. However, in contrast with arable crops, there are very few statistics available for grassland productivity. Proxies such as the animal stocking density per unit grassland/rangeland area can be calculated. Nevertheless, in intensive livestock breeding areas, the diet of domestic herbivores includes a large fraction of concentrate feeds and roughage, inputs that are purchased rather than produced on farm. Hence, there is no direct relationship between the animal stocking density, as estimated from regional statistics, and the actual grazing pressure (Soussana et al. 2004b). Given the numerous shortcomings of regional budget methods, an upscaling methodology based on models tested at benchmark sites is an attractive alternative.

Carefully parametrised and evaluated process-based models allow extrapolating flux data to the regional scale. However, there have been only few attempts to estimate the net radiative forcing fluxes of European grasslands by mechanistic modelling. Vuichard et al. (2007a, b) have used a process-oriented model (PaSim, Riedo et al. 1998) to account explicitly for edaphic, climatic and management variability over Europe. The distribution of management practices (grazing vs. cutting, cut events, animal density) is simulated to match biological constraints rather than economic ones. This study assumed that the management in each grid point reflects the capacity of grasslands to optimally sustain animals through grazing and feeding on locally harvested forage. In other words, macro- and microeconomical and cultural forces were assumed to be of second order compared to biological constraints. Despite these somewhat simplistic assumptions, rather realistic values of seasonal leaf area index, forage yields, livestock numbers, CH_4 and N_2O emissions were found.

The management scheme developed by Vuichard et al. (2007b) assumes an optimal grassland use, whereas in Europe, economic factors are quite strong (subsidies, cost of fertilisers, revenue of animals products …). For example, extensive use of grasslands which tends to be common in some mountain regions in Europe (Marriott et al. 2004) was not simulated. Simulation results have shown that the GHG budget may vary from a sink to a source for any N fertilisation scenario depending on the current compared to long-term equilibrium C stocks in the soil (Vuichard et al. 2007b). This result underlines that the changes in soil organic C content, as affected by global (elevated CO_2, climate change and N deposition) and local (fertiliser application, grazing and mowing patterns, irrigation) drivers are a key for the GHG balance of European grasslands.

Data from the National Soil Inventory of England and Wales obtained between 1978 and 2003 (Bellamy et al. 2005) show that carbon was lost from top soils across England and Wales over the survey period at a mean rate of 0.6% year^{-1} (relative to the existing soil carbon content). The relative rate of carbon loss increased with soil carbon content and was more than 2% year^{-1} in soils with carbon contents greater than $100\,g\ kg^{-1}$. The relationship between rate of carbon loss and carbon content is irrespective of land use, suggesting a link to climate change (Bellamy et al. 2005).

The possible implication of climate change was further studied by Smith et al. (2005) who calculated soil carbon change using the Rothamsted carbon model and using climate data from four global climate models implementing four IPCC emission scenarios (SRES). Changes in net primary production (NPP) were calculated by the Lund–Potsdam–Jena model. Land use change scenarios were used to project changes in cropland and grassland areas. Projections for 1990–2080 for mineral soil show that climate effects (soil temperature and moisture) will tend to speed decomposition and cause soil carbon stocks to decrease, whereas increases in carbon input because of increasing NPP will slow the loss. Technological improvement may further increase carbon inputs to the soil. When incorporating all factors, for grassland soils, Smith et al. (2005) have found a small increase (3–6 tC ha^{-1}) in soil carbon on a per area basis under future climate.

13.6 Conclusions

The role of global (i.e. global warming, atmospheric CO_2 rise and N deposition) environmental factors versus local land use and land management factors for C sequestration by European grasslands is still unclear. The development of a long-term monitoring strategy of these managed ecosystems is crucial in order to understand how climate variability, climatic and atmospheric changes affect ecosystem productivity, carbon stocks and GHG fluxes. Several criteria should be met for ensuring the success of such a monitoring strategy.

Apart from generic data quality and site representativity criteria, specific requirements for grassland sites concern:

1. Grassland management. Most grassland sites are managed for agricultural or nature conservation goals. While some flexibility is needed to cope with climate variability, clear guidelines need to be established to avoid possible confounding effects of management changes for soil pools and for CO_2 and GHG fluxes. Moreover, the management needs to reflect standard regional practices, while avoiding unnecessary complexity (e.g. mixing herbivore types, mixing different plots in the footprint of the mast, …) which may prevent model calibration and evaluation.

2. All components of the C cycle and of the GHG emissions need to be studied. The hidden parts of the C cycle (imports and exports of organic carbon, CH_4 emissions, DOC and DIC losses, …) need to receive due attention, since they have a key role for net carbon storage rate (NBP). While it may not be possible to study all components of the C cycle at the same time, the role of the missing components of the budget needs to be estimated preferably through short-term campaigns, or at least through literature checks to constrain the uncertainties. In the same way, major sources of non-CO_2 trace gases (N_2O, CH_4) need to be measured, and as far as possible other indirect sources (e.g. NO_3^-, NH_3, NO_x) should be estimated. This is especially important in intensively managed grasslands, given the major role of non-CO_2 emissions for the GHG balance of these systems.

3. Ecosystem monitoring. A number of biotic interactions (plant–plant, plant–soil biota, plant–herbivore, …) affect ecosystem processes and C cycling in grasslands. Carefully monitoring the major state variables of the vegetation and soil compartments is a pre-requisite for model evaluation. However, there are many practical problems associated with the monitoring of a grazed pasture, such as fast plant tissue turnover, large spatial heterogeneity of the soil and vegetation and large changes in phenology and herbage mass within a few weeks only. Moreover, the functional role of the plant species diversity (Loreau et al. 2001) and dynamics (Lavorel and Garnier 2002; Louault et al. 2005) has not yet been included in ecosystem models which limits their predictive ability. Advanced models linking community and ecosystem ecology will be needed for C cycle modelling in semi-natural grasslands. To test such models at long-term flux monitoring sites, plant community structure data will be required.

While the development of a strong grassland site network and the advancement of models are important pre-requisites for a full accounting of the GHG balance at European scale, a third major component, which is currently missing, concerns georeferenced data of grassland types and grassland management. Given the major role of organic C fluxes in livestock farms (exports of forages from grassland plots and imports of manures, purchase of concentrate feed and roughage), grassland and farm typologies should be linked in order to allow for the calculation of the C balance of livestock farms. This approach would allow keeping track of changes as influenced by socio-economic trends and would help designing mitigation options that are relevant for farmers.

References

Aeschlimann, U., Nösberger, J., Edwards, P.J., Schneider, M.K., Richter, M., and Blum, H. 2005. Responses of net ecosystem CO_2 exchange in managed grassland to long-term CO_2 enrichment, N fertilization and plant species. Plant, Cell and Environment 28:823–833.

Allard, V., Soussana, J.F., Falcimagne, R., Berbigier, P., Bonnefond, J.M., Ceschia, E., et al. 2007. The role of grazing management for the net biome productivity and greenhouse gas budget (CO_2, N_2O and CH_4) of semi-natural grassland. Agriculture, Ecosystems and Environment 121:47–58.

Baldocchi, D., and Meyers, T. 1998. On using eco-physiological, micrometeorological and biogeochemical theory to evaluate carbon dioxide, water vapor and trace gas fluxes over vegetation: A perspective. Agriculture and Forest Meteorology 90:1–25.

Baldocchi, D., Valentini, R., Running, S., Oechel, W., and Dahlman, R. 1996. Strategies for measuring and modelling carbon dioxide and water vapor fluxes over terrestrial ecosystems. Global Change Biology 2:159–168.

Balesdent, J., and Balabane, M. 1996. Major contribution of roots to soil carbon storage inferred from maize cultivated soils. Soil Biology and Biochemistry 28:1261–1263.

Bellamy, P.H., Loveland, P.J., Bradley, R.I., Lark, R.M., and Kirk, G.J.D. 2005. Carbon losses from all soils across England and Wales 1978–2003. Nature 437:245–248.

Bird, S.B., Herrick, J.E., Wander, M.M., and Wright, S.F. 2002. Spatial heterogeneity of aggregate stability and soil carbon in semi-arid rangeland. Environmental Pollution 116:445–455.

Blaxter, K.L., and Clapperton, J.L. 1965. Prediction of the amount of methane produced by ruminants. British Journal of Nutrition 19:511–522.

Boeckx, P., and Van Cleemput, O. 2001. Estimates of N_2O and CH_4 fluxes from agricultural land in various regions of Europe. Nutrient Cycling in Agroecosystems 60:35–47.

Bouwman, A.F. 1996. Direct emission of nitrous oxide from agricultural soils. Nutrient Cycling in Agroecosystems 46:53–70.

Cannell, M.G.R., Milne, R., et al. 1999. National inventories of terrestrial carbon sources and sinks, the UK experience. Climate Change 42.

CarboEurope IP 2003. Assessment of the European Terrestrial Carbon Balance. Integrated Project. Sixth Framework Programme. Priority 1.1.6.3 Global Change and Ecosystems.

Carlier, L., De Vliegher, A., van Cleemput, O., and Boeckx, P. 2004. Importance and functions of European grasslands. Proceedings of the COST Action 627 "Carbon storage in European Grasslands", Ghent, 3–6 June 2004, pp. 7–16.

Casella, E., and Soussana, J.F. 1997. Long-term effect of CO_2 enrichment and temperature increase on the carbon balance of a temperate grass sward. Journal of Experimental Botany 48:1309–1321.

Cernusca, A. (Project co-ordinator) 2004. Official project web site: CARBOMONT: http://botany. uibk.ac.at/forschung/forschungsprojekte/carbomont_ordner/carbomont/

Chapin, F.S. III, Matson, P.A., and Mooney, H.A. 2002. Principles of Terrestrial Ecosystem Ecology. Springer, New York.

Chevallier, T., Voltz, M., Blanchart, E., Chotte, J.L., Eschenbrenner, V., Mahieu, M., and Albrecht, A. 2000. Spatial and temporal changes of soil C after establishment of a pasture on a long-term cultivated vertisol (Martinique). Geoderma 94:43–58.

Ciais, P., Reichstein, M., Viovy, N., Granier, A., Ogee, J., Allard, V., Aubinet, M., Buchmann, N., et al. 2005. Europe-wide reduction in primary productivity caused by the heat and drought in 2003. Nature 437:529–533.

Clayton, H., McTaggart, I.P., Parker, J., Swan, L., and Smith, K.A. 1997. Nitrous oxide emissions from fertilised grassland: A 2-year study of the effects of N fertiliser form and environmental conditions. Biology and Fertility of Soils 25:252–260.

Conant, R.T., Paustian, K., and Elliott, E.T. 2001. Grassland management and conversion into grassland: Effects on soil carbon. Ecological Applications 11:343–355.

CORINE 1995. CORINE land cover–Part 1: Methodology. EEA technical report.

CORINE 2000. Addendum to the land cover technical guide. EEA technical report.

Delgado, C.L. 2005. Rising demand for meat and milk production in developing countries: Implications for grasslands-based livestock production. In: McGilloway (Ed.), Grassland: A Global Resource, pp. 29–41. Wageningen Acad. Publisher. ISBN907699871X.

De Mazancourt, C., Loreau, M., and Abbadie, L. 1998. Grazing optimization and nutrient cycling: When do herbivores enhance plant production? Ecology 79:2242–2252.

Dueck, T.A., de Visser, R., Poorter, H., Persijn, S., Gorissen, A., de Visser, W., et al. 2007. No evidence for substantial aerobic methane emission by terrestrial plants: A ^{13}C-labelling approach. New Phytologist, doi: 10.1111/j.1469-8137.2007.02103.x.

Hendriks, D.M.D., van Huissteden, J., Dolman, A.J., and van der Molen, M.K. 2007. The full greenhouse gas balance of an abandoned peat meadow. Biogeosciences 4:411–424.

European Environment Agency (EEA) 2005. The European Environment: State and outlook 2005. Part A. Integrated assessment 245 pp. EEA.

Food and Agricultural Organisation (FAO) 2004. FAOSTAT data, FAO.

Flechard, C.R., Neftel, A., Jocher, M., Ammann, C., and Fuhrer, J. 2005. Bi-directional soil/atmosphere N_2O exchange over two mown grassland systems with contrasting management practices. Global Change Biology 11:2114–2127.

Flechard, C.R., Ambus, P., Skiba, U., Rees, R.M., Hensen, A., van den Pol, A., Soussana, J.-F., et al. 2007. Effects of climate and management intensity on nitrous oxide emissions in grassland systems across Europe. Agriculture Ecosystems and Environment 121:135–152.

Follett, R.F. 2001. Organic carbon pools in grazing land soils. In: Follett, R.F., Kimble, J.M., and Lal, R. (Eds.), Potential of US Grazing Lands to Sequester Carbon and Mitigate the Greenhouse Effect, pp. 65–86. Lewis Publishers Inc., Boca Raton.

Freibauer, A., Rounsevell, M.D.A., Smith, P., and Verhagen, J. 2004. Carbon sequestration in the agricultural soils of Europe. Geoderma 122:1–23.

Giger-Reverdin, S., Sauvant, D., Vermorel, M., and Jouany, J.P. 2000. Modélisation empirique des facteurs de variation des rejets de méthane par les ruminants. Rencontre Recherche Ruminants 7:187–190.

Gilmanov, T.G., Tieszen, L.L., Wylie, B.K.,Flanagan, L.B., Frank, A.B., Haferkamp, M.R., Meyers, T.P., and Morgan, J.A. 2005. Integration of CO_2 flux and remotely-sensed data for primary production and ecosystem respiration analyses in the Northern Great Plains: Potential for quantitative spatial extrapolation. Global Ecology and Biogeography 14:271–292.

Gilmanov, T., Soussana, J.F., Aires, L., Allard, V., Ammann, C., Balzarolo, M., Barcza, Z., Bernhofer, C., Campbell, C.L., Cernusca, A., et al. 2007. Partitioning of the tower-based net CO_2 exchange in European grasslands into gross primary productivity and ecosystem respiration components using light response functions analysis 2007. Agriculture, Ecosystems and Environment 121:93–120.

INRA 2002. In: Arrouays, D., Balesdent, J., Germon, J.C., Jayet, P.A., Soussana, J.F., and Stengel, P. (Eds.), Contribution à la lutte contre l'effet de serre. Stocker du carbone dans les sols agricoles de France? Institut National de la Recherche Agronomique (INRA), 147 rue de l'Université, Paris. ISBN 2-7380-1054-7.

IPCC 1996a. Revised guidelines for national greenhouse gas inventories. Intergovernmental Panel on Climate Change, IPCC, Cambridge University Press.

IPCC 1996b. Climate change 1995. The science of climate change. Contribution of working group I to the 2nd assessment report of the IPCC. Intergovernmental Panel on Climate Change, Cambridge University Press.

IPCC 2001a. Climate change 2001: The scientific basis (contribution of working group I to the third assessment report of the Intergovernmental Panel on Climate Change). Intergovernmental Panel on Climate Change, Cambridge University Press.

IPCC 2001b. Good practice guidance and uncertainty management in national greenhouse gas inventories. Intergovernmental Panel on Climate Change (IPCC), Institute for Global Environmental Strategies, Tokyo, Japan.

IPCC 2004. Good practice guidance on land use change and forestry in national greenhouse gas inventories. Intergovernmental Panel on Climate Change (IPCC), Institute for Global Environmental Strategies, Tokyo, Japan.

Janssens, I.A., Freibauer, A., Ciais, P., Smith, P., Nabuurs, G.-J., Folberth, G., Schlamadinger, B., Hutjes, R.W.A., Ceulemans, R., Schulze, E.-D., Valentini, R., and Dolman, A.J., 2003. Europe's biosphere absorbs 7–12% of anthrogogenic carbon emissions. Science 300:1538–1542.

Johnson, K., Huyler, M., Westberg, H., Lamb, B., and Zimmerman, P. 1994. Measurement of methane emissions from ruminant livestock using a SF6 tracer technique. Environmental Science and Technology 28:359–362.

Jones, M.B., and Donnelly, A. 2004. Carbon sequestration in temperate grassland ecosystems and the influence of management, climate and elevated CO_2. New Phytologist 164:423–439.

Keppler, F., Hamilton, J.T.G., Brass, M., and Rockmann, T. 2006. Methane emissions from terrestrial plants under aerobic conditions. Nature 439:187–191.

Körner, C. 2003. Atmospheric science: Slow in, rapid out—Carbon flux studies and Kyoto targets. Science 300:1242–1243.

Lal, R. 2004. Soil carbon sequestration impacts on global climate change and food security. Science 304:1623–1627.

Lavorel, S., and Garnier, E. 2002. Predicting changes in community composition and ecosystem functioning from plant traits: Revisiting the Holy Grail. Functional Ecology 16:545–556.

Leahy, P., Kiely, G., and Scanlon, T.M. 2004. Managed grasslands: A greenhouse gas sink or source? Geophysical Research Letters 31, L20507, doi:10.1029/2004GL021161.

Lemaire, G., and Chapman, D. 1996. Tissue flows in grazed plant communities. In: Hodgson, J., Illius, A.W. (Eds.), The Ecology and Management of Grazing Systems. CABI, Wallingford.

Loreau, M., Naeem, S., Inchausti, P., Bengtsson, J., Grime, J.P., Hector, A., Hooper, D.U., Huston, M.A., Raffaelli, D., Schmid, B., Tilman, D., and Wardle, D.A. 2001. Biodiversity and ecosystem functioning: Current knowledge and future challenges. Science 294:804–808.

Louault, F., Pillar, V.D., Aufrere, J., Garnier, E., and Soussana, J.F. 2005. Plant traits functional types in response to reduced disturbance in a semi-natural grassland. Journal of Vegetation Science 16:151–160.

Lovett, D.K., Shalhoo, L., Dillon, P., and O'Mara, F.P. 2006. A systems approach to quantify greenhouse gas fluxes from pastoral dairy production as affected by management regimes. Agricultural Systems 88:156–179.

Machefert, S.E., Dise, N.B., Goulding, K.W.T., and Whitehead, P.G. 2002. Nitrous oxide emission from a range of land uses across Europe. Hydrology and Earth System Sciences 6:325–337.

Marriott, C.A., Fothergill, M., Jeangros, B., Scotton, M., and Louault, F. 2004. Long-term impacts of extensification of grassland management on biodiversity and productivity in upland areas. A review. Agronomie 24:447–462.

Novick, K., Stoy, P., Katul, G., Ellsworth, D., Siqueira, M., Juang, J., and Joren, R. 2004. Carbon dioxide and water vapor exchange in a warm temperate grassland. Oecologia 138:259–274.

Ogle, S.M., Conant, R.T., and Paustian, K. 2004. Deriving grassland management factors for a carbon accounting approach developed by the Intergovernmental Panel on Climate Change. Environmental Management 33:474–484.

Olesen, J.E., Schelde, K., Weiske, A., Weisbjerg, M.R., Asman, W.A.H., and Djurhuus, J. 2006. Modelling greenhouse gas emissions from European conventional and organic dairy farms. Agriculture, Ecosystems and Environment 112:207–220.

Paustian, K., Andrèn, O., Clarholm, M., Hansson, A.C., Johansson, G., Lagerlof, J., Lindberg, T., Pettersson, R., and Sohlenius, B. 1990. Carbon and nitrogen budgets of four agro-ecosystems with annual and perennial crops, with and without N fertilization. Journal of Applied Ecology 27:60–84.

Pinares-Patino, C.S., Baumont, R., and Martin, C. 2003. Methane emissions by Charolais cows grazing a monospecific pasture of timothy at four stages of maturity. Canadian Journal of Animal Science 83:769–777.

Pinares-Patino, C.S., Dhour, P., Jouany, J.-P., and Martin, C. 2007. Effects of stocking rate on methane and carbon dioxide production by grazing cattle. Agriculture, Ecosystems and Environment 121:30–46.

Reynolds, S.G., Batello, C., Baas, S., and Mack, S. 2005. Grasslands and forage to improve livelihoods and reduce poverty. In: McGilloway (Ed.), Grassland: A Global Resource, pp. 29–41. Wageningen Acad. Publisher. ISBN907699871X.

Riedo, M., Grub, A., Rosset, M., and Fuhrer, J. 1998. A pasture simulation model for dry matter production, and fluxes of carbon, nitrogen, water and energy. Ecological Modelling 105:141–183.

Robertson, G.P., Paul, E.A., and Harwood, R.R. 2000. Greenhouse gases in intensive agriculture: Contributions of individual gases to the radiative forcing of the atmosphere. Science 289:1922–1925.

Robles, M.D., and Burke, I.C. 1998. Soil organic matter recovery on Conservation Reserve Program fields in southeastern Wyoming. Soil Science Society of America Journal 62:725–730.

Rogiers, N., Eugster, W., Furger, M., and Siegwolf, R. 2005. Effect of land management on ecosystem carbon fluxes at a subalpine grassland site in the Swiss Alps. Theoretical and Applied Climatology 80:187–203.

Salètes, S., Fiorelli, J.L., Vuichard, N., Cambou, J., Olesen, J.E., Hacala, S., Sutton, M., Furhrer, J., and Soussana, J.F. 2004. Greenhouse gas balance of cattle breeding farms and assessment of mitigation option. In: Greenhouse Gas Emissions from Agriculture Conference. Leipzig, Germany 203-208 (10–12 February 2004).

Schils, R.L.M., Verhagen, A., Aarts, H.F.M., and Šebek, L.B.J. 2005. A farm level approach to define successful mitigation strategies for greenhouse gas emissions from ruminant livestock systems. Nutrient Cycling in Agroecosystems 71:163–175.

Schils, R.L.M., Olesen, J.E., del Prado, A., and Soussana, J.F. in press. A farm level modelling approach for mitigating greenhouse gas emissions from ruminant livestock systems. Livestock Science 112: 240–251.

Siemens, J., and Janssens, I.A. 2003. The European carbon budget: A gap. Science 302:1681.

Smith, P., Goulding, K.W.T., Smith, K.A., Powlson, D.S., Smith, J.U., Falloon, P.D., and Coleman, K. 2001. Enhancing the carbon sink in European agricultural soils: Including trace gas fluxes in estimates of carbon mitigation potential. Nutrient Cycling in Agroecosystems 60:237–252.

Smith, J., Smith, P., Wattenbach, M., Zaehle, Z., Hiederer, R., Jones, R.A., Montanarella, L., Rounsevell, M.D.A., Reginsters, I., and Ewert, F. 2005. Projected changes in mineral soil carbon of European croplands and grasslands, 1990–2080. Global Change Biology 11:2141–2152.

Sommer, S.G., Petersen, S.O., and Møller, H.B. 2004. Algorithms for calculating methane and nitrous oxide emissions from manure management. Nutrient Cycling in Agroecosystems 69:143–154.

Soussana, J.F., Loiseau, P., Vuichard, N., Ceschia, E., Balesdent, J., Chevallier, T., and Arrouays, D. 2004a. Carbon cycling and sequestration opportunities in temperate grasslands. Soil Use and Management 20:219–230.

Soussana, J.F., Salètes, S., Smith, P., Schils, R., and Ogle, S. 2004b. Greenhouse gas emissions from European grasslands. In: Sezzi, E., Valentini, R. (Eds.), Report 4/2004, Specific Study 3, CarboEurope GHG, Concerted Action, Synthesis of the European Greenhouse Gas Budget. University of Tuscia, Viterbo, Italy. ISSN1723-2236.

Soussana, J.F., Allard, V., Pilegaard, K., Ambus, P., Ammann, C., Campbell, C., Ceschia, E., Clifton-Brown, J., et al. 2007. Full accounting of the greenhouse gas (CO_2, N_2O, CH_4) budget of nine European grassland sites. Agriculture, Ecosystems and Environment 121:121–134.

Sozanska, M., Skiba, U., and Metcalfe, S. 2002. Developing an inventory of N_2O emissions from British soils. Atmospheric Environment 36:987–998.

Van Den Pol-Van Dasselaar, A. 1998. Methane emissions from grasslands. Ph.D. Thesis, 179 pp, Wageningen University.

Velthof, G.L., and Oenema, O. 1997. Nitrous oxide emission from dairy farming systems in the Netherlands. Netherlands Journal of Agricultural Science 45:347–360.

Vermorel, M. 1995. Prédictions gazeuses et thermiques résultant des fermentations digestives. In: Jarrige, R., Ruckebusch, Y., Demarquilly, C., Farce, M. H., Journet, M., (Eds.), Nutrition des Ruminants Domestiques—Ingestion et Digestion. INRA, Paris.

Vleeshouwers, L.M., and Verhagen, A. 2002. Carbon emission and sequestration by agricultural land use: A model study for Europe. Global Change Biology 8:519–530.

Vuichard, N., Soussana, J.F., Viovy, N., Calanca, P., Clifton-Brown, J., and Ciais, P. 2007a. Estimating the greenhouse gas fluxes of European grasslands with a process-based model: 1. Model evaluation from in situ measurements. Global Biogeochemical Cycles 21:14. GB1004. doi:10.1029/2005GB002611.

Vuichard, N., Ciais, P., Viovy, N., Calanca, P., and Soussana, J.F. 2007b. Estimating the greenhouse gas fluxes of European grasslands with a process-based model: 2. Simulations at the continental level. Global Biogeochemical Cycles 21:13. GB1005. doi:10.1029/2005GB002612.

Yazaki, Y., Mariko, S., and Koizumi, H. 2004. Carbon dynamics and budget in a *Miscanthus sinensis* grassland in Japan. Ecological Research 19:511–520.

Chapter 14
Regional Measurements and Modelling of Carbon Exchange

A. Johannes Dolman, Joel Noilhan, Lieselotte Tolk, Thomas Lauvaux,
Michiel van der Molen, Christoph Gerbig, Franco Miglietta,
and Gorka Pérez-Landa

14.1   Introduction

Atmospheric measurements of CO_2 mixing ratios at a number of locations around the globe have helped significantly to quantify the source–sink distribution of carbon at the global and sub-hemispheric scales (e.g. Rödenbeck et al. 2003, chapter 3 and 11). The techniques that achieve this (e.g. Gurney et al. 2002), use a globally distributed network of atmospheric concentration observations of CO_2 and other trace gasses together with an atmospheric transport model that back calculates an 'optimal' source–sink distribution. So far, this global inversion approach has yielded estimates of regional sinks and sources at scales of the order of a few hundreds of kilometres. For example, the distribution of the Northern Hemisphere carbon uptake in longitude between the oceans, North America, Europe and Asia is subject to many investigations but also to many uncertainties (e.g. Peylin et al. 2002; Fan et al. 1998). Stephens et al. (2007) question the common understanding of a large Northern Hemispheric sink, by comparing modelled profiles against observed ones from a few locations, and conclude that transport errors in the models may have contributed to putting too much of the global sink in the Northern Hemisphere transport.

At the local scale ($1 \, km^2$), direct flux measurements by the eddy covariance technique (Baldocchi et al. 2001; Valentini et al. 2000, chapter 11) constrain the net ecosystem exchange (NEE) to within 20%, comparable to the uncertainty estimated from inverse models (e.g. Janssens et al. 2003). In parallel, intensive field studies can determine the changes in vegetation and soil carbon stocks using biometric techniques, which allow independent quantification of the average carbon balance of ecosystems, albeit also with significant errors (Schulze et al. 2000; Wirth et al. 2002; Curtis et al. 2002). How the two scales, the global and local, interact at the regional level is unknown. It remains a major challenge to quantify the carbon balance at this 'missing scale'. Understanding the link between the local and global scale will add significant value to the existing local and global networks and will ultimately help to improve the constraints on the dynamics and vulnerability of the continental scale carbon balance.

The atmospheric boundary layer (ABL) dynamics play a crucial role in the transfer of CO_2 between the surface and the troposphere at the regional scale. Two aspects appear particularly relevant. First, the coupling between CO_2, radiation and turbulent fluxes at the surface has important implications for the surface energy balance. The cloudiness in the ABL controls the available solar radiation at the surface, which largely drives the CO_2 assimilation during daytime conditions. CO_2 assimilation is in turn strongly related to the vegetation transpiration through stomatal control, and hence directly impacts the magnitude of the Bowen ratio. For weak-to-moderate wind conditions, the surface sensible heat flux is the main driver of the development of the diurnal ABL and of the vertical mixing. This affects the CO_2 concentration through dilution. It is clearly important to know how strong this dilution effect is, when one is trying to relate atmospheric concentrations to surface sources (e.g. Karstens et al. 2006, chapter 4). The sequence of impacts and feedbacks shows the complexity of the processes involved in the interaction between the surface and the ABL. This complexity is comparable to that of soil moisture and the ABL (e.g. Van Ek and Holtslag 2004). Second, turbulence is the main process for upward transport of air as well as entrainment of tropospheric air at the top of the ABL. The turbulence in the ABL and the entrainment of clean and dry air from the upper free troposphere through the capping inversion are still an area of active research. Little is known of entrainment of CO_2 within the ABL and the values for entrainment rates (i.e. the ratio of the flux at the ABL top to the surface one), even for sensible heat flux remain highly speculative.

Recent experimental analyses show how entrainment significantly dries and dilutes the concentration of water vapour and CO_2 in the ABL (Vila-Guerau et al. 2004). At nights under weak wind conditions, the stable stratification of the ABL is responsible for the nocturnal accumulation of CO_2 close to the surface. Nocturnal inversion layers prevent the vertical mixing, and hence the concentration of CO_2 because of surface respiration may increase to high values within a thin atmospheric layer close to the surface (values as high as 10^3 ppmv can be observed in forests). This aspect is also very challenging for atmospheric modelling because it requires both refinements of turbulence schemes and a very high vertical resolution. The nocturnal storage of CO_2 close to the surface, however, is an initial condition for the CO_2 dynamics during the following day, and thus adequate assessment of this starting condition is key to a successful assessment of the daytime behaviour.

As an example of the complexity of boundary layer processes, Fig. 14.1 (Vila-Guerau et al. 2004) shows vertical profiles of fluxes of sensible and latent heat and carbon dioxide measured by an aircraft (Gioli et al. 2004). The graphs show the importance of the entrainment process. By calculating the ratio of the entrainment flux to the surface flux and linearly extrapolating the observed fluxes to the top of the ABL and on the basis of the data collected in the afternoon flights during an experimental campaign in the Netherlands, an indirect estimation of the ratio of the entrainment to the surface flux for the virtual potential temperature, specific humidity and carbon dioxide were obtained. Vila-Guerau et al. (2004) suggest for the case shown in Fig. 14.1 values of 0.6 for sensible heat flux, 0.25 for latent heat flux and 2.9 for the carbon dioxide flux. These values indicate that the turbulent eddies entrain warmer, drier and cleaner (lower concentration of CO_2) air from the free

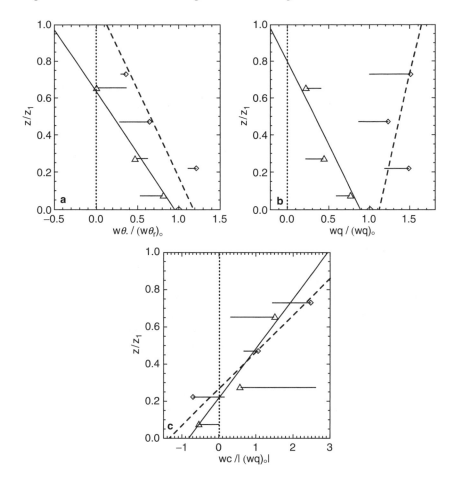

Fig. 14.1 Vertical profiles of the fluxes of (1.1.a) virtual potential temperature, (1.1.b) specific humidity and (1.1.c) carbon dioxide. The horizontal bars represent the standard deviation from the observations. The fluxes are made non-dimensional by their surface values. Diamonds represent observations in the morning, and triangles represent afternoon observations. The dashed and solid lines are linear fits to the observations which include the values of the fluxes at $z/z_i = 0$ (from Vila-Guerau et al. 2004).

troposphere down into the ABL. Importantly, Vila-Guerau et al. (2004) also define two regimes for the development of carbon dioxide in the boundary layer: an entrainment-diluting regime closely related to the rapid morning growth of the boundary layer, and a surface CO_2 assimilation regime which acts as a sink of CO_2 throughout the day but becomes dominant only in the afternoon hours. In the morning regime, the entrainment ratio is high (>1) and in the afternoon it declines as surface processes become more dominant in the evolution than entrainment processes at the top of the boundary layer.

At the regional scale, the 3-D structure of the ABL also becomes important. Mesoscale circulations generated by land surface heterogeneities like sea breezes, vegetation-gradient-induced breezes or regional circulations resulting from topography then start to play an important role in the spatio-temporal evolution of CO_2. Such circulations can be sensitive to surface heterogeneities at scales as small as a few kilometres. Therefore, high resolution (e. g. 1–2 km) and non-hydrostatic mesoscale modelling is necessary to simulate these circulations which could affect the whole ABL (Pérez-Landa et al. 2006a, b). Monitoring the CO_2 evolution at the regional scale would thus require simultaneous documentation of both the horizontal advection and the ABL dynamics during the complete diurnal cycle.

Gerbig et al. (2003a) indicate that a significant fraction of the information contained in the signature of boundary layer CO_2 is contained in relatively small spatial and temporal scales. This is largely caused by the scale of the surface and ABL heterogeneities. In Fig. 14.2, the representation error of a series of measurements over the North American continent during the COBRA experiment (CO_2 Budget and Rectification Airborne study in the Northern USA) is shown. This error is shown as the variance between samples and the difference in horizontal distance between them, combined with an uncertainty analysis of both, of which details can be found in Gerbig et al. (2003a). Figure 14.2 shows that for this experiment significant increase in representation errors occurs when horizontal grid size (distance) increases. At very small scales, the uncertainty approaches the measurement error of 0.2 ppm. At large grid sizes, the representation error becomes quickly large: at only 30 km it is of order twice the measurement error and the representation error dominates the measurement error; at about 150 km it is already 1 ppm. Given the grid size of atmospheric models used in current inversions (e.g. Gurney et al. 2002), 200–400 km, representation errors of 1–2 ppm can be expected. This suggests that to be able to use the variance of the observed signal during the COBRA study, analysis with grid cells of less than 30 km is required. This requires experimental

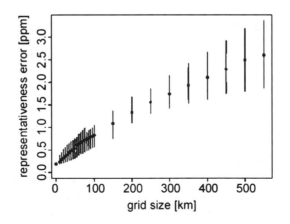

Fig. 14.2 Total representativeness error of mixed layer averaged CO_2 mixing ratios (combined measurement uncertainty and representativeness error) plotted against the horizontal dimension of the region. Vertical bars indicate the 5–95% range (from Gerbig et al. 2003a).

sampling and model development to resolve the diurnal timescale and to be appropriate at such small spatial scales.

The combination of 1-D and 3-D processes, interacting at the regional scale, makes observation and modelling of regional scale carbon exchange particularly challenging. In this chapter, we will briefly review the state of the art of observational and modelling techniques that are currently used in regional carbon exchange studies. We will also describe how these could be used strategically to bridge the gap in our understanding of local and large-scale carbon exchange. An example of such a focused study is given as illustration of the issues that are important at the regional scale (Dolman et al. 2006). We will close with an analysis of progress in inverse modelling at small scales and formulate some recommendations for future experiments.

14.2   State of the Art

14.2.1   Flux Measurement Aircraft

The possibility to measure directly the CO_2 flux of a region has recently been enhanced by the introduction of airborne eddy covariance from small environmental research aircraft (SERA, Isaac et al. 2004). These are able to measure spatially integrated surface properties, leading to reliable estimates of fluxes over different land uses (Gioli et al. 2004). The combination of flux measurements made from towers and from aircraft (Desjardins et al. 1997; Isaac et al. 2004) can be even more powerful as those two techniques can actually complement each other in terms of spatial and temporal integration of the measurements. The technique was pioneered more than 20 years ago by Lenschow et al. (1981) and Desjardins et al. (1982) and has since been applied in numerous field experiments. In particular, Crawford et al. (1996) and Brooks et al. (2001) have shown that aircraft flux measurements can be very accurate and reliable and still be executed on small lightweight planes.

However, there are still a number of unresolved issues that hinder full-scale application, and comparison between data from flux tower sites and aircraft has not unequivocally met with success. Two issues stand out. First, the difference in footprint between the fluxes measured from the aircraft and from the tower complicates a direct comparison. Particularly in heterogeneous terrain, this can be a problem, and the precise height of the flight track chosen may influence and almost certainly complicates such a comparison. Second, vertical flux divergence in the boundary layer implies that extrapolation to the tower fluxes is needed when aircraft fluxes measured at a certain height well above the surface are compared with measurements from a tower. This can be achieved by executing (nearly) simultaneous fights at several levels within the Planetary Boundary Layer (PBL).

The instruments that are now used in state-of-the-art aircraft have reached a high level of perfection and include high precision pressure spheres on the nose of aircraft, open-path infrared gas analysers for high frequency measurements of CO_2

and water, differential GPS for estimating the horizontal velocity relative to the ground and accelerometers to calculate 3-D pitch, roll and heading of the aircraft. An example of such a 'state of the art' plane is the Sky Arrow ERA described in Oechel et al. (1998) and in essence based on the Long-EZ developed by Crawford and Dobosy (1992). The Sky Arrow also carries on board Photosynthetically Active Radiation (PAR) and net radiation sensors and the possibility of a spectral video camera that allows classification of the underlying land surface through indices such as the Normalized Difference Vegetation Index (NDVI). This two-seater plane can fly at slow speed ($35\,m\,s^{-1}$) and at low altitudes up to $10\,m$ above the surface, so that meaningful comparisons can be made with tower flux measurements. Sampling at $50\,Hz$, it can detect eddies of sizes larger than $1.4\,m$ in no wind conditions (Gioli et al. 2004).

The combination of flux measurements made from towers and from aircraft (Desjardins et al. 1997; Isaac et al. 2004) can be even more powerful if those two techniques can actually complement each other in terms of spatial and temporal integration of the measurements. Miglietta et al. (2007) discuss measurements of net regional ecosystem exchange (NEE) that were made over a period of 21 days in the summer 2002 in the south-central part of the Netherlands using a combination of NEE data from a research flux aircraft (Sky Arrow ERA), half-hourly eddy covariance (EC) data from three towers, half-hourly weather data recorded by three weather stations and information on regional land use from digital maps. This combination allowed them to estimate the net contribution of terrestrial ecosystems to the overall regional carbon flux and to map dynamically the temporal and spatial variability of the fluxes, during the study period, albeit with uncertainty estimates of similar magnitude as the NEE of individual land use types.

14.2.2 ABL Concentration Measurements

ABL budgeting has been used to determine regional ecosystem fluxes during daytime in Central Siberia, the Amazon and Europe (Lloyd et al. 2001; Laubach and Fritsch 2002; Wofsy et al. 1988; Culf et al. 1997; Levy et al. 1999; Schmitgen et al. 2004), as well as to determine the regional discrimination of NEE against ^{13}C (Lloyd et al. 2001) with varying success. Both Lagrangian approaches, that track a column of air, as well as Eulerian approaches whereby advective terms need to be estimated have been used by various groups (Laubach and Fritsch 2002; Lin et al. 2004; Schmitgen et al. 2004). The latter approach can be applied for homogeneous surface fluxes, whereas Lagrangian experiments principally allow flux estimation also for heterogeneous areas (Chou et al. 2002). These studies require the use of aircraft for vertical profiling of CO_2 concentrations. For in situ measurements of CO_2 onboard aircraft, the Non-Dispersive InfraRed (NDIR) instrument CONDOR, developed at Laboratoire des Sciences du Climat et de l'Environment (LSCE), Paris, was used in a regional campaign in France (Dolman et al. 2006). Modified commercial CO_2 NDIR sensors can also be used. In situ measurement of CO in addition to CO_2 will help quantifying the fossil fuel contribution (Gerbig et al. 2003). Often, simultaneous flask sampling with subsequent analysis provides a consistent quality check on CO and CO_2, but, more importantly, provides important additional tracer information: carbon and oxygen isotope data can be helpful in separating carbon fluxes from different processes (assimilation vs respiration, heterotrophic vs autotrophic respiration); CH_4 and SF_6 provide additional constraints on the fossil fuel emission sources; the gases H_2 and N_2O are related to further biological processes in the soil.

Laubach and Fritsch (2002) used the profile information to calculate the fluxes of surface fluxes of CO_2 and sensible and latent heat using various assumptions about the state of the air column under consideration. Evaluating the budget for the assumption that the column has a fixed mass, rather than a variable one with the top fixed at the inversion, yielded errors of about 10–20% for the CO_2 and sensible heat flux and slightly higher ones for latent heat flux (20–30%). Their attempt to use the information of profiles just above the inversion to quantify advection met with variable success, and no clear physical explanation could be found to explain why in some cases it improved the estimates and in other cases in did not. This appears to be a recurring feature in ABL budgeting.

To overcome the problems with advection, Schmitgen et al. (2004) used a Lagrangian experimental approach whereby the aircraft follows a particular air mass. The Lagrangian experiment took place under favourable conditions over a productive, extended and spatially homogenous pine forest in South-West France in summer times, with flat terrain, and negligible anthropogenic sources that were checked in parallel by in situ measurements of CO, and simple boundary conditions given by homogenous maritime air moving over the forest. Yet, only for one flight out of four in total, the proper Lagrangian conditions were met with persistent low wind speeds ($6 km h^{-1}$) and a stable inversion layer at ca 700 m. Their results indicate that during its exposure to surface uptake, the air mass loses CO_2 at a rate of 0.3 ppm km^{-1}, which translates into an estimate for daytime NEE of 16 umolC $m^{-2} s^{-1}$ over the 100-km flight domain. This value compares well within its errors with the local scale measurements of the eddy covariance tower located further north of the flight tracks, and the results of a regional remote-sensing model based on 1-km resolution Moderate Resolution Imaging Spectroradiometer (MODIS) data.

Schmitgen et al. (2004) quantified instrumental errors because of possible drifts (±0.5 ppm) in the CO_2 sensor that would alias with the signal of fluxes (14%), the aggregation error as 22% as implied by solving for a mean flux for the whole domain, a bias of 18% because of the uncertainties in characterizing the true ABL mean value of $<s_i>$ and a systematic error that takes into account the lack of adequate entrainment data of 29%. Adding these error estimates suggests that the order of magnitude of the error is equal to the magnitude of the total flux (16 μmol $m^{-2} s^{-1}$). Schmitgen et al. (2004) argue convincingly that these errors can be drastically reduced by more adequate flight plans (both horizontal and vertical stacks) and better sampling and higher resolution measurements. Taking these arguments into account, they conclude that an overall error of 22% is reasonable. This may seem comparable to the estimates of Laubach and Fritsch (2002); however, these latter authors neglect to quantify the aggregation and representation errors involved, and refer primarily to the instrumental and method errors.

14.3 Models

Meteorological mesoscale models have proved to be important tools for helping to understand the temporal and spatial variations of CO_2 mixing ratios in the atmosphere. This variability sometimes shows large gradients associated with the different meteorological processes that determine the dispersion and transport conditions of the atmosphere and is arguably one of the causes for the unexplained failures of the ABL budgeting techniques. It is convenient to distinguish here between meso-α and meso-γ processes, the latter operating at scales that are resolvable by hydrostatic model (10-km grid distance), the former requiring non-hydrostatic high resolution (1–4-km grid distance) models.

Nicholls et al. (2004) employed the model Regional Atmospheric Modeling System (RAMS) coupled to the biosphere model SiB-2 to investigate some of the mechanisms leading to CO_2 variability in an area around the WLEF tower in Wisconsin (USA) with measurements available for 5 days in July 1997. The results were sensitive to different PBL parameterizations in the RAMS model and the cloud microphysics package. Their model reproduced lake breezes that created important CO_2 horizontal and vertical variability in a relatively small area. Lu et al. (2005) also found large spatial CO_2 gradients in a simulation with a similar modelling approach during a river-breeze event over the Tapajos River in Brazil. The large spatial and temporal gradients of CO_2 are not limited to summer conditions. For example, Hurwitz et al. (2004) analysed abrupt temporal changes in the CO_2 mixing ratio for the Wisconsin WLEF tall tower during four events under low pressure systems. The surface exchange processes, such as photosynthesis and respiration, were not able to produce these rapid changes, which were found to be mainly because of either vertical or horizontal CO_2 transport processes. The conclusion is that to properly interpret measurements, it is necessary to understand both the effects of the dynamical and biological processes and their coupling in the variability of CO_2. These processes involve a wide range of spatial and temporal scales that must be resolved.

Karstens et al. (2006) present simulations with two meso (γ) scale models over Europe and Siberia to assess the 'signal to noise' ratio of atmospheric CO_2 concentrations. They define this as the detectability of a specific CO_2 signal, given all the variability that is induced by all sources and transport processes. They find that large coherent spatial structures emerge from synoptic disturbances in Europe with typical scale lengths of 170–260 km. These disturbances are advected in the model domain despite slowly varying large-scale lateral boundary conditions and slowly varying surface fluxes. In Siberia with large-scale high pressure systems dominating the weather patterns, that distance may be larger, of order 380–490 km. They also note that predicting night-time concentrations involves high degrees of uncertainty because of the uncertainty involved in modelling nocturnal boundary layer processes that are responsible for large accumulation of CO_2 in the night.

Van der Molen and Dolman (2007) demonstrated for an area in Central Siberia around the Yenisey that relatively modest relief may cause horizontal carbon dioxide concentration gradients of 35 ppm in the boundary layer, that persisted for several hours. The advection of such gradients causes concentration changes in time, which disturb the link between surface fluxes and atmospheric concentration changes (Fig. 14.3a). Other causes of mesoscale variability in the atmosphere include variations in turbulence intensity, surface roughness, Bowen ratio, boundary layer height, subsidence velocity and/or entrainment flux. The CO_2 concentration gradient of 35 ppm is of similar strength as the mean diurnal cycle observed at flux tower sites. They found that the effect of advection of this concentration gradient on the boundary layer concentrations is 3–6 times larger than the effect of the surface uptake. They also calculated the representation errors as of the order of 1–5 ppm, comparable to the observed values of Gerbig et al. (2003). These are also large in comparison to the diurnal variation in boundary layer concentrations of 15 ppm that was observed on 15 July and 5 ppm on 16 July 1996 on that site (Lloyd et al. 2001). The fact that such strong concentration gradients appeared in the simulation while the topography differences are relatively modest made them suggest that mesoscale variability may significantly influence many CO_2 observation stations around the world.

In the Regional Carbon Balance Project (RECAB) within the frame of the CarboEurope cluster of projects, airborne measurements were carried out in two complex terrain regions in the Western Mediterranean Basin under summer conditions: on a valley site in Lazio (Italy) and on a coastal site in Valencia (Spain)

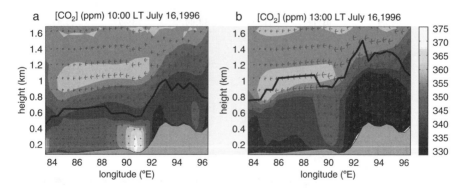

Fig. 14.3 Cross section of the simulated CO_2 concentration field of the coarsest grid at (a) 10:00 h LT and (b) 13:00 h LT on 16 July 1996. The colours and contours indicate concentration differences at 5 ppm intervals; the arrows indicate longitudinal and vertical wind velocity and the solid black line represents the boundary layer height. The surface topography is also shown. Note that the horizontal extent of the domain covers 864 km, so that the horizontal is much compressed relative to the vertical. The cross sections are drawn at 59.2 °N (*See Color Plate* 6).

(Gioli et al. 2004). In both regions and for most days of the campaigns, vertical profiling of the lower atmosphere (first 1000 m) showed a complex structure of CO_2 mixing ratio, with large vertical gradients changing with time. Figure 14.4 shows four different profiles measured on the 2nd of July 2001 in Valencia. The RAMS meteorological mesoscale model (Pielke et al. 1992) high resolution simulations (grid distances of 1.5 km), enabled Pérez-Landa et al. (2006a, b) to diagnose the presence of mesoscale circulations that led to the vertical injection of surface air masses and their sinking because of compensatory subsidence, and that resulted in a complex layering of the air masses in both regions.

To quantify the influence of these processes on the atmospheric CO_2 concentration, and their coupling with biological phenomena, they used RAMS simulations to run the Lagrangian Particle Dispersion Model (LPDM) Hybrid Particle and Concentration Transport Model (HYPACT) (Walko et al. 2001) on one of the Valencia summer campaign days. The time evolution of CO_2 emissions uptake was estimated by a simple empirical approach based on data sets from eddy covariance towers present in the region to parameterize the CO_2 fluxes of three main land categories representative of the region: rice, citrus and natural vegetation. The capability of HYPACT to define the shapes of the different sources was employed to estimate a realistic spatial distribution of these categories in the Valencia basin. Simulation results showed large vertical gradients changing with time. Figure 14.4 shows the contribution by source of the simulated net local CO_2 vertical distribution signal at the profiling site, at different times of the day. Direct comparison of simulated and observed profiles is not possible, but the simulation provides key process information to understand the observed profiles.

For example, the sea breeze cell that is confined in the lower 200 m regulates vertical mixing, as can be observed at 11:40, 13:00, 13:40 and 15:40 h. During this

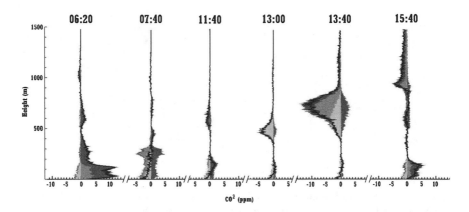

Fig. 14.4 Contribution to the net CO_2 profile at Lon = −0.37 and Lat = 39.37 at the corresponding time, classified by emission source. Light grey, grey and dark grey colours represent the contribution of the rice, citrus and natural sources, respectively. Solid black line represents the net CO_2 obtained by the addition of negative and positive signals.

time period, the vegetation uptake is maximum, which combined with the small dilution, permits the strong negative signals that can be observed at 13:00 and 13:40 h. The histories of the air masses in the same profile can be rather different as shown by the different origins of the signal and the different times since emission. This explains the strong positive and negative signals observed within only 1000 m at 15:40 h. The strong signals are maintained for a while by the stable atmosphere resulting from the compensatory subsidence above the breeze cell (see 13:00, 13:40 and 15:40 h profiles). The coupling between the diurnal cycles of both kinds of processes tends to separate positive and negative signals, except during the transition periods between meteorological and biological processes. Modelling assumptions could explain some of the strong differences with the observed profiles. For instance, the empirical approach of the CO_2 fluxes may be too coarse to represent the biological processes. Also as noted before, the vertical resolution could still be insufficient to solve layerings of tenths of metres. Another important source of error could be due to the limitations of mesoscale models in the parameterization of turbulence under these conditions, as only horizontal homogeneous steady-state boundary layer parameterizations are available for use in these models. A key problem in the performance of these models is the description of the boundary layer. Precise estimation of its height and the strength of the inversion gradient determines to a large extent the resulting dilution rates and concentration gradients. This requires very high vertical resolution in the boundary layer and near the inversion or very adequate parameterizations of Convective Boundary Layer (CBL) growth and entrainment to produce realistic simulations.

14.4   Case Study of a Regional Experiment: The CarboEurope Regional Experiment Strategy Les Landes

Determination of the CO_2 budget at the regional scale is based on the estimation of the various terms of the budget equation, which can be written (in a simplified form, where horizontal x axis is orientated along the mean wind) as:

$$\frac{\delta[CO_2]}{\delta t} + U\frac{\delta[CO_2]}{\delta x} + W\frac{\delta[CO_2]}{\delta z} + \frac{\delta\left[\overline{w'CO'_2}\right]}{\delta z} = 0 \qquad (14.1)$$

$$\text{(A)} \qquad\quad \text{(B)} \qquad\quad \text{(C)} \qquad\quad \text{(D)}$$

In this equation, U and W are the horizontal and vertical mean wind component, respectively. (A) represents the time evolution of the concentration, (B) and (C) the contributions due to horizontal and vertical advection, respectively, and (D) the contribution due to turbulent transfer of CO_2. The latter term is the vertical divergence of the turbulent flux, whose profile is controlled by both the surface flux $(W'CO_2')_0$ and the flux at the top of the boundary layer $(W'CO_2')_{Zi}$. The latter represents the exchanges between ABL and the free troposphere, whereas the former is the budget of the transfer through soil surface and vegetation. Since only a regional model describing atmospheric processes, as well as transfers into the soil and vegetation is able to monitor the CO_2 budget as described by the equation above, a realistic experimental strategy would try to estimate the various terms, either in their time or space variability, and thus to constrain the behaviour of the model.

In short-term intensive campaigns, the fixed network is extended with a rather large amount of additional measurements. Optimal use of these measurements requires a trade off between representation of a large area and detail, and between frequency and accuracy (Gerbig et al. 2003b). The characteristics of the most advantageous measurements are to a large extent determined by the question under study; apparently similar quantities might require very different strategies. Therefore, considerable specificity is required in defining the objectives for an optimization.

We show results from the CERES campaign (Dolman et al. 2006). The central methodology of CERES was to make concentration measurements both within and above the boundary layer and to couple those via a modelling/data assimilation framework to the flux measurements at the surface and within the boundary layer. To achieve this, a region near Les Landes Forest was instrumented with ground and air-based measurements at high spatial and temporal resolution. This area was chosen because of the wealth of supporting data that exist from the previous HAPEX–Mobilhy experiment (André et al. 1986) and the vicinity of Météo-France in Toulouse with state of the art forecasting tools. Such a multiple constraint approach had not been tried before at the regional level and the data were thought to give a better understanding of the surface source–sink distributions and allow for a better quantification of the representation and aggregation errors mentioned above.

We present some preliminary results from the second Intensive Observation Period (IOP) that lasted from 24 to 27 May (Dolman et al. 2006). During this period, there was a strong anti-cyclonic situation over the area with very weak variable winds. Very deep boundary layers were observed from the radio sounding performed above the forest near Cape Sud (Fig. 14.5). They were topped by a very sharp inversion, accompanied by a strong humidity gradient, where above 1.7 km a very dry layer of air had developed. This high final boundary layer depth appeared to be caused by a quick development through the residual boundary layer of the previous day in the early part of the afternoon.

Figure 14.6a and b shows CO_2 concentration and temperature images obtained from the flight of the Dimona research aircraft on the afternoon of 27 May. Figure 14.6c shows the flight plan of the aircraft. The aircraft tried to intercept a constant volume balloon that was released near Cape Sud, to obtain a perfect Lagrangian flight path. The main advantage of this type of sampling is that any changes in the airmass can, in principle, be attributed directly to fluxes encountered by the airmass from the ground or overlying air, as it travels along its path. CO_2 concentrations vary along the flight track between 375 and 385 ppm in the boundary layer, with the highest concentration close to the ocean surface. The lowest concentrations are observed over the agricultural area. Because the winds were weak and variable, most of this variation should be related to different uptake and emission regimes of the land surface. If this is true, the agricultural areas in the east part of the flight

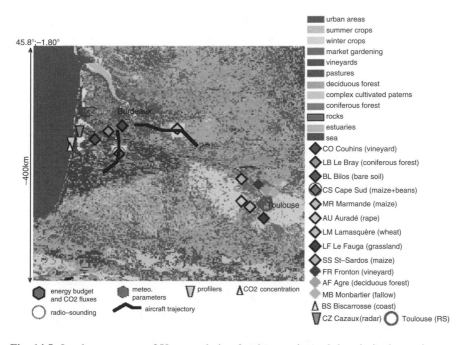

Fig. 14.5 Land cover map at 250-m resolution for the experimental domain in the south-west region of France showing the different location of summer and winter agricultural crops (classification by Champeaux et al. 2005). Also shown are the locations of the ground-based observation sites of surface fluxes and boundary layer. Flight tracks indicate the path flow by the Sky Arrow flux aircraft for agriculture and forested regions (*See Color Plate* 6).

would be a larger sink than the forest in the west. Surface flux measurements over Les Landes Forest for this day and the previous day, however, show a relatively high uptake rate, while the flux station in Marmande shows small net CO_2 fluxes (Dolman et al. 2006). These latter observations were obtained over an area of growing maize, that had large parts of soil exposed to the air and produced relatively high respiration. Crop sites towards the south-east showed indeed fluxes of comparable magnitude as the forest fluxes. In fact, analysis of the land cover (Fig. 14.5) shows that towards the south-east of the domain the crops are predominantly winter crops that already have high CO_2 uptake rates, while the summer crops are still having low rates.

The CO_2 fluxes from the Sky Arrow flux aircraft over the forests are much larger than in the agricultural areas, but comparable to those over the vineyards (Dolman et al. 2006). Thus, the observed high concentration of CO_2 over the forest is not directly correlated to the flux of the forest, suggesting that the air over the forest area would be more depleted in CO_2 than observed. Also, the areas planted with maize appear to act as a source of CO_2, as maize still is in the early stages of development at this time of the year.

Trajectories and footprints obtained by a Lagrangian dispersion model the Stochastic Time Inverted Lagrangian Transport model (STILT) (Lin et al. 2003), driven wind fields forecasted by the mesoscale model Aire Limitée Adaptation dynamique Développement InterNational (ALADIN) at 8-km resolution, provided some further insight into this apparent discrepancy between fluxes and concentrations. They showed that the footprint of concentration profile near Marmande was lying in a mostly westerly direction, in strong contrast to the Biscarrosse footprint that was strongly north-south oriented. In fact, the air sampled by the aircraft over Les Landes Forest suggests that CO_2 rich air was advected northward along the coast, while air sampled near Marmande was strongly depleted in CO_2 because it moved over active vegetation areas in the south-east. Taking either the concentration at Biscarrosse or Marmande as representative of local fluxes, or taking the flux measurements at those sites as representative of larger regions would thus lead to

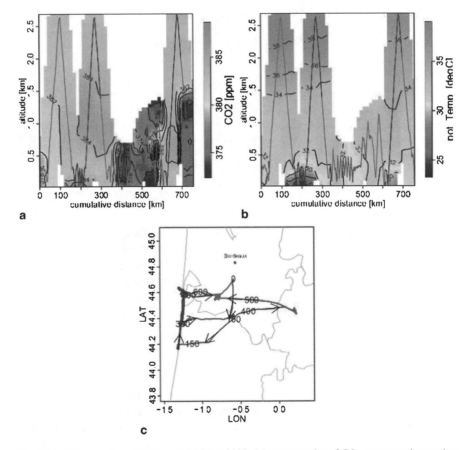

Fig. 14.6 Dimona aircraft flights on 27 May 2003. (**a**) contour plot of CO_2 concentration against distance flown, (**b**) contour plot of potential temperature against distance flown, (**c**) the flight track. Red colour codings indicate agricultural (summer crops) areas, black forested areas and blue the sand dunes of the coastline (*See Color Plate* 7).

incorrect conclusions. Clearly joint consideration of the synoptic and regional flow, fluxes and land surface is required for a correct interpretation. The significant difference of the Biscarrosse to the Marmande footprint, given that the distance is only 100 km, indicates a high variability of the synoptic flow that requires mesoscale modelling to properly resolve the observed tracer gradients.

CERES provided considerable insight into the large variability of spatial CO_2 fields (10–20 ppm over 200 km) and the strong diurnal range in the boundary layer that can approach 100 ppm. The scale of these variations is comparable to those simulated in the mesoscale models described above. Without appreciation of this 3-D context of the relation between surface fluxes and atmosphere, some of the patterns could not be understood. It is exactly this 3-D context that has limited the usefulness of 1-D inversions such as the convective boundary layer technique (Styles et al. 2002; Lloyd et al. 2001; Laubach and Fritsch 2002).

14.5 Inverse Modelling Approaches at the Regional Scale

Global inverse models have been widely used to infer global to regional sources of greenhouse gas (e.g. Gurney et al. 2002; Rödenbeck et al. 2003; Bousquet et al. 2000; Rayner and Law, 1999). However, they suffer from coarse resolutions and simple parameterization of PBL transport. The use of mesoscale models with refined physical parameterization and better resolution would help improving the estimation of biospheric and anthropogenic emissions. For example, Bakwin et al. (1998) suggested to use mesoscale models through regional inversions combined with continuous measurements from a high tower and high precision data on the structure of the convective boundary layer to study the interplay between biospheric sinks and industrial sources at the regional scale.

Regional scale inversions aim to constrain the surface CO_2 fluxes at a high spatial and temporal resolution (2–10 km grids, and day to month timescale). They make use of atmospheric CO_2 concentrations, a priori estimates of the source and sink strength and a regional transport model. Through Bayesian techniques, the uncertainties of the fluxes are then reduced and a posteriori fluxes obtained with error estimates. These show the reduction in uncertainty that is achieved through the inversion. At high spatial resolution, at the regional scale, the number of unknowns in the problem can quickly become very large. Thus, inversions at a fine scale require additional information compared to aggregated global scale inversions. Carouge (2006) showed that daily averages of data from the current European large tower network provide an optimal constraint on scales of about 500 km × 500 km and 8 days within Western Europe. At smaller scales, noise dominates the signal. This compares reasonably well with the results obtained by Karstens et al. (2006). Therefore, a dense measurement network is crucial to reduce the optimal aggregation area to the desired scale. The high temporal resolution of available sources and of the transport model hopefully improves our ability to simulate temporal variations in the data and so these variations are increasingly employed to constrain fluxes. In intensive campaigns like CERES (Dolman et al. 2006) such a dense measurement network is available. Sarrat et al. (2006) show the performance of the French non-hydrostatic mesoscale model Meso-NH, for simulating transport and boundary layer properties during the IOP mentioned above, to be quite good. The model was able to reproduce the horizontal and vertical gradients quite well.

Particularly in mesoscale inversions, limited domains make the model results sensitive to the precise formulation of the boundary conditions. The relevant importance of fluxes within the domain, advection of air from the boundaries and the initial condition is a function of domain size and the inversion time window. A case study with CERES data (Lauvaux et al. 2007) showed that during the 27th of May in Les Landes, even the low wind speed (around $5\,\mathrm{m\ s^{-1}}$ near the surface) allows particles to cross the entire domain within 24 h. Therefore, estimates of the initial value and the boundary conditions are required. For this particular simulation, values were extracted from global scale simulations. The sensitivity of CO_2 concentrations to the fluxes was computed in a backward in time mode, in an off-line coupling with a LPDM (Uliasz 2003). Initial concentrations, inflow fluxes or offsets can be added, describing the complete forcing for the concentration observations.

A high resolution inversion approach (top–down) will optimize more realistic flux dynamics, but will also lead to a weakly constrained inversion with a nearly infinite number of flux patterns as possible solutions. For example, a slightly higher observed concentration than expected from the priori fluxes may be explained by a forest fire far away or by a strong local source in the vicinity of the observation. Without additional constrains, the top–down approach adjusts the flux in a limited number of pixels surrounding the observation, which results in a flux uncertainty reduction in a small area and leaves most of the domain at a priori flux values. The inversion yields thus only 'islands of knowledge in a sea of ignorance'. In fact, without a priori constraints on fluxes the optimal solution that would be chosen might turn out to be a time-dependent source close to the intake of each sensor. A particular choice among the various flux patterns that can fit the observations is embodied in a priori smoothness constraints, usually expressed as spatial and temporal correlations. Spatial structures of the flux errors may be expected to be related to distance, meteorological systems and biome types, or a combination of those (e.g. Carouge 2006). The relevant correlations are not those of the fluxes themselves but rather of the differences between prior and true fluxes. Specifying appropriate correlations in prior errors is a requirement for unbiased inversion results, as discussed by Gerbig et al. (2006). In a comparison of modelled and observed fluxes, Chevallier et al. (2006) cast doubt both on the existence of such spatial correlations, at least in daily data, as well as the convenient Gaussian distribution of errors usually used in such inversions. Another strategy consists in treating the characteristics of the prior error covariances as statistical parameters that are optimized in the Bayesian inversion process (Michalak et al. 2004). This method, however, is still difficult to implement for large-dimension problems. Ensemble-based data assimilation methods, such as Kalman filtering (e.g. Peters et al. 2005) or Maximum Likelihood Ensemble Filters, reduce the uncertainty in certain locations, and thus create an error correlation pattern by updating analysis and forecast error covariance per assimilation cycle using information from the observations (Zupanski and Zupanski 2006).

Despite all these concerns, the regional scale inversion approach appears to be promising, as both ensemble and variational methods showed their ability to retrieve the synthetic data distributed in different zones. Dolman et al. (2005) evaluated the potential of the inverse modelling method for inferring the regional distribution of carbon sources and sinks from measurements taken at different heights along a tall tower within the CBL at Cabauw the Netherlands. When they assumed a perfect model, the standard deviation in the a posteriori surface fluxes was reduced significantly, both for the surface fluxes of the local vegetation type and for the dominant vegetation types within a $900\,km \times 900\,km$ domain. Also when they assumed that 20% of the standard deviation was due to model errors, significant reductions in the uncertainty of the a posteriori estimates of the surface fluxes were still achieved for the local vegetation type around the Cabauw tower and for crops/mixed vegetation, the dominant land use type within the $900\,km \times 900\,km$ domain. They concluded that inversion of regional transport models was possible. Thus, the continuous monitoring of carbon dioxide concentrations at towers could yield estimates of the regional distribution of sources and sinks of carbon dioxide. However, and arguably most important, they also concluded that applying and further developing of these inversion methods should be accompanied by major improvements in mesoscale atmospheric models. This holds particularly for the model's ability to simulate realistic night-time accumulation of carbon in the nocturnal boundary layer, as this presents the starting point of the daytime concentration profile development.

14.6 Conclusions and Recommendations

Discovering the 'missing scale' implies closing the gap between the local-based measurements such as eddy covariance towers and inventory studies, and the continental scale inversions. There are several reasons why this should happen and why regional studies provide added value to continental-scale carbon studies.

- First, since the local-scale measurements are selectively representative of a larger domain, upscaling this to larger areas needs development of upscaling techniques or methodology. Only then can they be robustly compared to independent estimates at the regional scale. Regional studies with their wealth of observations stimulate the development of such methods, and, in principle, provide a verification. Regional studies thus provide added value to global networks that suffer from inherent representation errors such as FLUXNET (chapter 11).
- Second, in large parts of the continents, human-induced land use change has resulted in a large variability in land cover. In these areas, global or even continental-scale inversions have to assume that the land cover is homogeneous. Only regional studies are able to provide the assessment of how this affects the mean concentrations and fluxes. This aspect is also referred to as aggregation error. In areas with high surface heterogeneity, regional or mesoscale information adds an

extra constraint on the surface fluxes and would thus be useful to apply in a large-scale multiple constraint approach.

- Third, land use activities typically take place at the regional scale, and any attempt to quantify these in the context of the Kyoto Protocol must necessarily take into account the relevant regional scales.
- Fourth, future data assimilation studies must necessarily take into account at what level the information of, for instance, flux towers or inventories is useful and can be ingested in the system. Insight into the variability of fluxes at subgrid level is then a prerequisite.

An example of the potential for errors involved in this was given by Van der Molen and Dolman (2007) who calculated the potential for disturbance by topography for 27 GlobalView stations. They concluded that only five were potentially free from mesoscale disturbances and that the remaining stations could suffer from a variable degree of topography-induced mesoscale effects similar to those found for the Central Siberia case described in this chapter and in Van der Molen and Dolman (2007).

A well-designed regional experiment consists of an adequate mix of surface and air-based instrumentation, preferably complemented by some space borne or satellite measurements of radiometric properties. There are few guidelines as to what exactly constitutes 'an adequate mix', but some recommendations can nevertheless be made. Surface fluxes of the dominant land cover types are a prerequisite, as shown by the CERES campaign, where eddy correlation measurements of winter and summer crops were essential to appreciate the regional CO_2 balance. Repeated transects with flux aircraft proved helpful in delineating spatial variability within land cover types and are also essential for upscaling (Miglietta et al. 2007). Concentration measurements in the ABL are essential to provide constraints on the surface fluxes. It appears from the little experience that it may be more important to fly regular vertical profiles than to do long horizontal transects. Obviously when funding permits it, a combination is preferable. In fact, which patterns should be flown depends very much on the questions asked: is a grid average value required then long horizontal transects may be appropriate, whereas when spatial heterogeneity is the issue, repeated vertical ABL profiling may be more helpful (e.g. Gerbig et al. 2003b). The distinction is, however, rather vague, and it may be useful in cases where there is no heterogeneity, to also invest in vertical profiles. Modelling with LPDMs may be used to more accurately determine a flight pattern. Such LPDMs should preferably be run with the output of mesoscale models to obtain the necessary spatial resolution.

Most regional experiments, and particularly those discussed in this chapter, have focused on relatively short campaigns. It remains a major challenge how to scale this information in the temporal domain to longer timescales. This is particularly important in the carbon cycle where the slow accumulation processes may take tens of years and ultimately determine emission rates during short-term experiments. It is clear that merging of regional, soil and forest inventory data with bottom–up and top–down models may provide the tool to achieve this. When regional Carbon Data

Assimilation systems are in place, this may happen on a more regular basis. At this moment such systems are in dire need to be developed and it is unclear how they should operate precisely. There are several attempts underway to provide seasonal regional carbon balances as part of the continental-scale carbon programmes. Longer-term observations with tall towers and surface fluxes would at least provide part of the essential data. Network design studies are urgently required to address the question of what constitutes an optimal configuration. The answer to this question is likely to be strongly dependent on the specific scientific questions asked: Does the regional carbon budget need to be closed? Or is information on land cover types required? These two questions would require rather different setups of the design study (chapter 3). Network design studies require an appropriate regional-scale data assimilation system to be available, including appropriate formulation of error covariances. Otherwise the network will be only ideal for the idealized world for which it was developed.

Equally important is accurate information on the anthropogenic CO_2 fluxes. These are generally available at large spatial resolution only, and need to be downscaled temporally also, to be of use in mesoscale models. Downscaling algorithms may exist, but validity remains questionable unless they are verified. There is some hope that CO may be useful as a tracer for fossil fuel emissions of CO_2 (chapter 4).

Inverse modelling at the regional (particularly, meso-γ) scale is still in its infancy. Estimating the various errors, from both the model and measurements in Bayesian inversions at the regional scale is complicated and some attempts (Gerbig et al. 2003a,b) have shown that they may be large. It is still unclear whether the 'noise' generated by mesoscale processes dominates the signal and inversions become like the proverbial searching for the needle in the haystack. Data from high intensity campaigns, such as COBRA (Gerbig et al. 2003) and CERES (Dolman et al. 2006), provide an excellent test bed for such studies. Most studies at the regional scale so far point to the direction that improvements in inverse modelling need to go parallel with improvements in our capability to model the development and spatial heterogeneity of the ABL. Sarrat et al. (2006) show how a regional model can capture the essential dynamics and spatial patterns.

References

André, J.-C., Goutorbe, J.-P., and Perrier, A. 1986. HAPEX-MOBILHY: A hydrologic atmospheric experiment for the study of water budget and evaporation flux at the climatic scale. Bulletin of the American Meteorological Society 67, 138–144.

Bakwin, P. S., Tans, P. P., Hurst, D. F., and Zhao, C. 1998. Measurements of carbon dioxide on very tall towers: Results of the NOAA/CMDL program. Tellus 50B, 401–415.

Baldocchi, D. D., Falge, E., Gu, L., Olson, R., Hollinger, D., Running, S., Anthoni, P., ChBernhofer, Davis, K., Fuentes, J., Goldstein, A., Katul, G., Law, B., Lee, X., Malhi, Y., Meyers, T., Munger, J. W., Oechel, W., Pilegaard, K., Schmid, H. P., Valentini, R., Verma, S., Vesala, T., Wilson, K., and Wofsy, S. 2001. FLUXNET: A new tool to study the temporal and spatial variability of ecosystem-scale carbon dioxide, water vapor and energy flux densities. Bulletin of the American Meteorological Society 82, 2415–2435.

Bousquet, P., Peylin, P., Ciais, P., Le Quéré, C., Friedlingstein, P., and Tans, P. 2000. Regional changes in carbon dioxide fluxes of land and oceans since 1980. Science 290, 1342–1346.

Brooks, S. B., Dumas, E. J., and Verfaillie, J. 2001. Development of the SkyArrow surface/atmosphere flux aircraft for global ecosystem research. American Institute of Aeronautics and Astronautics Journal and Proceedings, 39th Aerospace Sciences Meeting & Exhibit, January 8–11, 2001, Reno Nevada, 10 pp.

Carouge, C. 2006. Vers une estimation des flux de CO_2 journaliers européens à haute résolution par inversion du transport atmosphérique. PhD thesis University Paris 6.

Champeaux, J.-L., Fortin, H., and Han, K.-S., 2005. Spatio-temporal characterization of biomes over SW France using SPOT/VEGETATION and Corinne Land Cover datasets. IGARRS proceedings.

Chevallier, F., Viovy, N., Reichstein, M., and Ciais, P. 2006. On the assignment of prior errors in Bayesian inversions of CO_2 surface fluxes. Geophysical Research Letters 33.

Chou, W. W., Wofsy, S. C., Harriss, R. C., Lin, J. C., Gerbig, C., and Sachse, G. W., 2002. Net fluxes of CO_2 in Amazonia derived from aircraft observations. Journal of Geophysical Research-Atmospheres 107 (D22), 4614, doi: 10.1029/2001JD001295.

Crawford, T. L., and Dobosy, R. J. 1992. A sensitive fast response probe to measure turbulence and heat flux from any airplane. Boundary Layer Meteorology 59, 257–278.

Crawford, T. L., Dobosy, R. J., McMillen, R. T., Vogel, C. A., and Hicks, B. B. 1996. Air-surface exchange measurement in heterogeneous regions: Extending tower observations with spatial structure observed from small aircraft. Global Change Biology 2, 275–285.

Culf, A. D., Fiosch, G., Mahli, Y., and Nobre, C. A. 1997. The influence of the atmospheric boundary layer on carbon dioxide concentrations over a tropical rain forest. Agricultural and Forest Meteorology 85, 149–158.

Curtis, P. S., Hanson, P. J., Bolstad, P., Barford, C., Randolf, J. C., Scmid, H. P., and Wilson, K. B. 2002. Biomertix and eddy covariance based estimates of annual carbon storage in five eastern North American deciduous forests. Agricultural and Forest Meteorology 113, 3–19.

Desjardins, R. L., Brach, E. J., Alno, P., and Schuepp, P. H. 1982. Aircraft monitoring of surface carbon dioxide exchange. Science 216, 733–735.

Desjardins, R. L., MacPherson, J. I., Mahrt, L., et al. 1997. Scaling up flux measurements for the boreal forest using aircraft-tower combinations. Journal of Geophysical Research-Atmospheres 102 (D24), 29125–29133.

Dolman A.J., Ronda, R., Miglietta, F., and Ciais, P. 2005. Regional measurement and modelling of carbon balances. In Griffith H and Jarvis, P.G. (Eds). The carbon balance of forest biomes. Taylor Francis, Abingdon UK., p 93–108.

Dolman, A. J., Noilhan, J., Durand, P., Sarrat, C., Brut, A., Piguet, B., Butet, A., Jarosz, N., Brunet, Y., Loustau, D., Lamaud, E., Tolk, L., Ronda, R., Miglietta, F., Gioli, B., Magliulo, V., Esposito, M., Gerbig, C., Körner, S, Galdemard, P., Ramonet, M., Ciais, P., Neininger, B., Hutjes, R. W. A., Elbers, J. A., Macatangay, R., Schrems, O., Pérez-Landa, G., Sanz, M. J., Scholz, Y., Facon, G., Ceschia, E., and Beziat, P. 2006. CERES, the CarboEurope regional experiment strategy in Les Landes, South West France, May–June 2005. Bulletin of the American Meteorological Society 87 (10), 1367–1379.

Fan, S., Gloor, M., Mahlman, J., Pacala, S., Sarmiento, J., Takahashi, T., and Tans, P. 1998. A large terrestrial carbon sink in North America implied by atmospheric and oceanic carbon dioxide and models. Science 282, 53–74.

Gerbig, C., Lin, J. C., Wofsy, S. C., Daube, B. C., Andrews, A. E., Stephens, B. B., Bakwin, P. S., and Grainger, C. A. 2003a. Toward constraining regional-scale fluxes of CO_2 with atmospheric observations over a continent: 1. Observed spatial variability from airborne platforms. Journal of Geophysical Research 108 (D24), 4756, doi:10.1029/2002JD003018.

Gerbig, C., Lin, J. C., Wofsy, S. C., Daube, B. C., Andrews, A. E., Stephens, B. B., Bakwin, P. S., and Grainger, C. A. 2003b. Toward constraining regional-scale fluxes of CO_2 with atmospheric observations over a continent: 2. Analysis of COBRA data using a receptor-oriented framework. Journal of Geophysical Research 108 (D24), 4757, doi:10.1029/2003JD003770.

Gerbig, C., Lin, J. C., Munger, J. W., and Wofsy, S. C. 2006. What can tracer observations in the continental boundary layer tell us about surface-atmosphere fluxes? Atmospheric Chemistry and Physics 6, 539–554.

Gioli, B., Miglietta, F., De Martino, B., Hutjes, R. W. A., Dolman, A. J., Lindroth, A., Lloyd, J., Sanz, M. J., Valentini, R., and Dumas, E. 2004. Comparison of tower and aircraft-based eddy correlation fluxes at five sites in Europe. Agricultural and Forest Meteorology 127, 1–16.

Gurney, K., Law, R. M., Denning, A. S., et al. 2002. Towards robust regional estimates of CO_2 sources and sinks using atmospheric transport models. Nature 415, 626–630.

Hurwitz, M. D., Ricciuto, D. M., Davis, K. J., Wang, W., Yi, C., Butler, M. P., and Bakwin, P. S. 2004. Advection of carbon dioxide in the presence of storm systems over a northern Wisconsin forest. Journal of the Atmospheric Sciences 61, 607–618.

Isaac, P. R., McAneney, J., Coppin, P., and Hacker, J. 2004. Comparison of aircraft and ground-based flux measurements during OASIS95. Boundary Layer Meteorology 110, 39–67.

Janssens, I. A., Freibauer, A., Ciais, P., Smith, P., Nabuurs, G.-J., Folberth, G., Schlamadinger, B., Hutjes, R. W. A., Ceulemans, R., Schulze, E.-D., Valentini, R., and Dolman, A. J. 2003. Europe's biosphere absorbs 7–12% of anthropogenic carbon emissions. Science 300, 1538–1542.

Karstens U., Gloor, M., Heimann, M., and Rödenbeck, C. 2006. Insights from simulations with high-resolution transport and process models on sampling of the atmosphere for constraining midlatitude land carbon sinks. Journal of Geophysical Research 111, D12301, doi:10.1029/2005JD006278.

Laubach, J., and Fritsch, H. 2002. Convective boundary layer budgets derived from aircraft data. Agricultural and Forest Meteorology 111, 237–263.

Lauvaux, T., Uliasz, M., Sarrat, C., Chevallier, F., Bousquet, P., Lac, C., Davis, K. J., Ciais, P., Denning, A. S., and Rayner, P. J. 2007. Mesoscale inversion: First results from the CERES campaign with synthetic data, ACPD submitted.

Lenschow, D. H., Pearson, Jr. R., and Stankov, B. B. 1981. Estimating the ozone budget in the boundary layer by use of aircraft measurements of ozone eddy flux and mean concentration. Journal of Geophysical Research 86, 7291–7297.

Levy, P., et al. 1999. Regional scale CO_2 fluxes over central Sweden by a boundary layer budget method. Agricultural and Forest Meteorology 99, 159–167.

Lin, J. C., Gerbig, C., Wofsy, S. C., Andrews, A. E., Daube, B. C., Grainger, C. A., Stephens, B. B., Bakwin, P. S., and Hollinger, D. Y. 2004. Measuring fluxes of trace gases at regional scales by Lagrangian observations: Application to the CO_2 Budget and Rectification Airborne (COBRA) study. Journal of Geophysical Research 109, D15304, doi:10.1029/2004 JD004754.

Lin, J. C., Gerbig, C., Wofsy, S. C., Daube, B. C., Andrews, A. E., Bakwin, P. S., Davis, K. J., Stith, J., and Grainger, A. 2003. A near-field tool for simulating the upstream influence of atmospheric observations: The Stochastic Time-Inverted Lagrangian Transport (STILT) model. Journal of Geophysical Research 108 (D16), 4493, doi:10.1029/2002JD003161.

Lloyd, J., Francey, R., Mollicone, D., Raupach, M. R., Sogachev, A., Arneth, A., Byers, J. N., Kelliher, F. M., Rebmann, C., Valentini, R., Chin-Wong, S., Bauer, G., and Schulze, E.-D. 2001. Vertical profiles, boundary layer budgets, and regional flux estimates for CO_2 and its $^{13}C/^{12}C$ ratio and for water vapor above a forest/bog mosaic in central Siberia. Global Biogeochemical Cycles 15, 267–284.

Lu, L., Denning, A. S., Silva-Dias, M. A., Silva-Dias, P., Longo, M., Freitas, S. R., and Saatchi, S. 2005. Mesoscale circulations and atmospheric CO_2 variations in the Tapajós Region, Pará, Brazil. Journal of Geophysical Research, in press.

Michalak, A. M., Bruhwiler, L., and Tans, P. P. 2004. A geostatistical approach to surface flux estimation of atmospheric trace gases. Journal of Geophysical Research-Atmospheres 109, D14109, doi:10.1029/2003JD004422.

Miglietta, F., Gioli, B., Hutjes, R. W. A., and Reichstein, M. 2007. Net regional ecosystem CO_2 exchange from airborne and ground-based eddy covariance, land-use maps and weather observations. Global Change Biology 13, 548–550. doi:10.1111/j./1365–2486.2006.01219.x.

Nicholls, M. E., Denning, A. S., Prihodko, L., Vidale, P.-L., Davis, K., and Bakwin, P. 2004. A multiple-scale simulation of variations in atmospheric carbon dioxide using a coupled biosphere-atmospheric model. Journal of Geophysical Research, 109, D18117, doi:10.1029/2003JD004482.

Oechel, W. C., Vourlitis, G. L., Brooks, S. B., Crawford, T. L., and Dumas, E. J. 1998. Intercomparison between chamber, tower, and aircraft net CO_2 exchange and energy fluxes measured during the Arctic system sciences land-atmosphere-ice interaction (ARCSS-LAII) flux study. Journal of Geophysical Research 103, 28993–29003.

Pérez-Landa, G., Ciais, P., Sanz, M. J., Gioli, B., Miglietta, F., Palau, J. L., Gangoiti, G., and Millán, M. M. 2006a. Mesoscale circulations over complex terrain in the Valencia coastal region, Spain, Part 1: Simulation of diurnal circulation regimes. Atmospheric Chemistry and Physics Discussions 6, 2809–2852.

Pérez-Landa, G., Ciais, P., Gangoiti, G., Palau, J. L., Carrara, A., Gioli, B., Miglietta, F., Schumacher, M., Millán, M. M., and Sanz, M. J. 2006b. Mesoscale circulations over complex terrain in the Valencia coastal region, Spain, Part 2: Linking CO_2 surface fluxes with observed concentrations. Atmospheric Chemistry and Physics Discussions 6, 2853–2895.

Peters, W., Miller, J. B., Whitaker, J., Denning, A. S., Hirsch, A., Krol, M. C., Zupanski, D., Bruhwiler, L., and Tans, P. P. 2005. An ensemble data assimilation system to estimate CO_2 surface fluxes from atmospheric trace gas observations. Journal of Geophysical Research-Atmospheres 110, D24304, doi:10.1029/2005JD006157.

Peylin, P., Baker, D., Sarmiento, J., Cias, P., and Bousquet, P. 2002. Influence of transport uncertainty on annual mean and seasonal inversions of atmospheric CO_2 data. Journal of Geophysical Research 107, 10.1029/2001JD000 857.

Pielke R. A., Cotton, W. R., Walko, R. L., Tremback, C. J., Lyons, W. A., Grasso, L. D., Nicholis, M. E., Moran, M. D., Wesley, D. A., Lee, T. J., and Copeland, J. H. 1992. A comprehensive meteorological modelling system-RAMS. Meteorology and Atmospheric Physics 49, 69–91.

Rayner, P. J., and Law, R. M. 1999. The interannual variability of the global carbon cycle. Tellus Series B-Chemical and Physical Meteorology 51, 210–212.

Rödenbeck, C., Houweling, S., Gloor, M., and Heimann, M. 2003. CO_2 flux history 1982–2001 inferred from atmospheric data using a global inversion of atmospheric transport. Atmospheric Chemistry and Physics 3, 1919–1964.

Sarrat, C., et al. (2007). Atmospheric CO_2 modeling at the regional scale: Application to the CarboEurope Regional Experiment. *J. Geophys. Res.*, 112, D12105, doi:10.1029/2006JD008107.

Schmitgen, S., Ciais, P., Geiss, H., Kley, D., Voz-Thomas, A., Neiniger, B., Baeumle, M., and Brunet, Y. 2004. Carbon dioxide uptake of a forested region in southwest France derived from airborne CO_2 and CO observations in a Lagrangian budget approach. Journal of Geophysical Research 109, D14, doi:10.1029/2003JD004335.

Schulze, E. D., Hoegberg, P., Van Oene, H., Persson, T., Harrison, A. F., Read, D., Kjoller, A., and Matteucci, G. 2000. Interactions between the carbon and nitrogen cycles and the role of biodiversity: A synopsis of a study along a north-south transect through Europe. In: Carbon and Nitrogen Cycling in European Forest Ecosytems, pp 468–491.

Stephens, B., et al. 2007. Weak northern and strong tropical land carbon uptake from vertical profiles of atmospheric CO2. Science 316, 1732–1735.

Styles, J.M., Lloyd, Zolotukhin, D., Lawton, K.A., Tchebakova, N., Francey, R.J., Arneth, A., Salamakho, D., Kolle, O., Schulze, E-D., 2002. Estimates of regional surface carbon dioxide exchange and carbon and oxygen isotope discrimination during photosynthesis from concentration profiles in the atmospheric boundary layer. Tellus B 54:5 768.

Uliasz, M. 2003. A modelling framework to evaluate feasibility of deriving mesoscale surface fluxes of trace gases from concentration data. Available at http://biocycle.atmos.colostate.edu/marek.mesoinversion7c.pdf.

Valentini, R., Matteucci, G., Dolman, A. J., et al. 2000. Respiration as the main determinant of carbon balance in European forests. Nature 404, 861–865.

Van der Molen, M. K., and Dolman, A. J. 2007. Regional carbon fluxes and the effect of topography on the variability of atmospheric CO_2. Journal of Geophysical Research-Atmosphères 112, D01104. doi:10.1029/2006JD007649. http://dx.doi.org/10.1029/2006JD007649

Van Ek, M. B., and Holtslag, A. A. M. 2004. Influence of soil moisture on boundary layer cloud development. Journal of Hydrometeorology 5, 86–99.

Vila-Guerau de Arellano, J., Gioli, B., Miglietta, F., Jonker, H. J. J., Klein Baltink, H., Hutjes, R. W. A., and Holtslag, A.A.M. 2004. Entrainment process of carbon dioxide in the atmospheric boundary layer. Journal of Geophysical Research 109D, 18110, doi:10.1029/2004JD004725.

Walko,R. L.,Tremback, C. J., and Bell, M. J. 2001. HYPACT Hybrid Particle and Concentration Transport Model, Users Guide. Mission Research Corporation, Fort Collins, CO.

Wirth, C., Czimczik, C. I., and Schulze, E.-D. 2002. Beyond annual budgets: Carbon flux at different temporal scales in fire-prone Siberian S cots pine forests. Tellus, 54 B, 611–630.

Wofsy, S. C., Harriss, R. C., and Kaplan, W. A. 1988. Carbon dioxide in the atmosphere over the Amazon basin. Journal of Geophysical Research-Atmospheres 93, 1377–1387.

Zupanski, D., and Zupanski, M. 2006. Model error estimation employing an ensemble data assimilation approach. Monthly Weather Review 134, 1337–1354.

Chapter 15
Using Satellite Observations in Regional Scale Calculations of Carbon Exchange

**Shaun Quegan, Philip Lewis, Tristan Quaife, Gareth Roberts,
Martin Wooster, and Mathias Disney**

15.1 Introduction: How Satellite Data Can Interact with Carbon Exchange Calculations

Estimates of carbon exchange at regional to global scale have been achieved by essentially three means: (1) use of atmospheric concentration measurements with models of atmospheric transport to infer carbon sources and sinks (atmospheric inversion), (2) inventory and (3) process modelling. These approaches can be applied independently, with results that are rarely consistent even amongst approaches of the same type (e.g. Janssens et al. 2003), and each has particular strengths and weaknesses. The first two simply involve conservation of mass and are based on measurements (although these are normally highly undersampled in space). They have little power to explain the observed fluxes or predict their future behaviour. In contrast, the third approach has both explanatory and predictive power, but is normally poorly constrained by measurements. These complementary strengths motivate current efforts to combine the three approaches in model–data fusion schemes (Ciais et al. 2003). Satellite data can play an important part in such schemes, since satellite sensors provide the only means of making global, frequently repeated measurements of the Earth's surface and atmosphere. However, capitalising on these measurements to make improved carbon exchange estimates requires that they be brought into the same framework as the models that actually perform the estimates. This interface depends on both the measurements and the type of model.

As is the case for ground-based flask data, satellite measurements of atmospheric gas concentrations (typically of column content, though with possibilities to obtain height-resolved values) provide no direct information on carbon exchange processes, but can constrain atmospheric transport models (or transport-chemistry models for reactive gases) whose boundary conditions are the surface fluxes we wish to estimate. In contrast, satellite measurements can provide information on relevant surface properties and processes, but the connection to carbon exchange estimates is normally indirect, and can only be achieved by interfacing them with an ecosystem model. Effective use of the data requires that this interface be well founded, and most of this chapter will be devoted to exploring how this can be achieved. The later part of the chapter will deal with atmospheric data and the structure of a complete model-data fusion system.

A context for the discussion of satellite measurements of land surface properties is provided by the generic form of dynamic vegetation model (DVM) illustrated in Fig. 15.1. Here, the surface state in a model grid-cell centred on position **x** at time t is described by a state vector $\mathbf{S}_t(\mathbf{x})$; this holds such information as the proportion of each land-cover type in the grid-cell, the size of the carbon pools, the soil moisture and such like. Using climate data, the current state can be updated to the new state $\mathbf{S}_{t+1}$ on the basis of the parameterised processes that make up the model. For data from satellites or elsewhere to affect these calculations, they must modify the state vector (including helping to initialise it) or provide information on the parameters or processes of the model. Data can also be used to test the model if the value of an observation can be calculated from the model state; this requires an **observation operator** (sometimes called a forward operator) which may be very simple (e.g. if the proportion of forest is a state variable, the observation operator would just be the area itself) or very complicated (e.g. a non-linear radiative transfer calculation).

The concept of an observation operator is fundamental to meaningful use of data, as it tells us what should be observed if the system is in a particular state. Depending on the uncertainty associated with both the measurements and the model, we can then tell if they are mutually consistent. If not, either our estimates of the error characteristics are wrong or the model state is wrong. From this, we can

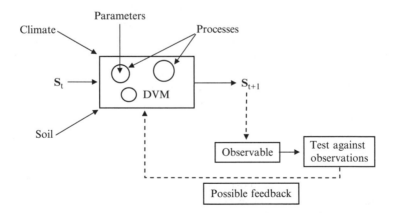

Fig. 15.1 The upper part of this diagram shows the essential structure of a dynamic vegetation model (DVM), containing several linked processes, many of which involve parameters. The current state is used, together with input data on climate and soil texture, to calculate the state at the next time step. One of the key ways to test the model, indicated by dashed lines, is to predict an observable from the state vector and compare it with observations; data assimilation involves feeding this information back into the model to improve the estimate of the state vector.

learn about the performance of the model. Even better, for models with simple enough structure, the existence of an observation operator may allow the model state or parameters to be optimised in a model-data fusion or data assimilation approach.

In contrast with atmospheric observations, where dedicated sensors are typically used to make specific measurements, land surface properties are usually derived from multi-purpose sensors. The primary measurement is received power above the atmosphere (although some microwave sensors, such as synthetic aperture radar, also measure phase). Going beyond this to measure properties of the surface is rarely straightforward, and involves techniques ranging from pure correlation to physical modelling. A further contrast between use of satellite data for meteorology and for land process analysis is that lower-level data are preferred in the former to avoid introducing artefacts, whereas much effort by the land remote sensing community is devoted to deriving high-level products. A key argument of this chapter is that the development of model-data fusion provides strong reasons to adopt the meteorological approach in many aspects of carbon exchange. An exhaustive survey of carbon-relevant measurements (e.g. which would include moisture variables) is not attempted here. Instead, in Sect. 15.2, we focus on four high-level derived products (land cover, fraction of Absorbed Photosynthetically Active Radiation, (fAPAR), fire and biomass), together with reflectance, while Sect. 15.3 deals with satellite measurements of atmospheric trace gases.

15.2 Bio-Geophysical Variables Derived from Satellite Data

The land surface products discussed in this section are chosen both for their importance and because they illustrate the issues in trying to influence carbon exchange calculations with satellite data. Each subsection seeks to describe concisely the availability and characteristics of each product, together with the pros and cons of attempting to use them in carbon exchange models. Although the arguments are intended to be general, many of the calculations rely on specific biospheric models, in particular, the Sheffield dynamic global vegetation model (SDGVM) (Woodward and Lomas 2004) and the data assimilation-linked ecosystem carbon (DALEC) model (Williams et al. 2005).

15.2.1 Land Cover

Knowledge of land cover is crucial for many purposes, not least as a basic control on carbon exchange. Satellite data provide the only means of providing continually updated information on land cover and land-cover change over large regions, so that much effort has been devoted to exploiting them in land-cover classification. The methods used are almost all statistical in nature, using ground data to either train a classification algorithm (supervised classification) or to interpret the clusters found by a clustering algorithm (unsupervised classification). Whatever approach is taken, a core issue is whether the spectral and/or temporal signatures of different types of land cover allow them to be reliably distinguished. For carbon calculations, this simplifies to distinguishing plant functional types (PFTs), which are the functional unit of most biospheric models. In carbon terms, only misclassifications at the PFT level matter, not misclassification of land covers within the same PFT.

Although land-cover maps produced from high-resolution data are likely to be more accurate, they are available for only a few regions in the world. Regional scale carbon modelling must, therefore, often rely on the freely available global maps produced from moderate resolution (around 1 km) satellite data, of which the best known are Global Land Cover 2000 (GLC 2000) (Fritz et al. 2003) and a range of land-cover products derived from Moderate-Resolution Imaging Spectroradiometer (MODIS) data. These include the International Biosphere-Geosphere Programme (IGBP) land-cover product (Loveland and Belward 1997); the University of Maryland (UMD) modification of the IGBP product (Hansen et al. 2000); land cover optimised for the MODIS leaf area index (LAI)/fAPAR product (Lotsch et al. 2003); land cover optimised for the biogeochemical cycle (BGC) model-derived MODIS net primary production (NPP) product (Running et al. 1999); and a PFT classification for use with climate models (Bonan et al. 2002). Each of these MODIS products is derived with a single globally applied algorithm. In contrast, GLC2000 was produced from SPOT-4 VEGETATION data by a set of teams that each covered a different region of the world; they used their own algorithms and local knowledge to optimise the algorithms for their own region.

Two very important questions for regional carbon calculations are (1) do the different land-cover products produce significantly different estimates of carbon fluxes and (2) can we obtain any insight into the accuracy likely with any of these products? The first question is relatively easy to answer, but the second is much harder, since data to assess accuracy are not generally available. However, the UK has a recent high-resolution land-cover map, LCM2000 (Fuller et al. 2002), providing a reference against which to assess carbon calculations based on medium-resolution products. This allows us to perform calculations that address both questions.

An example comparing MODIS-IGBP with LCM2000 for Great Britain is shown in Fig. 15.2. Here, data are represented on a 1/6th of a degree grid (the grid size is determined by the spacing of climate data used for the model calculations shown later in this section) and the grey scale indicates the proportion of each of the C3 grass, crop, deciduous broadleaf and evergreen needleleaf PFTs in each grid-cell; these are the four dominant PFTs in Great Britain. The most striking feature of this comparison is that much of the grassland in the West seen in the reference land cover is misclassified as crops in the MODIS product, although there are also significant differences in the forest assignments.

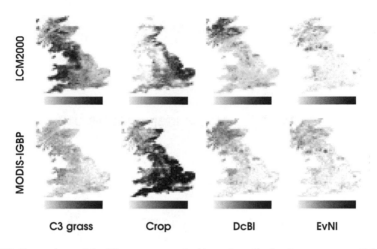

Fig. 15.2 Proportions of the C3 grass, crops, deciduous broadleaf and evergreen needleleaf plant Functional Types in each 1/6ᵗʰ degree grid-cell across Great Britain according to the LCM2000 (top) and MODIS–IGBP (bottom) land cover maps. The scale runs from 0% (black) to 100% (white).

Table 15.1 Mean fluxes in gC m⁻² for the year 2000 from Sheffield Dynamic Global Vegetation Model calculations in which Great Britain is covered by a single Plant Functional Type

	C3 grasses	Crops	Deciduous broadleaf	Evergreen needleleaf
GPP	1657	1196	1382	1714
NPP	936	1012	771	921
NEP	93	223	160	215

More important than the misclassifications themselves are their effects on carbon fluxes. These depend on which fluxes are being considered, as can be illustrated by SDGVM simulations in which Great Britain is completely covered by a single PFT. The resulting mean values of gross primary production (GPP), NPP and net ecosystem production (NEP) (shown in Table 15.1) indicate, for example, that misclassifying C3 grasses as crops has drastic effects on GPP and NEP, but comparatively little effect on NPP.

Table 15.1, when combined with information such as that shown in Fig. 15.2, helps to interpret differences between flux maps derived from LCM2000 and the other land-cover products. Figure 15.3 shows these differences for NEP for each of

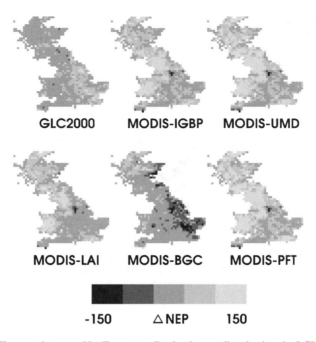

Fig. 15.3 Differences between Net Ecosystem Production predicted using the LCM2000 and the six moderate-resolution land-cover products. Dark and light areas respectively, indicate under-prediction and over-prediction compared to LCM2000.

the six moderate-resolution land-cover products considered. Four of the MODIS products show similar patterns of difference, improving in accuracy in the sequence UMD, PFT, IGBP and LAI, largely because of decreases in the amount of C3 grassland assigned to crops. MODIS–BGC appears to perform well, but this hides very large errors in the underlying GPP calculations that are compensated for in the associated soil respiration. Overall, GLC2000 gives the best performance, but over-estimates the aggregated NEP of Great Britain by 16%.

The calculations above are for Great Britain, but several general features emerge from them. Firstly, different land-cover products yield very different carbon flux estimates: for the land-cover products considered, errors in the aggregated Great Britain carbon flux relative to estimates based on a high-resolution land-cover map ranged from −82 to 91 gC m^{-2} y^{-1} (−8.5 to 9.4%) in GPP; −10 to 92 gC m^{-2} y^{-1} (−1.8 to 16.91%) in NPP and −3 to 34 gC m^{-2} y^{-1} (−5.2 to 58.6%) in NEP. Secondly, some systematic types of misclassification have big effects. For example, several of the products have problems in separating crops and grasses; this is likely to have significant effects on carbon budget (also methane and N$_2$O) calculations at regional to global scales because large parts of the globe are given over to cropping and grasslands, often in proximity, and the carbon fluxes from these two cover types are quite different. Similar remarks apply to confusion between moorland/upland heath and forest, which is likely to be very important at higher latitudes. Thirdly, positive biases in all flux estimates occur when using land cover derived from any moderate-resolution satellite sensor, because small non-productive areas (e.g. small urban areas) are missed. Great Britain is particularly heterogeneous and consequently such biases are fairly strong. More generally, biases are likely in regions, such as Europe or south-east Asia, which contain complex, heterogeneous mosaics of agriculture, woodland and urban areas. For more homogeneous regions, such as Siberian boreal forests, such biases will tend to be smaller.

15.2.2   Photosynthesis: APAR and fAPAR

Vegetation growth is driven by the interception of solar radiation, mainly by green leaves, and the use of this energy in photosynthesis to convert carbon dioxide and water into carbohydrates. This process can be limited by factors, such as temperature and the availability of nutrients and water. The leaf pigments involved (typically chlorophyll) absorb radiation only at short wavelengths (400–700 nm), and the incoming radiation in this region is generally known as photosynthetically active radiation (PAR), which is typically expressed in units of mol $m^{-2}s^{-1}$ or MJ m^{-2}.

Solar irradiance at the Earth surface can vary significantly in directional and wavelength distribution with varying solar zenith angle (diurnal, latitudinal and seasonal effects) and atmospheric composition. Despite PAR being an integrated quantity over incidence angle and wavelength, the proportions of direct and diffuse radiation are sensitive to these terms, and both the proportion of diffuse radiation and the total PAR are significantly affected by cloud cover. In addition, the PAR available to the leaves on an individual plant will be affected by the spatial arrangement of leaves, stems, branches and such like, in the canopy as well as factors such as topography.

An important quantity when assessing light use by vegetation is fAPAR, which we here treat as the fraction of PAR that is absorbed by chlorophyll in the leaves (i.e. omitting the portion absorbed by non-photosynthetic material, such as senescent material, branches and stems; Zhang et al. 2005). This is often approximated by the average proportion of radiation *intercepted* by the canopy because absorption of PAR by green leaves is usually greater than 80% (Monteith and Unsworth 1990). It is possible to measure fAPAR by measuring the bi-hemispherical reflectance and transmittance of the canopy over the PAR region, and subtracting them from unity. In such measurements, it is common to treat the leaves and soil as totally absorbing, so that reflectance is zero and transmittance is unity minus canopy interception, though this will tend to overestimate fAPAR. Further, such measurements do not measure the proportion available for photosynthesis.

Monteith (1972, 1977) used the above concepts in a framework known as the production efficiency model, in which the GPP, the quantity of assimilate produced by photosynthesis, can be written:

$$GPP = \varepsilon \times fAPAR \times PAR \qquad (15.1)$$

Here, the coefficient ε is known as the light-use efficiency. The NPP is

$$NPP = GPP - R_a \qquad (15.2)$$

where R_a is the autotrophic respiration from living plant material, which is often considered to be a proportion of GPP. This model has been extended and applied to satellite data by various authors, including Prince and Goward (1995), Running et al. (2000) and Veroustraete et al. (2002). These latter papers and subsequent articles express ε as the product of maximum efficiency and limiting factors (such as CO_2, temperature and vapour pressure deficit).

A criticism that can be levelled at the production efficiency approach is that it greatly oversimplifies the process of light use by vegetation. In fact, fAPAR is a complex function of the spectral and directional illumination conditions and the arrangement of photosynthesising elements within the canopy. The penetration of diffuse radiation into a canopy is typically significantly greater than that of the direct solar beam, so the fAPAR under overcast conditions is likely to be greater than under clear sky conditions. Also, the photosynthetic response of a leaf to light intensity is non-linear, having an initial linear portion followed by saturation at a certain light level. Since different leaves will be subject to different light levels (the starkest contrast being between sunlit and shaded leaves), this non-linearity complicates the calculation of total absorbed PAR. It is therefore preferable to calculate the proportion of sunlit and shaded leaves before aggregating to canopy photosynthesis. Alton et al. (in press) found that diffuse illumination enhanced light-use efficiency by between 6% and 33% for various FLUXNET forest test sites. This is due to a relatively even distribution of light levels over the canopy in comparison to the direct sunlit case. For direct sunlight, there is more variation in the radiation levels on leaves, with a significant proportion of sunlit leaves being light saturated and shaded leaves receiving too little light to photosynthesise. Other papers, such as Gu et al. (2002), predict an even stronger influence of diffuse illumination, of up to approximately a factor of 2. Another complicating factor is that leaves have varying levels of chlorophyll. When this is considered in models, it is generally assumed that chlorophyll (or nitrogen) levels are proportional to PAR availability (Drouet and Bonhomme 2004).

However, the main motivation for calculating photosynthesis with a production efficiency model is that direct estimates of canopy photosynthesis and production are laborious and can only be conducted at relatively small test plots. When estimating the uptake of carbon at larger scales in space ($1\,km^2$ or greater) and time (weekly to annual), the averaging involved may reduce the impact of many of the local effects. Since the only available data for wider area (regional or global) studies are those derived from satellite observations, production efficiency models represent a pragmatic approach to calculating GPP and NPP, in which major departures from expected or observed behaviour (such as diffuse light effects) can be dealt with by introducing additional multiplicative terms to ε.

One reason why such models are particularly attractive for modelling photosynthesis is that there appears to be a roughly linear relationship between fAPAR and the widely-used normalised difference vegetation index (NDVI), a normalised difference ratio between satellite-measured reflectance at red and near-infrared wavelengths. NDVI is simple to calculate from satellite data, although the relationship is known to be sensitive to many external factors, including viewing and illumination angles and soil spectral properties (e.g. Fensholt et al. 2004). Additionally, its formulation as a ratio helps to reduce systematic calibration errors and it is continuously available from the beginning of the advanced very high-resolution radiometer (AVHRR) satellite sensor era in 1982, making it the longest satellite time series giving global information on land processes. The last point makes the NDVI particularly important in global change studies and has motivated major efforts to reprocess the AVHRR dataset in order to reduce its known deficiencies (http://ltdr. nascom.nasa.gov/ltdr/ltdr.html). However, there are some intrinsic weaknesses in satellite-derived fAPAR values for carbon exchange estimates. For example, at given latitude, polar-orbiting satellites (the main sources of fAPAR data) are constrained to a particular local time because of orbital constraints. In addition, most available fAPAR products involve procedures that combine data from several days to mitigate the effects of clouds.

Various indices other than NDVI have been derived to relate satellite observations to fAPAR, such as those of Gobron et al. (2000). This approach acknowledges the shortcomings in a simple global index like NDVI, and replaces it with indices that are designed to have maximum sensitivity to the parameter of interest and minimum sensitivity to extraneous factors. The formulation of the indices is arrived at through sensitivity analyses using canopy reflectance and atmospheric radiative transfer models. Since sensors vary, these indices need to be separately optimised for different sensors, ensuring consistent values for different satellites. Ruimy et al. (1999) noted that fAPAR derived from AVHRR data was negatively biased relative to predictions by ecosystem models, but Gobron et al. (2006) claim uncertainty in the fAPAR derived from the Sea-viewing Wide Field-of-view Sensor (SeaWiFS) to be better than 0.1 in comparison to ground measurements, with little apparent bias.

Instead of using indices, a physically based model of canopy reflectance can be inverted to recover fAPAR. Canopy reflectance is sensitive to several factors other than those directly affecting fAPAR (such as soil reflectance) and there may be insufficient information in the data alone to account for them all, but the resulting uncertainties can be described through the use of such models. This approach is adopted in the MODIS LAI/fAPAR algorithms (Knyazikhin et al. 1998); these are continually being developed, and the latest version (collection 5) of the LAI/fAPAR product was released in 2007.

From an ecological point of view, a more valuable quantity than fAPAR is LAI, since leaves are the functional units of photosynthesis, rain interception and water and gas exchange. Hence, fAPAR is not an explicit quantity in many carbon models, while LAI is. They can be connected if the biophysical model implies some description of how light is absorbed in the canopy. The simplest and most widely used such description is derived from Beer's law for the proportion of radiation intercepted by the canopy, i_0:

$$i_0 = 1 - e^{-k \times \text{LAI}} \tag{15.3}$$

If leaves are assumed perfectly absorbing, i_0 is equivalent to the fAPAR. Here, the extinction coefficient, k, is equivalent to an attenuation coefficient for black leaves, and is a function of the leaf angle distribution, multi-scale clumping in the canopy and the angle of incident radiation. It is commonly taken to have the value 0.66 for diffuse illumination and $0.5/\cos \theta$ for the direct solar beam (where θ is the illumination zenith angle), this being the value for a spherical leaf angle distribution in a homogeneous canopy. Note that Eq. 15.3 actually describes radiation *interception* by the canopy, rather than fAPAR. It takes no account of leaf scattering (rather than absorption), and does not consider canopy clumping, non-photosynthetic canopy elements, varying levels of chlorophyll or non-linear leaf response to light levels. Interestingly, the LAI of sunlit foliage is given by Monteith and Unsworth (1990) as $(1 - e^{-k \times \text{LAI}})/k$. This is used in some ecological models to upscale leaf-level photosynthesis from direct illumination to the canopy level (e.g. Best 2005).

As with fAPAR, satellite products of LAI are available (Knyazikhin et al. 1998), but there are major difficulties in its estimation because (1) optical sensor response saturates at high LAI (5 or 6); (2) optical reflectance is essentially only sensitive to the product $k \times \text{LAI}$, and k must be prescribed to estimate true LAI. These issues may have little effect on estimates of radiation interception or fAPAR, but are important when relating satellite-derived LAI to values from an ecosystem model (note that ground-based optical methods for measuring LAI suffer from the same problems). Validation for collection 4 of the MODIS LAI/fAPAR product currently shows an average overestimate in LAI of around 12% compared to scaled field data (Yang et al. 2006), with the majority of this bias coming from high LAI values, particularly for grasses and cereals (their Fig. 11). The bias is small up to LAI values of 3, with an uncertainty of less than 0.5.

An alternative to the production efficiency approach for predicting terrestrial carbon flux is to use DVMs, as described in Sect. 15.1. This type of model does not require fAPAR data and the environmental limitations on photosynthesis are more comprehensively modelled. Typically such models deal with light interception in a more sophisticated way by calculating vertical profiles of light absorption throughout the canopy and taking into account the proportions of sunlit and shaded leaves. Nonetheless, they still contain an implicit concept of fAPAR, so can be compared with satellite data. In the simplest case, this would entail inserting values of LAI calculated by the model into Eq. 15.3 to calculate fAPAR, but ideally fAPAR would be calculated by considering the absorbed radiation at each level in the canopy.

Figure 15.4 shows fAPAR data for May and August 1999 over Europe derived from SeaWiFS and calculated from the SDGVM. A major difference between the two datasets, clear from the scatter plots, is a strong positive bias in the model values relative to the satellite product. This is partly explained by the different spatial patterns in the two datasets. For example, the model predicts much higher fAPAR in northern than mid-latitude Europe, whereas SeaWiFS shows the opposite. It is

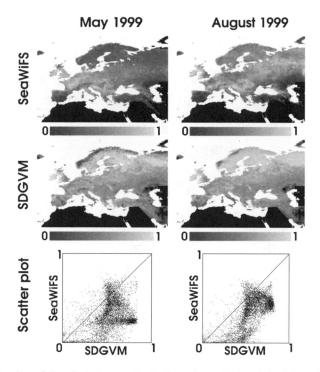

Fig. 15.4 Fraction of absorbed photosynthetically active radiation derived from SeaWiFS data (top) and estimated by the Sheffield Dynamic Global Vegetation Model (middle) for (left) May, (right) August 1999. Scatterplots of the two datasets are shown at the bottom, with SDGVM values along the *x*-axis (fAPAR supplied by the Joint Research Centre of the EU, Ispra).

not known whether the satellite data or the model represents these spatial patterns better. The difference is unlikely to be due to errors in land cover since the spatial distribution of vegetation in the model is given by satellite-derived land-cover maps. However, a possible factor is that the light interception routines in the SDGVM do not account for canopy heterogeneity. Two forest areas with the same LAI may have very different stem densities, resulting in different values of the extinction coefficient *k* in Eq. 15.3, and hence different fAPARs. The sparser the tree crowns, the lower the fAPAR. Ignoring this will lead the SDGVM to overestimate fAPAR in open canopies, such as are typical in northern forests.

The satellite-derived fAPAR can be used to predict GPP via a production efficiency model. Figure 15.5 shows the GPP that directly corresponds to the fAPARs in Fig. 15.4. To characterise the efficiency, ε, the algorithm of Veroustraete et al. (2002) was used, and to ensure as much comparability as possible between the models, the same climate data sets were used to drive both. Although there are still marked differences in spatial pattern, the general trends across Europe show much better agreement. To some extent, this reflects the fact that any trend in the common PAR data will affect both estimates. However, the agreement in *overall* magnitude

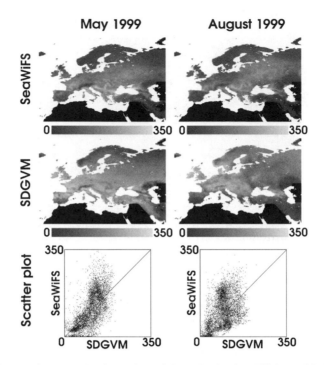

Fig. 15.5 Gross Primary Production estimated by a Production Efficiency Model driven by SeaWiFS fraction of absorbed photosynthetically active radiation (top) and by the Sheffield Dynamic Global Vegetation Model (middle) for May (left) and August 1999 (right). Scatterplots of the two datasets are shown at the bottom, with SDGVM values along the *x*-axis.

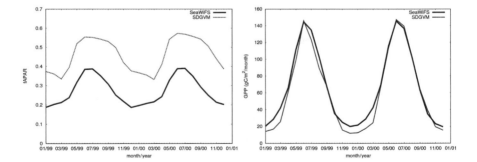

Fig. 15.6 Temporal trend of fraction of absorbed photosynthetically active radiation (left) and Gross Primary Production (right) for 1999 and 2000, averaged across the whole of Europe. The solid and dashed lines show SeaWiFS data and Sheffield Dynamic Global Vegetation Model calculations, respectively.

is still much better than would be expected from the large differences between model and measured fAPAR shown in Fig. 15.4. The implication is that the efficiency used in the production efficiency model compensates for the differences in fAPAR to yield the same effective light use as the SDGVM.

Figure 15.6 shows the temporal trajectories of fAPAR and GPP from the sources used in Figs. 15.4 and 15.5, averaged over the whole of Europe for 1999 and 2000. The modelled and measured values of fAPAR display similar temporal behaviour, but there is a strong bias between the two datasets. Remarkably, the GPPs are very similar in magnitude and behaviour across the entire time span, demonstrating again that the efficiency used in the production efficiency model is in effect tuned to yield the same light use as the DVM.

In summary, fAPAR should, in principle, provide a firm point of comparison between satellite measurements and DVMs:

- It is a radiometric property and hence more closely related to the satellite signal than parameters such as LAI;
- Most DVMs contain some concept of radiation absorption and hence it is possible to derive a value of fAPAR, even if it is not explicitly calculated within the model itself.

However, the brief results above indicate some of the difficulties in comparing models with satellite-derived fAPAR data, and using these data for regional carbon flux estimates. A valuable step forward would be to enhance the way PAR absorption is represented in the DVMs, to account more accurately for factors that control the satellite signal, such as crown level clumping, heterogeneity at the landscape level, the spectral and angular distribution of the PAR irradiance and multiple scattering within the canopy. Greater confidence is also needed in the accuracy of the satellite products, which are mostly validated at only a small number of sites. This should include a comprehensive assessment of their reliability under a range of conditions and for different biomes. Such work is underway by several teams but progress is slow. Perhaps most important is for the model and remote-sensing communities to commonly recognise that developments are needed in both domains in order to combine the two sources of information, and hence improve our estimates of carbon flux.

15.2.3   *Fire*

Globally, on average, vegetation fires generate carbon emissions equivalent to between one third and one half of those from fossil fuels. Emissions from fires exhibit great variation in time and space, have different properties (e.g. emission rate, chemistry, injection height) in different biomes and are affected in complex ways by human behaviour. The sporadic, patchy nature of fires, their large scale and their frequent occurrence in regions with few systematic observations make data from satellites one of the most important sources for their quantification. Two key fire phenomena can be identified: actively burning fires and their radiated power, and mapping of post-fire burnt area. In addition, enhanced concentrations of fire-emitted trace gases, such as CO and methane, and detection of smoke aerosols can be identified and measured from space, and it is becoming possible to detect the associated perturbations of atmospheric CO_2 concentration.

Active fires (hot spots): Detection of active fires is provided by instruments measuring in the mid-to-thermal infrared (~3–14 µm), such as AVHRR, MODIS, the Advanced Along-Track Scanning Radiometer (AATSR) and certain of the recent geostationary satellites (e.g. Meteosat-8). The great intensity of thermal radiant energy emission within the shorter wavelength range of this spectral region ensures that unobscured fires covering even a very small fraction (10^{-4} to 10^{-3}) of the sensor instantaneous field of view are detectable. Systematic global monthly maps of nighttime active fire counts from November 1995 to the present are provided by the European Space Agency (ESA) in the World Fire Atlas (http://dup.esrin.esa.it/ionia/wfa/index.asp), based on data from the ATSR-2 on ERS-2 and AATSR on Envisat (see also http://earth.esa.int/rootcollection/eeo9/earth_esa.html). Whilst unprecedented in their duration, there are several reasons why these data are relatively limited as regards inferring atmospheric emissions from fires, including

- In regions dominated by grassland fires, such as the African savannah, the number of fires may be an order of magnitude or more lower at night than in the day (see Fig. 15.8). Similar, but less exaggerated, differences occur during forest fire episodes. Data on the typical extent of the tropical fire diurnal cycle is provided by active fire detections from the Tropical Rainfall Measuring Mission (Giglio et al. 2003).
- Active fire detections provide no information on fire intensity, so that relatively small, weakly burning fires are not distinguished from much more intense events covering far greater proportions of the sensor instantaneous field of view, yet the latter will clearly emit much greater amounts of material to the atmosphere.
- Cloud hinders detection of underlying fires, but the amount of cloud cover is not provided. Further, orbit characteristics mean that the satellite may not view an area when a fire occurs or a fire-front passes through an area. Absence of fire detections may, therefore, not imply absence of fires.
- Fire counts are processed per scene, and orbital convergence means that there are more viewing opportunities at higher latitudes, so the data suffer from a latitudinal bias.

Nonetheless, active fire counts are an invaluable source of information on the seasonal and geographic variations in fire occurrence, and provide some indication of levels of inter-annual fire activity variation that can be used to scale emissions data from other sources (Schultz 2002). They can also be of great help when validating satellite-derived estimates of burnt area, and cumulative active fire counts have even been used as a proxy for burnt area when alternative data are unavailable (Soja et al. 2004).

Burnt area: Satellite measurements in the near and short wave infrared are sensitive to changes in surface reflectance caused by burning, and mapping of burnt area mostly relies on detecting changes in indices based on these wavelengths. Currently, the most detailed and wide-ranging products of this type were produced in the Global Burnt Area 2000 (GBA2000) project (http://www-tem.jrc.it/Disturbance_by_fire/products/burnt_areas/global2000/global2000.htm), which provides maps of monthly burnt area at 1-km resolution, together with statistics by country and by region, based on a series of different ecosystem-dependent 'fire-scar' detection algorithms devised in collaboration with local teams. Unfortunately, GBA2000 provides data only for the year 2000. However, in late 2007, the ESA Globcarbon project will release monthly composites of global burned area for 1998–2003 at ½°, ¼° and 10-km resolutions, together with some regional products at 1-km resolution.

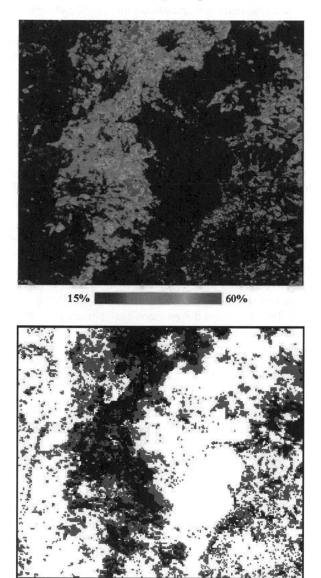

Fig. 15.7 (a) Percentage change in NIR reflectance on day of burn, measured from MODIS reflectance data. This will usually be larger for higher intensity burns or where a bigger proportion of the pixel is burned. (b) Burnt areas detected by both MODIS and GBA 2000 are marked in black, those by only MODIS in red. The MODIS product detects a significantly larger burnt area than GBA2000, although the latter successfully detects the large fires (*See Color Plate* 8).

An important methodological advance was made by Roy et al. (2001), who used a model of reflectance to account for day-to-day variations in sensor viewing and illumination geometries. This is more powerful than previous empirical approaches, and is flexible enough to be extended to multiple instruments. It provides a much firmer basis for burnt area detection and allows more subtle changes to be detected (Fig. 15.7). A major enhancement in information on burnt area comes with the recent release of monthly 500-m resolution global burnt area maps derived from the full MODIS dataset (1999 to the present) using the Roy et al. (2001, 2005a) method, and supported by robust validation protocols (Roy et al. 2005b). This dataset also includes the approximate date of burning.

A valuable interim step taken by Giglio et al. (2006) prior to the new MODIS dataset was to establish regional correlations between MODIS active fire counts and burnt area estimates derived from selected MODIS 500-m imagery. This allowed the production of $1° \times 1°$ resolution global monthly burnt area maps for the period from late 2000 to mid-2005. Van der Werf et al. (2006) have used this information, together with model-based estimates of fuel-load and combustion completeness, to derive a widely used monthly global biomass burning emissions database.

At the European scale, annual satellite-based mapping of burnt areas larger than 50 ha has been carried out since 2000 within the European Forest Fire Information System (EFFIS; http://effis.jrc.it/Home/). The area limitation derives from the use of the 250-m spatial resolution MODIS optical channels, but is believed to be adequate since fires of this size generate 75% of the total area burnt in Europe (Joint Research Centre 2006). The burned area maps can be readily combined with land-cover data, for example, from the European Comission Programme for Coordination of Information on the Environment (CORINE) system, to provide a breakdown of the areas of each land-cover type affected by fire; for recent years, this is a routine part of the EFFIS system on a per country basis. As an example, Table 15.2 gives the values for Portugal for 2005 (Joint Research Centre 2006).

Fire radiative power: Converting burned area into estimates of fuel consumption and carbon release requires information on the amount of biomass per unit area available to be burnt, and the fraction of that fuel which actually burns. Information on these biomass density and combustion completeness parameters is not easy to

Table 15.2 Totals of burnt areas of at least 50 ha and their distribution among land-cover classes for Portugal in 2005 (source: Joint Research Centre, 2006)

Land cover	Area burned (ha)	% of total area burned
Forest land	248,906	90.3
Agriculture	25,191	9.1
Wetlands	1,118	0.4
Artificial surfaces	408	0.2
Total	275,622	100.0

acquire, although there are prospects for deriving biomass from space (see Sect. 15.4). Hence, more direct methods have been sought, of which by far the most promising is the use of fire radiative power observations (Wooster et al. 2005; Roberts et al. 2005). The key relationship involved is that the biomass consumed is proportional to the integrated total radiant energy emitted by the fire. Instantaneous radiative power measures can be calculated from estimates of the fire effective temperature and area, derived from high-spatial resolution, multi-spectral data such as that provided by the Bi-spectral Infrared Detection (BIRD) micro-satellite mission (Wooster et al. 2003; Zhukov et al. 2005), or from lower spatial resolution systems through an approach based on quantifying the fire's spectral radiant energy emissions in the middle infrared (Wooster et al. 2005). Observations at such middle infrared wavelengths are available from several satellite sensors, but to measure the integrated energy requires repeated measurements not too widely spaced in time, together with a sensor whose dynamic range is sufficient not to saturate over the more intense events. A near-ideal sensor for this purpose is the Spinning Enhanced Visible and Infrared Imager (SEVIRI) instrument onboard the Meteosat-8 geostationary satellite, which can provide radiative power measurements every 15 min over the African continent at a 3 km sub-satellite sampling distance. An example of the power of this technique is seen in Fig. 15.8, which shows the total biomass

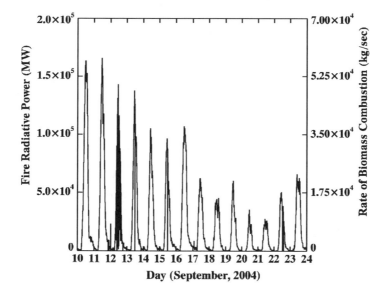

Fig. 15.8 Fire radiative power (FRP) observations made by the SEVIRI sensor on Meteosat-8 over southern Africa for 2 weeks in September, 2004. Data are cumulative FRP from all detected fires at 15-min temporal resolution, adjusted for cloud-cover variations on a 1° grid-cell basis with the methodology in Roberts et al. (2005). The equivalent biomass combustion rate is shown on the RH y-axis. Total fire-radiated energy over the period is 2.6×10^{10} MJ, equivalent to consuming 9.6 Tg of fuel. Data dropouts affect some of the measurements, particularly on 12 September.

combustion over southern Africa for 2 weeks in September 2004. The strong
diurnal variation, peaking in the early afternoon, is very obvious (and illustrates the
weakness of relying only on night-time observations, such as those used by ATSR
for active fire mapping).

The relationship between integrated radiant energy emission and fuel consumption has been tested across a range of combustion conditions and fuel types
(Wooster et al. 2005), and matches between instantaneous radiative power observations made by high and low-spatial resolution systems can be used to adjust the
geostationary-derived estimates for bias caused by their inability to detect the
smallest/least intense fires. The total biomass consumed during the 2-week period
covered by Fig. 15.8 is estimated as 9.7 Tg, not including adjustment for fuel moisture content (which is expected to be low at this stage of the dry season).

Up to now, the fire radiative power method has only been applied to SEVIRI, but
other geostationary satellites with potentially suitable spectral channels include
GOES-E and GOES-W, MTSAT, FY2C and INSAT3D. Together with the future
Russian geostationary systems, these systems should offer near-global coverage of
fire-affected zones below 50°–55° latitude, thus including all of southern Europe.
Above 55°, the obliqueness of the geostationary view limits the accuracy of this
approach, though at higher latitudes, polar orbiting data from sensors such as
MODIS can be used, since their revisit frequency maximises in such regions. Even
at lower latitudes, regional-scale radiative power estimation may be possible by
integrating data from several polar-orbiting satellite overpasses, as demonstrated by
Roberts et al. (2005).

In addition to their successes, it is important to highlight some of the limitations
of current methods for estimating burned area and radiative power from remote
sensing. One important limitation, common to both techniques, is that a higher
proportion of surface fires burning under dense forest canopies will be missed than
will fires in more open ecosystems. An example is rainforest fires, though these are
usually confined to areas already partly impacted by human activity, hence tend to
occur in areas where the canopy has already been disturbed. At present, there is no
well-validated method of quantifying the true number or extent of such below-canopy
fires and unadjusted emissions estimates derived from such regions are, therefore,
likely to be underestimates.

Development of validation procedures for fuel consumption/carbon release estimates made by the burned area and radiative power approaches is continuing, with
the likelihood that a combination of high spatial resolution burned area maps and
field sampling offers the best way to validate emissions from individual fire events.
At regional to global scales, probably the best datasets for comparison are top–down
source estimates derived from inversion of atmospheric trace gas concentration
measurements, currently predominantly CO measurements from the Measurement
of Pollution in the Troposphere (MOPITT) sensor (Arellano et al. 2006). However,
such inversion-derived estimates are not without their ambiguities; for example, they
rely on assumptions about the injection height of emissions and the emissions
factor relating the amount of biomass burnt to the amount of CO released. The possibilities of future remotely derived estimates of emission injection height (e.g. from

multi-view angle data) and for concentration measurements of increased numbers of trace gas species (e.g. from the Scanning Imaging Absorption Spectrometer for Atmospheric Chartography, SCIAMACHY, the Orbital Carbon Observatory, OCO and the Greenhouse Gases Observing Satellite, GOSAT), therefore offer considerable potential for further improvement.

15.2.4 Forest Biomass

One of the biggest information gaps in carbon cycle studies at regional to global scale is the distribution of biomass. Available data on above-ground biomass mainly comes from inventories, which form the backbone of national reporting on forest area and biomass to the United Nations Framework Convention on Climate Change (UNFCCC). Global and national summaries are provided by the FAO in their forest resources assessments at ~5-year intervals, but the data are not gridded, and are subject to errors that differ widely between countries. The only available global gridded biomass data are based on old country and sample plot statistics (Olson et al. 1983), with some modifications based on land cover (World Resources Institute 2000); they almost certainly contain very large, geographically varying errors. This has motivated efforts to recover biomass from remote sensing. Because optical data primarily provide information about biochemical properties of the tree elements (mainly leaves), much of this effort has focussed on the use of radar or lidar.

Radar waves interact with a forest canopy through scattering and attenuation by the tree elements making up its foliar or woody biomass and by scattering from the soil. Since scatterers that are small relative to the wavelength make an insignificant contribution to the backscatter, the tree elements playing the major role in these interactions change with wavelength. Theoretical studies indicate that, in general, most of the scattering is from leaves or needles at the shortest wavelength used for vegetation remote sensing (about 3 cm), while trunk scattering dominates at the longest wavelengths (several metres).

Scattering increases with increasing volume of free water (and hence biomass) in the living vegetation components (leaf, branch or trunk), so the canopy backscatter increases with biomass. However, our ability to exploit this increase is limited by two factors. Firstly, the total backscatter from the forest consists of both canopy and soil returns, since microwaves can penetrate the canopy, and the canopy term needs to be isolated if we are to recover biomass. This mixing of signals is most important for young or sparse forest, since, as biomass increases, so does signal attenuation, causing the soil return to decrease. Secondly, the canopy backscatter saturates as biomass increases, which imposes an upper limit on the range of biomass that can be measured. Hence existing C-band radar data from the European Remote Sensing Satellite (ERS), the Canadian Radarsat and the European Environmental Satellite Advanced Synthetic Aperture Radar (Envisat ASAR) are of little value for measuring biomass, since saturation occurs at around 20–30 t/ha and the return is significantly affected by soil scatter below this value. The recently launched Japanese Advanced Land Observing Satellite–Phased Array type L-band Synthetic Aperture Radar (ALOS–PALSAR) L-band (24-cm wavelength) radar will do better, since at this wavelength, the soil effects are weaker, while saturation occurs at around 50 t/ha (we comment further on this saturation value below). In addition, ALOS has an image acquisition plan that will systematically provide annual (or better) global coverage of the main forest biomes and wetlands.

Although C-band backscatter is not useful for measuring biomass, another radar technique known as radar interferometry has been used to retrieve biomass up to 30–50 t/ha from C-band satellite data. This was possible when the two ERS satellites were in nearly identical orbits that allowed images of the same region on the ground to be acquired at intervals separated by 1 day. Such tandem data were only available up to 1998 and covered only part of the Earth's surface. However, there was good coverage of Europe, and a special tandem campaign in 1997–1998 imaged a 750,000 km^2 region in central Siberia. The latter allowed production of biomass maps (96,100 × 100 km^2 maps were mosaiced to form a consistent overall map with 50-m pixel size) which brought out strongly the importance of low-biomass forest and non-forest areas in this region. Overall, 4.5% of the area is occupied by water (including Lake Baikal), 5.4% by agriculture, bogs and grasslands, 7.0% by open forests due to disturbance (fire, logging, deforestation and windthrow), 15.1% by low-biomass forests (10–50 t/ha) and 40.6% by forests with biomass exceeding 50 t/ha. Post-disturbance forest areas, corresponding to open and low-biomass forests, account for 22.1% of the region, with a total area of ~165,700 km^2 (Le Toan et al. 2004). Failure to include such areas means that both the Olson et al. (1983) data and model calculations greatly overestimate the vegetation carbon stocks in the region (Le Toan et al. 2004). Such a large proportion indicates that it is crucial to understand the role of low-biomass areas in the carbon balance of the region. It also shows that saturation at 50 t/ha still permits very useful information on forest dynamics to be inferred; the global low-biomass data being acquired by ALOS should, therefore, add significantly to our knowledge on deforestation, forest degradation, low-productivity forest and forest regrowth in the tropics.

Attempts to increase the measurable range of biomass are following two paths. The first is to build sensors operating at longer wavelengths, and Fig. 15.9 indicates why this is desirable. It shows measurements from airborne radar systems operating at L-band (24-cm wavelength), P-band (68 cm) and very high frequency (VHF) (several metres). Saturation can clearly be seen for L- and P-band, with P-band exhibiting a higher level of saturation biomass, but VHF shows no sign of saturation. Unfortunately, the refraction of long wavelengths by the ionosphere means that currently the longest wavelength feasible for a space-based system is at P-band. Such a system (the BIOMASS mission) is in fact under consideration by the ESA as the next in line for their Earth Explorer series (but would not go into orbit before 2014).

The second approach to measure biomass is instead to measure vegetation height by either lidar or radar methods, and to recover biomass from regional allometric equations. This, in principle, removes the saturation limit, allowing full global recovery of biomass, and has been demonstrated with airborne systems, including radar systems operating at both L- and P-band (Garestier et al. 2007;

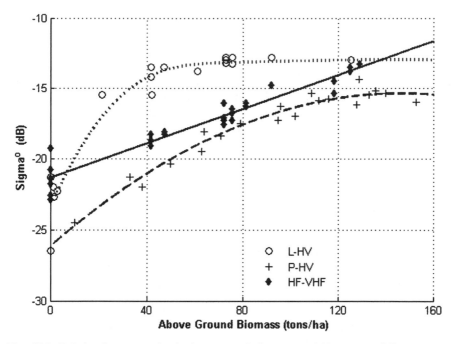

Fig. 15.9 Relation between radar backscatter and above-ground biomass at different wavelengths, derived from airborne measurements gathered over the Landes forest, France by three radar systems: the French RAMSES at L-band, the NASA AirSAR at P-band and the Swedish Defence CARABAS at very high frequency (VHF).

Kugler et al. 2007). Radar systems capable of performing this measurement must have both polarimetric and interferometric capabilities, and these form part of the specification of the BIOMASS mission. A spaceborne vegetation lidar (A-SCOPE) is also being considered by ESA as an Earth Explorer mission.

15.2.5 Low-Level Products: Optical Radiance and Microwave Backscatter

Sections 15.2.1–15.2.5 dealt with geophysical products relevant to carbon flux calculations that can be derived from satellite data. However, as discussed in Sect. 15.1 and illustrated in Fig. 15.1, these products have to be brought together with other information in some sort of model if a full carbon calculation is to be made. A fundamental issue in doing this effectively, and one that is often poorly understood, is to ensure that the measurements and the model representation of processes are conceptually consistent. As an example, many models represent fire as a parameterised stochastic process driven by probabilities derived from calculated fuel load, soil moisture, climate and such like. It would, therefore, be incorrect to expect the fires actually seen by a satellite to be predicted by the model; instead, conceptual consistency between the model and the data implies that the temporal and spatial statistics of the observed and modelled fire behaviour should agree, at some scale. If this modelling approach is adopted, then the role of satellite data is in process parameterisation, requiring long time series and global coverage. (Other ways to use fire observations in models are clearly possible. This is an area where considerable development can be expected, in part because of the new observations described in Sect. 15.3, and in part because of new approaches to physical and statistical modelling of fire.)

A more subtle form of potential inconsistency between carbon models and satellite data is when the remote-sensing product is derived using a model for the scattering or reflecting medium (e.g. vegetation may be represented as uniform slab of randomly oriented leaves). This leads to two possible sources of inconsistency:

(1) the model for the medium used to derive the remote-sensing product may be (and usually is) different from that used in the carbon model;
(2) the model for the medium may not capture the realities of the measurement (e.g. in the real world the vegetation may be clumped).

Inconsistencies of type (2) arise from unconsidered factors that may bias the measurement and widen its uncertainty relative to the modelled values. Inconsistencies of type (1) are more fundamental, since the quantities derived from the measurement process and from the carbon model are essentially different, and hence cannot be directly compared. A good example is shown in Fig. 15.4, where different models of the vegetation–soil system are used to derive the fAPAR values from satellite data and from the SDGVM. Hence, differences in absolute value and even in spatio-temporal pattern cannot be readily interpreted.

Avoiding such fundamental inconsistency between derived and modelled values requires both to be based on the same assumptions about the soil–vegetation medium. Unfortunately, this is usually hard to realise in practice, since the products are developed by groups that are disconnected from the carbon modellers. This problem can, however, be reduced by dealing with low-level products that make few assumptions about the medium (and which, as a further advantage, usually have well-characterised measurement errors). A second requirement for achieving consistency is that there must be an observation operator that predicts the measured quantity from the state vector of the model; the model and measurement can then be directly compared, with no internal inconsistency. This provides a secure basis for combining model and measurement information. Such an approach has become fundamental for atmospheric science but is still poorly developed for the land surface. However, it is likely to form a major future direction for robust use of satellite data in carbon models.

The use of low-level products suggests that one would ideally work in terms of the measured radiance or backscatter. However, two qualifications are needed. Firstly, satellite measurements are made above the atmosphere, whose properties are not normally part of the state vector, so that atmospheric correction based on external information is often needed. Hence, we have to accept some degree of modification of the primary measurements to yield the products actually used. Secondly, inconsistencies of type (1) between the model representation and the measurement may still exist and must be characterised if differences between the model and data are to be properly interpreted.

Use of the system state vector to predict the observed signal measured at a satellite can be applied at both microwave and optical frequencies. For example, in a study of the use of microwave coherence to estimate the age of plantation forests in the UK, Drezet and Quegan (2006) used the soil–plant–atmosphere model (Williams et al. 1996) to predict the coherence measured jointly by the two ERS satellites when they were operating in tandem mode. Driving the model by weather data produced time-varying estimates of canopy and soil moisture, which were then used to drive a simplified electromagnetic model for coherence. This allowed a consistent test of the model, and the predicted coherence values were indeed found to fit reasonably well to observations. However, despite giving insight into the factors affecting coherence, the conclusion of this study was that operational methods to infer forest age from coherence would still have to rely on semi-empirical methods.

The most powerful application of an observation operator is when the difference between the values predicted from the model state vector and a time series of measurements is fed back to improve the estimate of the state vector. This sequential updating of the state vector using data constraints is known as data assimilation. It requires not just a state space model, observation operator and measurements, but also characterisation of the error covariances in all three components, from which the uncertainty in the predicted state vector can also be quantified (Raupach et al. 2005).

This approach is becoming increasingly important in carbon budget modelling, and has been partially implemented by several researchers. For example, Knorr and Lakshmi (2001) predict fAPAR from the Biosphere Energy-Transfer Hydrology (BETHY) state vector using a canopy reflectance model derived from Gobron et al. (1997), and use it to optimise the parameters in the BETHY model (an approach that is perhaps more accurately described as model calibration). This has been subsumed into the Carbon Cycle Data Assimilation System (CCDAS: Rayner et al. 2005). However, uncertainty in the predicted fAPAR is not accounted for and the measured fAPAR is not based on the same assumptions about the soil–vegetation medium as those used by BETHY. Other papers developing observation operator and data assimilation concepts include Nouvellon et al. (2001) and Vivoy et al. (2001).

Use of data assimilation to merge satellite reflectance data with a carbon flux model is described by Quaife et al. (2007), based on previous successful assimilation of ground and weather data into the DALEC ecosystem model (Williams et al. 2005). In Quaife et al. (2007), the top-of-canopy reflectance is both predicted from the state vector of DALEC and derived from MODIS data. These values are then combined in an ensemble Kalman filter to improve the estimate of the state vector and reduce its uncertainty. A notable technical issue is that the radiative transfer calculation used to infer reflectance from the state vector is non-linear, requiring an augmented state vector approach. Figure 15.10 indicates the power of this technique, with the upper and lower panels respectively showing DALEC predictions

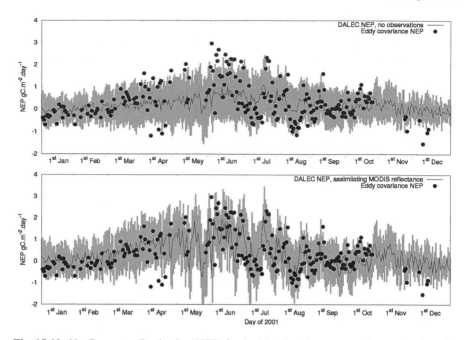

Fig. 15.10 Net Ecosystem Production (NEP) for the Metolius, Oregon, test-site, predicted by the Data Assimilation-Linked Ecosystem Carbon (DALEC) model with (bottom) and without (top) assimilating MODIS reflectance data. The light grey zone around the ensemble mean is one standard deviation of the ensemble. The black dots are data acquired by the eddy covariance tower.

of NEP before and after assimilating MODIS data for the Metolius, Oregon, test site (Law et al. 2001). Observations from the eddy covariance tower at the site are marked by dots. The improved correspondence between modelled and observed values after data assimilation is obvious.

15.3   Satellite Measurements of Atmospheric Trace Gases

Section 15.2 dealt with satellite measurements of surface properties and the issues involved in exploiting them in carbon exchange calculations. None of the measured quantities are direct carbon variables, so their use relies on the context provided by biospheric models. In contrast, several satellites make direct measurements of the concentrations of atmospheric trace gases, such as carbon dioxide, methane, nitrogen dioxide, ozone, formaldehyde and carbon monoxide (e.g. Bovensmann et al. 1999; Burrows et al. 1999). The step from these concentrations to surface fluxes is still indirect and relies on atmospheric transport models (and chemistry models for the shorter-lived species). Biospheric models can be exploited in this inversion problem both to provide prior information on the fluxes, and to help interpret the fluxes in terms of underlying processes. (Both issues also arise when using the *in situ* surface CO_2 network to infer surface fluxes.)

An additional complication when using space observations is that the concentrations are inferred from measurements of radiation that has traversed the whole atmospheric column and that may be sensitive to particular altitude regions of the atmosphere. Most current sensors relevant to carbon cycle estimates do not provide height profiles but instead measure a weighted integral of the atmospheric column concentration; the weighting depends on the technique used. To recover information on the location of trace gas sources and sinks, it is preferable to have the greatest sensitivity in the lower troposphere, otherwise interpretation of the data relies on models for vertical gas transport. For example, the SCIAMACHY instrument uses wavelengths in the near infrared and is most sensitive to CO_2 concentrations in the planetary boundary layer, whereas the sensitivity of the Atmospheric Infrared Sounder (AIRS) instrument, which uses wavelengths in the thermal infrared, peaks at altitudes around 12 km in the tropics and 7–9 km at mid-latitudes. Consequently, there are large temporal differences in the phase of the CO_2 anomaly relative to the monthly mean column volume mixing ratio measured by each instrument, and the anomalies are significantly smaller in the AIRS data (Barkley et al. 2006a), making the measurements harder to interpret in terms of surface fluxes.

In general terms, satellite observations of CO_2 provide much denser spatial coverage than the surface network, but have poorer temporal sampling and measurement precision, and their value has to be assessed within the context of this trade-off. A target for satellite measurements of column content CO_2 is that a precision of 1% within an 8° × 10° footprint is necessary to improve on the existing surface network (Rayner and O'Brien 2001). Current measurements from SCIAMACHY approach but apparently do not meet this stringent requirement, yet nonetheless show significant structure at resolutions of 1° × 1° that are absent from model simulations (Barkley et al. 2006b).

Two new satellite missions dedicated to measuring atmospheric CO_2 are likely to improve significantly on the performance of SCIAMACHY. These are the NASA Orbiting Carbon Observatory (OCO) and the Japanese GOSAT, both due for launch in late 2008. The precision of OCO is expected to be better than 1 ppmV (0.3%) for monthly averages over spatial scales around 1000 × 1000 km^2 (Crisp et al. 2004), which is well within the target quoted above. The target precision for GOSAT is 0.3% at best under clear sky conditions (Center for Global Environmental Research 2006). OCO, in particular, has been the subject of detailed performance assessments as regards recovering surface CO_2 fluxes over both land and ocean. With reasonable assumptions on measurement and model uncertainties, these lead to the conclusion that, at the 3.75° × 2.5° spatial grid scale of an atmospheric transport model, assimilation of OCO data can give up to 15–45% error reduction at a weekly timescale over land. The error in the monthly carbon budget over Europe is reduced by 70%, leaving a residual uncertainty of 0.16 Gt C. However, undetected small regional biases in the CO_2 retrieval (e.g. owing to aerosols) can adversely affect the regional flux estimates, leading to biases that for Europe are larger than the root mean square error (RMSE) of the prior estimates (Chevallier et al. 2007). Hence, OCO (and GOSAT) will find their greatest value in conjunction with other CO_2 observing systems.

Methods developed for satellite CO_2 data could equally be applied to methane and other trace gases. For example, lower tropospheric observations of methane by SCIAMACHY have been compared with predictions from the TM3 chemistry-transport model. The major discrepancy between the two estimates originates in the tropical forests, and suggests that a currently undefined methane source appears to be operating in these forests (Frankenberg et al. 2005).

A comparatively undeveloped area where satellite observations of the atmosphere and land surface can be combined is in achieving a better description of the emissions because of fire. In particular, observations of fire radiative power (Sect. 15.3) allow emissions of trace gases and aerosols to be predicted using emission factors such as those in Andreae and Merlet (2001). Within the constraints of a chemistry-transport model, these should be consistent with atmospheric observations. Hence, by exploiting the existing structures developed to infer CO_2 sources and sinks from atmospheric concentrations, we should be able to derive information on one of the more poorly described source terms, with associated improvements in the whole inversion procedure.

15.4 Conclusions

The recent comparison of coupled climate–carbon cycle models by Friedlingstein et al. (2006) revealed large uncertainties in describing the land carbon cycle. A crucial factor is the lack of strong data constraints on the models. Satellite data can play a major role in addressing this problem through advanced data assimilation techniques combined with observation operators that consistently link ecosystem state to measurements of the land surface from space. Up to now, the emphasis has been on using derived products (e.g. LAI/fAPAR) with carbon models. Whilst this approach can be of great value, and will continue to be developed, it has important weaknesses: (1) similar products from different sources are inconsistent; (2) product derivation uses different assumptions from those in ecosystem models; (3) the error properties of the products are poorly known, reducing their value in data assimilation. These problems can be avoided by schemes that use observation operators based on radiative transfer to link model state with low-level satellite measurements, such as radiance, backscatter and coherence, in a fully consistent manner. Successful schemes of this type have been implemented and point the way to much more effective use of remote-sensing data in improving carbon calculations.

Improved use of satellite measurements of surface properties provides one avenue to better carbon flux estimates, but major new insights should also follow from the new generation of satellite observations of column CH_4 and CO_2. While not being continuous or as precise as ground-based atmospheric concentration measurements, the satellite data should provide repetitive global coverage with sufficient precision to improve the flux inversions from the sparse ground network, particularly when used as an additional data stream. The OCO and GOSAT should also have sufficient spatial resolution to improve flux estimates at regional scales (~100 km).

A major challenge in the next few years is to combine surface and atmospheric measurements from satellites with ground measurements in a unified data assimilation approach to recover both surface fluxes and values of the parameters controlling these fluxes. This overall concept is set out in the International Global Carbon Observations strategy (Ciais et al. 2003). This paper indicates that many of the elements needed to make this happen are in place or well developed, and we are close to moving from concepts to practice.

Acknowledgements We would like to thank Paul Monks (University of Leicester) and Thuy Le Toan (CESBIO Toulouse) for comments and improvement of the text.

References

Alton, P. B., North, P. R., and Los, S. O. (in press). The impact of diffuse sunlight on canopy light-use efficiency, gross photosynthetic product and net ecosystem exchange in three forest biomes. *Global Change Biol.*, online: 10-Jan-2007 doi: 10.1111/j.1365–2486.2007.01316.x

Andreae, M. O., and Merlet, P. (2001). Emission of trace gases and aerosols from biomass burning. *Global Biogeochem. Cycles*, **15**(4), 955–966.

Arellano, A. F., Kasibhatla, P. S., Giglio, L., van der Werf, G. R., Randerson, J. T., and Collatz, G. J. (2006). Time-dependent inversion estimates of global biomass-burning CO emissions using Measurement of Pollution in the Troposphere (MOPITT) measurements. *J. Geophys. Res.*, **111**, D09303, doi:10.1029/2005JD006613.

Barkley, M. P., Monks, P. S., and Engelen, R. J. (2006a). Comparison of SCIAMACHY and AIRS CO_2 measurements over North America during the summer and autumn of 2003. *Geophys. Res. Letts.*, **33**, L20805, doi:10.1029/2006GL026807.

Barkley, M. P., Monks, P. S., Friess, U., Mittermeier, R. L., Fast, H., Korner, S., and Heimann, M. (2006b). Comparisons between SCIAMACHY atmospheric CO_2 retrieved using (FSI) WFM-DOAS to ground based FTIR data and the TM3 chemistry transport model. *Atmos. Chem. Phys.*, **6**, 4483–4498.

Best, M. (2005). *JULES Technical Documentation*, Met Office, Joint Centre for Hydro-Meteorological Research, Wallingford, UK. (based on Hadley Centre Technical Notes 24 (TRIFFID) and 30 (MOSES 2.2)).

Bonan, G. B., Levis, S., Kergoat, L., and Oleson, K. W. (2002). Landscapes as patches of plant functional types: An integrated concept for climate and ecosystem models. *Global Biogeochem. Cycles*, **16**(2), 1021–1051.

Bovensmann, H., Burrows, J. P., Buchwitz, M., Frerick, J., Noel, S., Rozanov, V. V., Chance, K. V., and Goede, A. P. H. (1999). SCIAMACHY: Mission objectives and measurement modes. *J. Atmos. Sci.*, **56**, 127–150.

Burrows, J. P., Weber, M., Buchwitz, M., Rozanov, V., Ladstatter-Weissenmayer, A., Richter, A., DeBeek, R., Hoogen, R., Bramstedt, K., Eichmann, K. U., and Eisinger, M. (1999). The global ozone monitoring experiment (GOME): Mission concept and first scientific results. *J. Atmos. Sci.*, **56**, 151–175.

Center for Global Environmental Research (2006). *GOSAT: Greenhouse Gases Observing Satellite* (http://www-cger.nies.go.jp/gosat/gosat060609_e.pdf) National Institute for Environmental Studies, Japan.

336 S. Quegan et al.

Chevallier, F., Breon, F.-M., and Rayner, P. J. (2007). The contribution of the Orbiting Carbon Observatory to the estimation of CO_2 sources and sinks: Theoretical study in a variational data assimilation framework, *J. Geophys. Res.* **112**, D09307, doi:10.1029/2006JD007375.

Ciais, P., Moore, B., Steffen, W., Hood, M., Quegan, S., Cihlar, J., Raupach, M., Rasool, I., Doney, S., Heinze, C., Sabine, C., Hibbard, K., Schulze, D., Heimann, M., Chédin, A., Monfray, P., Watson, A., LeQuéré, C., Tans, P., Dolman, A.J., Valentini, R., Arino, O., Townshend, J., Seufert, G., Field, C., Chu, I., Goodale, C., Nobre, A., Inoue, G., Crisp, D., Baldocchi, D., Tschirley, J., Denning, S., Cramer, W., and Francey, R. (2003). *Final Report on the IGOS-P Carbon Theme. The Integrated Global Carbon Observing Strategy: A Strategy to Build a Coordinated Operational Observing System of the Carbon Cycle and Its Future Trends*, FAO.

Crisp, D., Atlas, R. M., Breon, F.-M., Brown, L. R., Burrows, J. P., Ciais, P., Connor, B. J., Doney, S. C., Fung, I., Y., Jacob, D. J., Miller, C. E., O'Brien, D., Pawson, S., Randerson, J. T., Rayner, P. J., Salawitch, R. J., Sander, S. P., Sen, B., Stephens, G. L., Tans, P. P., Toon, G. C., Wennberg, P. O., Wofsy, S. C., Yung, Y. L., Kuang, Z., Chudasama, B., Sprague, G., Weiss, B., Pollock, P., Kenyon, D., and Schroll, S. (2004). The Orbiting Carbon Observatory (OCO) mission. *Adv. Space Sci.*, **34**(4), 700–709.

Drezet, P. M. L., and Quegan, S. (2006). Environmental effects on the interferometric repeat-pass coherence of forests. *IEEE Trans. Geosci. Remote Sens.*, **44**(4), 825–837.

Drouet, J.-L., and Bonhomme, R. (2004). Effect of 3D nitrogen, dry mass per area and local irradiance on canopy photosynthesis within leaves of contrasted heterogeneous maize crops. *Ann. Bot.* **93**(6), 699–710.

Fensholt, R., Sandholt, I., and Schultz Rasmussen, M. (2004). Evaluation of MODIS LAI, fAPAR and the relation between fAPAR and NDVI in a semi-arid environment using *in situ* measurements. *Remote Sens. Env.*, **91**(3–4), 490–507.

Frankenberg, C., Meirink, J. F., van Weele, M., Platt, U., and Wagner, T. (2005). Assessing methane emissions from global space-borne observations, *Science*, **308**, 1010–1014.

Friedlingstein, P., Cox, P., Betts, R., Bopp, L., von Bloh, W., Brovkin, V., Cadule, P., Doney, S., Eby, M., Fung, I., Bala, G., John, J., Jones, C., Joos, F., Kato, T., Kawamiya, M., Knorr, W., Lindsay, K., Matthews, H. D., Raddatz, T., Rayner, P., Reick, C., Roeckner, E., Schnitzler, K.-G., Schnur, R., Strassmann, K., Weaver, A. J., Yoshikawa, C., and Zend, N. (2006). Climate-carbon cycle feedback analysis: results from the C4MIP model intercomparison. *J. Clim.*, **19**, 3337–3353.

Fritz, S., Bartholomé, E., Belward, A., Hartley, A., Stibig, H.-J., Eva, H., Mayaux, P., Bartalev, S., Latifovic, R., Kolmert, S., Roy, P., Agrawal, S., Bingfang, W., Wenting, X., Ledwith, M., Pekel, J.-F., Giri, C., Mücher, S., de Badts, E., Tateishi, R., Champeaux, J.-L., and Defourny, P. (2003). *Harmonisation, mosaicing and production of the Global Land Cover 2000 database*, Beta Version (2003). Publication of the European Commission EUR 20849.

Fuller, R., Smith, G., Sanderson, J., Hill, R., Thomson, A., Cox, R., N. J., B., Clarke, R., Rothery, P., and Gerard, F. (2002). *Land Cover Map 2000, Final Report 7*. Centre for Ecology and Hydrology, Cambridgeshire.

Garestier, F., Le Toan, T., and Dubois-Fernandez, P. (2007). Forest height estimation using P-band Pol-InSAR data. *Proc. Polinsar 2007*, ESA.

Giglio, L., Kendall, J. D., and Mack, R. (2003). A multi-year active fire data set for the tropics derived from the TRMM VIRS. *Int. J. Remote Sens.*, **24**, 4505–4525.

Giglio, L., van der Werf, G. R., Randerson, J. T., Collatz, G. J., and Kasibhatia, P. (2006). Global estimation of burned area using MODIS active fire observations. *Atmos. Chem. Phys.*, **6**, 957–974.

Gobron, N., Pinty, B., Verstraete, M. M., and Govaerts, Y. (1997). A semi-discrete model for the scattering of light by vegetation. *J. Geophys. Res.*, **102**, 9431–9446.

Gobron, N., Pinty, B., Verstraete, M. M., and Widlowski, J.-L. (2000). Advanced vegetation indices optimized for up-coming sensors: Design, performance, and applications. *IEEE Trans. Geosci. and Remote Sens.*, **38**(6), 2489–2505.

Gobron, N., Pinty, B., Aussedat, O., Chen, J. M., Cohen, W. B., Fensholt, R., Gond, V., Huemmrich, K. F., Lavergne, T., Melin, F., Privette, J. L., Sandholt, I., Taberner, M., Turner,

D. P., Verstraete, M. M., and Widlowski, J.-L. (2006). Evaluation of fraction of absorbed photosynthetically active radiation products for different canopy radiation transfer regimes: Methodology and results using Joint Research Center products derived from SeaWiFS against ground-based estimations. *J. Geophys. Res.*, **111**, D13110, doi:10.1029/2005JD006511.

Gu, L., Baldocchi, D., Verma, S. B., Black, T. A., Vesala, T., Falge, E. M., and Dowty, P. R. (2002). Advantages of diffuse radiation for terrestrial ecosystem productivity. *J. Geophys. Res.*, **107**(D6), 4050, doi:10.1029/2001JD001242.

Hansen, M. C., DeFries, R. S., Townshend, J. R. G., and Sohlberg, R. (2000). Global land cover classification at 1 km spatial resolution using a classification tree approach. *Int. J. Remote Sens.*, **21**, 1331–1364.

Janssens, I. A., Freibauer, A., Ciais, P., Smith, P., Nabuurs, G.-J., Folberth, G., Schlamadinger, B., Hutjes, R. W. A., Ceulemans, R., Schulze, E. -D., Valentini, R., and Dolman., A. J. (2003). Europe's terrestrial biosphere absorbs 7 to 12% of European anthropogenic CO_2 emissions. *Science*, **300**, 1538–1542.

Joint Research Centre (2006). *Forest Fires in Europe 2005*, EUR 22312 EN, European Commission, 54 pp.

Knorr, W., and Lakshmi, V. (2001). Assimilation of fAPAR and Surface Temperature into a Land Surface and Vegetation Model. In: Lakshmi, V., Albertson, J., and Schaake, J. (eds.), *Land Surface Hydrology, Meteorology and Climate: Observations and Modelling*, AGU: Washington, 177–200.

Knyazikhin, Y., Martonchik, J. V., Myneni, R. B., Diner, D. J., and Running, S. W. (1998). Synergistic algorithm for estimating vegetation canopy leaf area index and fraction of absorbed photosynthetically active radiation from MODIS and MISR data. *J. Geophys. Res.*, **103**, 32,257–32,276.

Kugler, F., Papathanassiou, K., and Hajnsek, I. (2007). Forest parameter estimation in tropical forests by means of Pol-InSAR: Evaluation of the Indrex-II campaign. *Proc. Polinsar 2007*, ESA.

Law, B. E., Thornton, P., Irvine, J., Van Tuyl, S., and Anthoni, P. (2001). Carbon storage and fluxes in ponderosa pine forests at different developmental stages. *Global Change Biol.*, **7**, 755–777.

Le Toan, T., Quegan, S., Woodward, I., Lomas, M., Delbart, N., and Picard, G. (2004). Relating radar remote sensing of biomass to modelling of forest carbon budgets, *J. Clim. Change*, **67**(2), 379–402.

Lotsch, A., Tian, Y., Friedl, M. A., and Myneni, R. B. (2003). Land cover mapping in support of LAI and FPAR retrievals from EOS-MODIS and MISR: Classification methods and sensitivities to errors. *Int. J. Remote Sens.*, **24**(10), 1997–2016.

Loveland, T. R. and Belward, A. S. (1997). The IGBP-DIS global 1 km land cover data set, DIScover: First results, *Int. J. Remote Sens.*, **65**(9), 1021–1031.

Monteith, J. L. (1977). Climate and the efficiency of crop production in Britain. *Phil. Trans. Roy. Soc. Lond.*, B **281**, 277–294.

Monteith, J. L. (1972). Solar radiation and productivity in tropical ecosystems. *J. Appl. Ecol.*, **9**, 747–766.

Monteith, J. L., and Unsworth, M. (1990). *Principles of Environmental Physics*, 2nd edition, Arnold, London.

Nouvellon, Y., Moran, M. S., Seen, D. L., Bryant, B., Rambal, S., Ni, W., Bégué, A., Emmerich, B., Heilman, P., and Qi, J. (2001). Coupling a grassland ecosystem model with landsat imagery for a 10-year simulation of carbon and water budgets. *Remote Sens. Env.*, **78**, 131–149.

Olson, J. S., Watts, J. A., and Allison, L. J. (1983). *Carbon in Live Vegetation of Major World Ecosystems*. Report ORNL-5862. Oak Ridge National Laboratory, Oak Ridge, Tennessee, 164 pp.

Prince, S. D., and Goward, S. N. (1995). Global primary production: A remote sensing approach. *J. Biogeogr.*, **22**(4/5), 815–835.

Quaife, T., Lewis, P., De Kauwe, M., Williams, M., Law, B. E., Disney, M., and Bowyer, P. (2007). Assimilating canopy reflectance data into an ecosystem model with an Ensemble Kalman Filter. *Remote Sens. Env.* (in press).

Raupach, M. R., Rayner, P., Barrett, D. J., DeFries, R., Heimann, M., Ojima, D., Quegan, S., and Schmullius, C. (2005). Model-data synthesis in terrestrial carbon observation: Methods, data requirements and data uncertainty specifications. *Global Change Biol.,* **11**, 378–397.

Rayner, P. J., and O'Brien, D. M. (2001). The utility of remotely sensed CO_2 concentration data in surface source inversions. *Geophys. Res. Lett.,* **28**, 175–178.

Rayner, P. J., Scholze, M., Knorr, W., Kaminski, T., Giering, R., and Widmann, H. (2005). Two decades of terrestrial carbon fluxes from a carbon cycle data assimilation system (CCDAS). *Global Biogeochem. Cycles,* **19**, doi:10.1029/2004GB002254.

Roberts, G., Wooster, M. J., Perry, G. L. W., Drake, N., Rebelo, L-M. and Dipotso, F. (2005). Retrieval of biomass combustion rates and totals from fire radiative power observations: application to southern Africa using geostationary SEVIRI Imagery. *J. Geophys. Res.,* **110**, D21111: doi:10.1029/2005JD006018

Roy, D. P., Lewis, P., and Justice, C. O. (2001). Burned area mapping using multi-temporal moderate spatial resolution data—A bi-directional reflectance model-based expectation approach. *Remote Sens. Env.,* **83**, 263–286.

Roy, D. P., Jin, Y., Lewis, P. E., and Justice, C. O. (2005a). Prototyping a global algorithm for systematic fire-affected area mapping using MODIS time series data. *Remote Sens. Env.,* **97**, 137–162.

Roy, D. P., Frost, P. G. H., Justice, C. O., Landmann, T., le Roux, J. L., Gumbo, K., Makungwa, S., Dunham, K., du Toit, R., Mhwandagaraii, K., Zacarias, A., Tacheba, B., Dube, O. P., Pereira, J. M. C., Mushove, P., Morisette, J. T., Santhana Vannan, S. K., and Davies, D. (2005b). The Southern Africa Fire Network (SAFNet) regional burned-area product-validation protocol. *Int. J. Remote Sens.,* **26**(19), 4265–4292.

Ruimy, A., Kergoat, L., Bondeau, A., and the participants of the Potsdam NPP model intercomparison (1999). Comparing global models of terrestrial net primary productivity (NPP): Analysis of differences in light absorption and light use efficiency. *Global Change Biol.,* **5**(suppl. 1), 56–64.

Running, S., Nemani, R., Glassy, J. M., and Thornton, P. (1999). *MODIS Daily Photosynthesis (PSN) and Annual Net Primary Production (NPP) Product (MOD17)*, Algorithm Theoretical Basis Document, version 3.0, April 29 1999.

Running, S. W., Thornton, P. E., Nemani, R. R., and Glassy, J. M. (2000). Global Terrestrial Gross and Net Primary Productivity from the Earth Observing System. In: Sala, O., Jackson, R., and Mooney, H. (Eds.), *Methods in Ecosystem Science*, Springer-Verlag: New York.

Schultz, M. (2002). On the use of ATSR fire count data to estimate the seasonal and interannual variability of vegetation fire emissions. *Atmos. Chem. Phys.,* **2**, 387–395.

Soja, A. J., Sukhinin, A. I., Cahoon Jr, D. R., Shugart, H. H., and Stackhouse Jr, P. W. (2004). AVHRR-derived fire frequency, distribution and area burned in Siberia. *Int. J. Remote Sens.,* **25**, 1939–1960.

van der Werf, G. R., Randerson, J. T., Giglio, L., Collatz, G. J., Kasibhatla, P. S., and Arellano, A. F. (2006). Interannual variability in global biomass burning emissions from 1997 to 2004. *Atmos. Chem. Phys.,* **6**, 3423–3441.

Veroustraete, F., Sabbe, H., and Eerens, H. (2002). Estimation of carbon mass fluxes over Europe using the C-fix model and Euroflux data. *Remote Sens. Environ.,* **83**, 376–399.

Vivoy, N., Francois, C., Bondeau, A., Krinner, G., Polcher, J., Kergoat, L., Dedieu, G., De Noblet, N., Ciais, P., and Friedlingstein, P. (2001). Assimilation of remote sensing measurements into the ORCHIDEE/STOMATE DGVM biosphere model. *Proceedings of 8th International Symposium Physical Measurements & Signatures in Remote Sensing.* Aussois, France, 713–718.

Williams, M., Rastetter, E. B., Fernandes, D. N., Goulden, M. L., Wofsy, S. C., Shaver, G. R., Melillo, J. M., Munger, J. W., Fan, S.-M., and Nadelhoffer, K. J. (1996). Modelling the soil-plant-atmosphere continuum in a Quercus-Acer stand at Harvard Forest: The regulation of stomatal conductance by light, nitrogen and soil/plant hydraulic properties. *Plant Cell Environ.,* **19**, 911–927.

Williams, M., Schwarz, P. A., Law, B. E., Irvine, J., and Kurpius, M. (2005). An improved analysis of forest carbon dynamics using data assimilation. *Global Change Biol.,* **11**, 89–105.

Woodward, F. I., and Lomas, M. R. (2004). Vegetation dynamics—Simulating responses to climatic change. *Biol. Rev.*, **79**(3), 643–670.

Wooster, M. J., Roberts, G., Perry, G., and Kaufman, Y. J. (2005). Retrieval of biomass combustion rates and totals from fire radiative power observations: calibration relationships between biomass consumption and fire radiative energy release. *J. Geophys. Res.*, **110**, D21111: doi:10.1029/2005JD006318.

Wooster, M. J., Zhukov, B., and Oertel, D. (2003). Fire radiative energy for quantitative study of biomass burning: Derivation from the BIRD experimental satellite and comparison to MODIS fire products. *Remote Sens. Env.*, **86**, 83–107.

World Resources Institute (2000). http://earthtrends.wri.org/maps_spatial/index. php?p=2&theme=9, PAGE, 2000.

Yang, W., Tan, B., Huang, D., Rautiainen, M., Shabanov, N. V., Wang, Y., Privette, J. L., Huemmrich, K. F., Fensholt, R., Sandholt, I., Weiss, M., Ahl, D. E., Gower, S. T., Nemani, R. R., Knyazikhin, Y., and Myneni, R. B. (2006). MODIS leaf area index products: From validation to algorithm improvement. *IEEE Trans. Geosci. Remote Sens.*, **44**(7), 1885–1898.

Zhang, Q., Xiao, X., Braswell, B., Linder, E., Baret, F., and Moore, B. (2005). Estimating light absorption by chlorophyll, leaf and canopy in a deciduous broadleaf forest using MODIS data and a radiative transfer model. *Remote Sens. Env.*, **99**, 357–371.

Zhukov, B., Lorenz, E., Oertel, D., Wooster, M. J., and Roberts, G. (2005). Spaceborne detection and characterization of fires during the Bi-spectral Infrared Detection (BIRD) experimental small satellite mission (2001–2004). *Remote Sens. Env.*, **100**, 29–51.

Chapter 16
The Lateral Carbon Pump, and the European Carbon Balance

Philippe Ciais, Alberto V. Borges, Gwenael Abril, Michel Meybeck, Gerd Folberth, Didier Hauglustaine, and Ivan A. Janssens

16.1 Introduction

Comparing atmospheric inversion estimates of the carbon fluxes of continents with bottom-up estimates (Pacala et al. 2001; Janssens et al. 2003; Peylin et al. 2005) is no easy task because (1) inversion fluxes always contain a certain amount of a priori information from bottom-up studies, so that the two approaches are not independent, (2) the time period for which inversion models and bottom-up estimates are produced is generally not the same, in the presence of substantial interannual variability, and (3) lateral carbon displacement makes some bottom-up estimates differ from inversions. Lateral displacement processes form a "carbon pump" which moves carbon away from the area where CO_2 was fixed from the atmosphere by photosynthesis with a very small additional sink from rock weathering. Lateral pumping of carbon implies that regional changes in *carbon storage* must differ from regional mean CO_2 fluxes (Tans et al. 1995; Sarmiento and Sundquist 1992). In presence of the lateral transport, one can write that

$$\text{Ecosystem } CO_2 \text{ budget} = \text{Ecosystem carbon budget} \quad (16.1)$$
$$+ \text{Lateral carbon flux}$$

Bottom-up inventory methods measure the carbon budget (first right hand term), while eddy covariance directly measure locally CO_2 fluxes (left hand term, excluding advection). Atmospheric inversions also measure CO_2 fluxes at continental level. This chapter has three major goals. The first objective is to analyze the lateral carbon flux underlying processes, and their implications for understanding regional carbon fluxes, and for designing an observing system to measure the European carbon balance. The second objective is to quantify the carbon pump of carbon within and across the borders of the European continent. The third goal is to produce spatially explicit maps of the land–atmosphere CO_2 fluxes associated with the lateral carbon pump. Janssens et al. (2003, 2005) accounted for some lateral carbon fluxes in their analysis of the European ecosystem carbon budget, but not for all of them, and their analysis did not map spatial distributions, so this study completes and enhances their analysis.

We consider the three mechanisms linking ecosystem–atmosphere CO_2 fluxes to the horizontal "carbon pump":

1. the trade of food, feedstuff, and wood products harvested,
2. the emissions of reduced carbon atmospheric compounds,
3. the river transport of dissolved inorganic carbon (DIC) and dissolved organic carbon (DOC).

In addition, CO_2 fluxes in coastal seas (Borges et al. 2006) are analyzed. This flux does not strictly correspond to an horizontal displacement from land to sea, but it is addressed here, as one possible flux needed to reconcile large-scale atmospheric inversion results with bottom-up estimates. Note that we do not treat the lateral atmospheric transport fluxes of CO_2. The first four sections treat each lateral transport process separately, and associated estimates of CO_2 fluxes. Sect. 16.6 summarizes the contribution of the lateral carbon pump to the carbon balance of Europe.

16.2 Crop and Forest Products Trade

16.2.1 Crop Products

Cultivated ecosystems remove CO_2 from the atmosphere during their growing season. Carbon incorporated into the biomass of crops is harvested and displaced to supply human and (some) animal consumption (Imhoff et al. 2004). After displacement, the consumption of food/feed products releases CO_2 back to the atmosphere. Over the globe, the annual displacement of harvested carbon remains approximately neutral for the atmosphere, given that the storage of food and feedstuff is negligible compared to harvested amounts. Locally, croplands are net annual CO_2 sinks (e.g., Anthoni et al. 2004). Conversely, populated areas where food and feedstuff is consumed are net sources of CO_2. This is illustrated in Fig. 16.1. International crop product trade transports carbon across the border of Europe, therefore, affecting the net carbon balance of the continent. Carbon exported corresponds to a CO_2 sink and carbon imported corresponds to a CO_2 source. Intra-European trade has similar consequences, and acts to redistribute carbon and generate CO_2 fluxes to and from the atmosphere within the continent, but with a zero effect on the net carbon balance of the continent.

We analyzed agrocultural trade statistics from FAO (2004). Cereals, essentially maize, wheat, and barley, explain nearly all the CO_2 sink matching the grain export. The corresponding source, on the other hand, is explained by the use (oxidation) of a mix of diverse crop products. On the perspective of individual countries, the situation becomes much more contrasted (Fig. 16.2). The largest CO_2 sink from the trade of crop products is France ($9\,TgC\ year^{-1}$), accounting for 90% of the total European sink. The largest CO_2 sources are Portugal, Belgium, Netherlands, Italy, and Spain (altogether more than $22\,TgC\ year^{-1}$). No significant correlation was

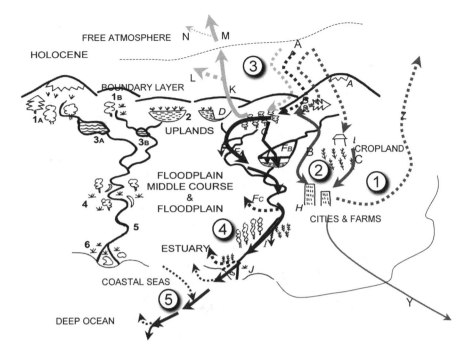

Fig. 16.1 Carbon cycle branches involving some lateral transport of carbon. Associated fluxes of atmospheric CO_2 are in dotted lines, fluxes of transported carbon in solid lines. **Green**. Cycle associated to photosynthesis (A), harvest of wood and crop products, transport by domestic (B, C) and international (Y) trade circuits, and consumption (Z) of carbon in crop products (1) and forest wood products (2). **Brown**. Cycle associated to photosynthesis (A), emissions and atmospheric transport in the boundary layer (K) and transport in free atmosphere (M), and oxidation in the boundary layer (L) and in the free atmosphere (N) of chemically reactive reduced carbon compounds (RCCs). **Blue**. Transport of carbon of atmospheric origin as dissolved inorganic carbon (DIC), dissolved organic carbon (DOC), and particulate organic carbon (POC) by river systems from river uplands to inner estuaries (see text for letters explanation) (4). **Purple**. Fluxes of carbon and CO_2 from coastal seas (5) (*See Color Plate* 9).

found between either the harvest (production) or the exported flux and the net carbon balance of each country.

Overall, Europe imports more food and feed carbon than it exports, thus being a net source to the atmosphere of $24\,TgC\ year^{-1}$. This flux represents 13% of the annual crop harvest and is equivalent to about 1% of the fossil fuel emissions.

The geographic pattern of CO_2 fluxes is calculated using geospatial information on (1) the main crop types (Ramankutty and Foley 1998), (2) human population distribution, and (3) housed poultry, pigs, and cattle populations, using the same methodology as in Ciais et al. (2006). Figure 16.3 shows the resulting geographic distribution of CO_2 fluxes. Major agricultural plains with intensive cultivation of cereals in northern France and southern England, Hungary, Po valley, and Denmark

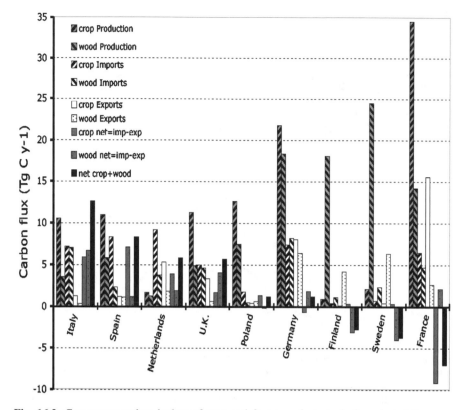

Fig. 16.2 Component carbon budget of crop and forest products trade for selected European countries, those with the largest such fluxes. Yellow is the production, equating the CO_2 uptake necessary to form crop and forest product biomass. Green are imports from outside Europe, oxidized into CO_2 within Europe and yielding a net CO_2 source. Red are exported products by Europe, yielding a net CO_2 sink. Magenta is the net balance resulting of import and export. Hatched bars are for forest products and plain colors for crop products.

are annual net CO_2 sinks, with uptake flux reaching up to $100\,gC\ m^{-2}\ year^{-1}$ locally. This uptake flux is large, equivalent to the mean sink of forests [$96\,gC\ m^{-2}\ year^{-1}$ after Janssens et al. (2003)]. This reflects the higher productivity and harvest index [ratio of yield to net primary production (NPP)] of crops compared with trees. In Fig 16.3, regions with dense population and intensive animal farming emit CO_2 to the atmosphere, up to $50\,gC\ m^{-2}\ year^{-1}$.

There are some uncertainties on these maps. Using statistics on country totals may smooth the fields. For instance, the feedstuff consumption by farmed animals is distributed equally in the same country on the basis of the animal density, while in reality, animals have regionally different reliance on feedstuff (e.g., dairy cows receive more feedstuff complements than do grazing cattle). Similarly, in large countries, crop harvest is distributed evenly on cropland area, neglecting regions where lower soil fertility or defavorable climate conditions induce lower yields.

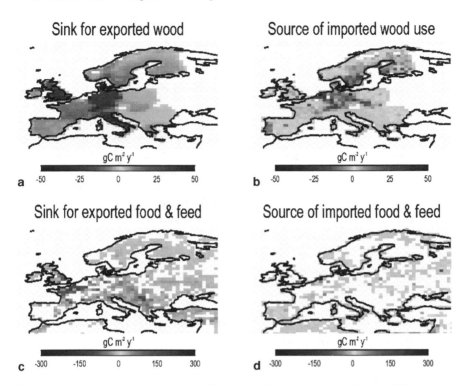

Fig. 16.3 Spatial patterns of trade-induced CO_2 fluxes with the atmosphere. Sources of CO_2 are positive and sinks of CO_2 are negative. **A** and **B** Sinks and sources of CO_2 associated to harvest, trade and consumption of crop products. **C** and **D** Same for forest products CO_2 fluxes (*See Color Plate* 10).

Finally, the release of consumed food is assumed to have the geographic distribution of human population, neglecting transport of organic carbon into dejections to sewage water and rivers.

16.2.2 Forest Products

We also analyzed wood and forest products carbon displacement. In the analysis, we include coniferous and nonconiferous industrial roundwood, sawn wood, wood panels, and paper listed in the FAO (2004) database. Wood harvest data in volume units are converted to mass of carbon by using a mean wood density of 0.5, and a 0.45 carbon fraction in dry biomass. The FAO (2004) statistics indicate that Sweden and Finland export more wood carbon than they import, thus being net sinks of atmospheric CO_2. Portugal and the Czech Republic are also net sinks, but of smaller magnitude. All other countries are CO_2 sources. The largest sources are Italy, Spain, the Netherlands, and the UK. When comparing crop and forest product lateral carbon pumps, it is striking to see that countries which export food products generally also export wood products (except Nordic countries). Conversely, countries which import food products (e.g., Mediterranean regions) import forest products as well.

To map the geographic distribution of CO_2 fluxes associated to forest product trade, we distributed the FAO (2004) country-level data using NPP fields from Lafont et al. (2002) and a forest cover map (CORINE Land cover 2000) at $1° \times 1°$ resolution. The results are given in Fig. 16.2. The decomposition or combustion of forest products, releasing CO_2, is determined by population density. This assumes that landfills are part of the population density. There are large uncertainties around these maps. First, the forests grid points with the largest NPP locally do not necessarily have the largest harvest, although one analysis (Myneni et al. 2001) shows a significant correlation between Normalized Difference Vegetation Index (NDVI) and biomass stocks in European countries. Furthermore, the areas emitting CO_2 by wood product decay may differ in their geographic location from the population density (e.g., depending on regional practice for using wood as a construction material).

16.3 Reduced Carbon Compounds Atmospheric Transport and Chemistry

Ecosystems and anthropogenic activities emit non-CO_2 reactive carbon compounds (RCCs). The RCCs are reduced species which include CO, CH_4, biogenic volatile organic compounds (BVOCs, such as isoprene, terpene), and anthropogenic volatile organic compounds (VOCs). If the objective of a study is to determine the *CO_2 budget* of Europe by inverse modeling of CO_2 concentration data, then the RCC emissions can rightfully be ignored. On the contrary, if the objective is to determine the *carbon budget* of Europe, the RCC fluxes must be added to the CO_2 fluxes (Eq. 16.1).

In the atmosphere, the RCCs are oxidized with various lifetimes. From ~9 years for CH_4 and few months for CO, the lifetime of RCCs drops to a mere few hours in the case of terpenes. The oxidation sequence of RCCs' species can be complex, but their main end product is a volume source of CO_2 in the atmosphere. The surface emission of RCC is small compared to gross ecosystem fluxes, but significant compared to net CO_2 fluxes. The global biogenic organic carbon emission to the atmosphere was estimated to amount to ~1% of gross primary production globally (Kesselmeier 2005).

Because the lifetime of atmospheric RCCs with respect to their chemical sink in the atmosphere can easily exceed typical atmospheric transport time scales, the carbon contained into RCCs is displaced away by atmospheric transport and eventually released as CO_2. Globally, if there is no perturbation of that branch of the carbon cycle, the atmospheric cycle RCC compounds should be carbon neutral.

To quantify how much carbon is displaced by atmospheric transport of RCCs, we made a European inventory of these compounds. Table 16.1 shows that the total

Table 16.1 Component budget of atmospheric RCCs over the European continent, an area bounded by 32°N and 73°N in latitude and −10°W and 40°E in longitude. Sources to the atmosphere are counted positive and sinks negative

RCC emissions over Europe (TgC year⁻¹)	
Methane	60.0
CO	82.0
BVOCs	27.0
Other VOCs	15.5
Total	184.5
CO_2 production from RCC atmospheric oxidation over Europe (TgC year⁻¹)	
Boundary layer	25.7
Free troposphere	19.3
Total	45.3
Carbon deposition in Europe (GtC year⁻¹)	
Surface dry deposition	12.0
Wet deposition	9.6
Total C deposited	21.6

RCC reactive carbon compound, *BVOC* biogenic volatile organic compound, *VOC* volatile organic compound

RCC emission over Europe is 185 TgC year⁻¹. In this total, the biogenic source of 87 TgC year⁻¹ must be matched by a CO_2 photosynthetic sink of equal magnitude.

The RCC emitted to the atmosphere react into CO_2. We estimated the atmospheric oxidation rates using a 3-D chemistry transport model (Folberth et al. 2005; Hauglustaine et al. 2004). The model explicitly accounts for two major oxidation channels of RCC into CO_2: (1) the oxidation of CO produced either by combustions, or by oxidation of CH_4 and VOCs, (2) the reaction of peroxy-radicals yielding CO_2. A third and minor channel is the direct ozonolysis of alkenoid compounds, which also produces CO_2. The total chemical CO_2 production in the volume of the atmosphere over Europe is given in Table 16.2. The modeled production of "photochemical" CO_2 in the European air shed was found to equal 45 TgC year⁻¹, among which 26 TgC year⁻¹ in the boundary layer (PBL) (Fig 16.4.). If "Europe" is defined as the continent and its PBL, then the net carbon loss is 45 − 26 = 19 TgC year⁻¹ (Table 16.3).

Opposing these CO_2 losses, a sink of carbon of 22 TgC year⁻¹ is caused by dry deposition and wet scavenging of RCC over the continent (Table 16.3). The net effect of the RCC cycle on the European continent carbon balance (Europe = continent + its PBL; source ≥ 0 flux) is given by

$$\text{Net carbon budget} = \text{RCC anthropogenic emissions} - \quad (16.2)$$
$$\text{photochemical } CO_2 \text{ in PBL} - \text{deposition sink}$$

This assumes that the biogenic RCC emissions are neutralized by a CO_2 photosynthetic sink of equal magnitude everywhere. Equation (16.2) gives a net carbon loss of 76 TgC year⁻¹, available as CO_2 for long-range transport outside the European domain.

Table 16.2 Summary of the lateral transport fluxes and associated atmospheric CO_2 flux for the European continent. The European continent includes its boundary layer and its inner estuary regions. Sources to the atmosphere are counted positive and sinks negative (all units in TgC year[-1])

	Estimated flux	Quality index of the estimate
Crop and forest products trade		
Crop and forest product import and use	110	+++
Crop and forest product export	−81	+++
Total CO_2 flux[b]	29	+++
Total carbon flux[b]	29	++
Atmospheric RCCs		
RCC total emissions (1)	185	++
RCC emissions of biogenic origin (2)[a]	87	++
CO_2 source from RCC oxidized in boundary layer (3)	26	++
Carbon sink from wet + dry deposition of RCC (4)	−22	+
Total net CO_2 flux = (3) − (2)	−61	+
Total net carbon flux[a] = (1) − (4) − (2)	76	++
River carbon fluxes[d]		
CO_2 outgassing in river systems	90	++
Carbon burial in lakes, dams, and estuarine sediments	33	++
Export to estuaries	53	+++
CO_2 outgassing in estuaries	10–20	+
CO_2 uptake by coastal seas	−68	+++
Total net carbon flux including coastal seas[c]	−113	+
Total net CO_2 flux including coastal seas[c]	−60 to −50	+
Total net CO_2 flux	−160 to −150	+
Total net carbon flux	−76	+

RCC reactive carbon compound

[a]Assumes that 100% of RCC biogenic sources originates from photosynthetic uptake of atmospheric CO_2

[b]Assumes that consumption of crop and forest products produces 100% of CO_2 released to the atmosphere

[c]Assumes that 90% of the DIC transport is a carbon sink is of atmospheric origin equal to −160 TgC year[-1] (see text), and accounting for the atmospheric share only of carbon flux exported

[d]Established for 8.16 Mkm[2], including Barentz and Black Sea river catchments

16.4   Rivers and Estuaries Carbon Transport

River systems (streams, lakes, river main stems, and floodplains and estuaries) transport carbon laterally, from the land to the ocean, and vertically as CO_2 degassing to the atmosphere and as carbon burial in sediments. Export to the ocean and burial in sediments is a net sink of atmospheric carbon, whereas degassing in aquatic systems is a source (Fig. 16.1).

Rivers and freshwater systems transport carbon originating from land ecosystems to estuaries under dissolved and particulate organic forms [DOC and particulate

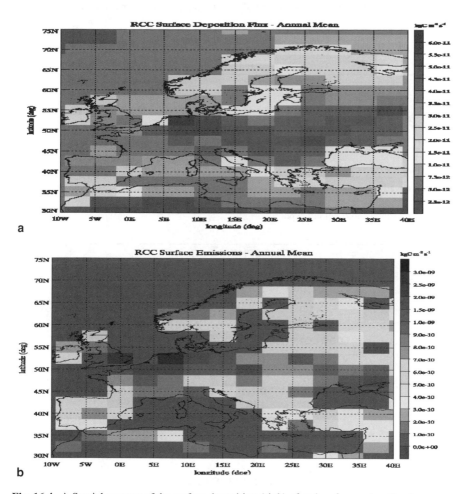

Fig. 16.4 A Spatial patterns of the surface deposition (sink) of carbon from reduced carbon compounds. **B** Patterns of reduced carbon compounds emissions to the atmosphere from antropogenic and biospheric sources (*See Color Plate* 11).

organic carbon (POC)] and under inorganic forms [DIC, particulate inorganic carbon (PIC), and CO_2]. Sources and sinks of river carbon in natural conditions are schematized in Fig. 16.1. Sources include: (1) wetlands and peatbog drains (Fig. 16.5A; A), (2) soil leaching and erosion (B), and (3) chemical weathering of soil minerals (C and D). This form of carbon is originally taken up from the atmospheric reservoir via photosynthesis (as followed by $CO_2 + H_2O \rightarrow C_{org} + O_2$), carbonate rock weathering ($CO_2 + H_2O + MCO_3 \rightarrow 2HCO_3^- + M^{2+}$), and by silicate rock weathering via the reaction ($2CO_2 + H_2O + MSiO_3 \rightarrow 2HCO_3^- + M^{2+} + SiO_2$). During weathering of noncarbonated rocks, 100% of river DIC originates from the atmosphere. In contrast, during weathering of carbonated sedimentary rocks, which

Table 16.3 Fluxes of carbon in TgC year^{-1} and origins of net river carbon reaching the continental shelf after estuarine filters (Abril and Meybeck, in preparation). Irrigation not taken into account

	Drainage area (10^3 km^2)	Water flow (km^3 year^{-1})	River carbon (TgC year^{-1})	Carbon yield (gC m^{-2} year^{-1})	%DOC	%POC	%DIC[a]
Northern Europe	2,528	806	13.6	5.4	54.3	4.4	41.1
Temperate Europe	4,699	1,188	24.5	5.2	23.3	9.0	67.6
Southern Europe	936	360	10.2	10.8	9.2	11.5	79.2
Total Europe	8,163	2,355	48.3	5.9	29.1	8.3	62.6

DOC dissolved organic carbon, *POC* particulate organic carbon, *DIC* dissolved inorganic carbon

[a] In percent of total carbon of atmospheric origin

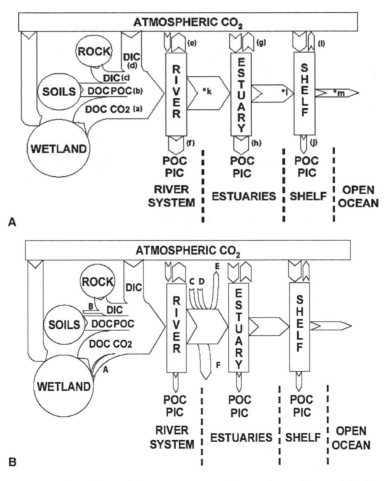

Fig. 16.5 Schematic transport of river carbon along the aquatic continuum. **A** Under natural conditions. **B** With multiple human impacts. See text for indications.

results in important fluxes of DIC, usually as bicarbonates, half of the DIC is taken up from the atmosphere, and half derives from fossil carbonates stored in rocks over 10 millions to 100 millions years. Thus, carbonated sedimentary rocks represent an important source of DIC, usually in the form of bicarbonates. Therefore, the nature and ages of river carbon species is very different (Meybeck 1993, 2005). The PIC is derived from the mechanical erosion. While being transported downstream to the coast, it is gradually trapped in lowlands, floodplains, lakes, estuaries, and on the continental shelf [Fig. 16.1. F_A, F_B, I, and J; Fig. 16.5A (f) and (h)]. This relocation of the rock PIC generally does not affect the CO_2 cycle. Under the specific climate and ecological conditions (such as, arid conditions, high river pH), some DIC may precipitate as calcite and can thus be stored in soils and in sediment "traps" on their way from headwaters to the coastal areas.

Factors controlling the river export of atmospheric carbon, that is, DOC + POC + atmospheric derived DIC, are first river runoff, then rock type (occurrence of carbonate rocks), and finally the presence of wetlands. Large lakes are also efficient filters of dissolved carbon fluxes (Fig. 16.1). A preliminary comparison of river carbon fluxes in Northern, Central, and Southern Europe shows major regional contrasts. Northern European catchments are characterized by high DOC export (most POC is trapped in lakes, which cover 5–20% of these river basins), derived from wetlands and peatbogs, the age of which probably ranges from 100 to 6,000 years. In Southern catchments, DIC is the dominant form of river carbon and nearly half of it originates from carbonate rock dissolution with carbon aged of 10–100 million years, the other half from contemporary soil and atmospheric CO_2. Central Europe catchments are intermediate, with carbon fluxes depending on river runoff and rock type. From mass-balance considerations, we estimated that the river transport of DIC corresponds to an ecosystem CO_2 sink of 160 TgC year^{-1} (see Table 16.3 and below).

Direct human pressures on river catchments may greatly modify carbon fluxes, transfers, and sinks (Fig. 16.5B). The exploitation of peatbogs generally increases DOC contents in headwaters (Fig. 16.1, D; Fig. 16.5B, A). Increased soil erosion and agricultural practices may result in soil POC inputs. Untreated organic waste water (Fig. 16.1, H; Fig. 16.5B, C) and eutrophication (Fig. 16.5B, D) of river and lakes are additional sources of very labile DOC and POC. It must be noted that the ages of these carbon species are also highly variable and range from a few days for river algal carbon to a thousand of years in the case of, for example, peat DOC. The CO_2 evasion (outgassing) from river, lakes, and reservoirs greatly depends on the reactivity of this organic carbon carried by rivers. During the 1970s, when rivers were accepting large quantities of untreated urban and industrial waste waters, river respiration (R) was exceeding river production (P) related to eutrophication: net CO_2 evasion was observed as in the Rhine River (Kempe 1984). Present conditions are changing because of waste treatment and the same river may have multiple changes of P/R ratio from headwaters to estuary as is the case for the Scheldt and Seine Rivers.

In addition to new carbon input to river system, river damming and irrigation control net river carbon fluxes to oceans. Reservoirs store up to 99% of particulate river material (Vörösmarty et al. 2003), including POC, and may degrade DOC and retain part of DIC as calcite precipitation. Irrigation canals continuously transfer river carbon to agricultural soils and the irrigation residues do not balance these fluxes. In Southern Europe, most river water flows to the ocean have been decreased, up to 90% (as for the Nile), although this impact on net river fluxes to the Mediterranean Sea or the Portuguese Coast is unknown as the last gauging and water quality stations are located upstream of the major water withdrawal for deltas irrigation (e.g., Ebro, Rhone, Axios, Nile) (Ludwig et al. 2003).

Continental aquatic surfaces including streams, lakes, river main stems, and floodplains and estuaries are sources of CO_2 to the atmosphere (Sobek et al. 2003). Except for a few cases occurring seasonally, CO_2 supersaturation episodes in the water prevail in streams (Hope et al. 2001; Billett et al. 2004), lakes (Cole et al. 1994), rivers (Kempe 1982; Jones and Mulholland 1998; Abril et al. 2000; Cole and Caraco 2001), and estuaries (Frankignoulle et al. 1998; Abril and Borges 2004). The high CO_2 concentrations in continental waters can have two distinct origins and are either produced on land by soil respiration, followed by surface runoff and riparian transport, or result from the oxidation of terrestrial organic carbon in the aquatic system itself, by microbial respiration and photochemistry (Graneli et al. 1996; Jones and Mulholland 1998; Abril and Borges 2004; Gazeau et al. 2005). Temperate rivers in Western Europe show a positive pCO_2 versus DOC relationship (Fig. 16.6), as a result of anthropogenic (mainly domestic) loads, that increase the DOC, followed by aquatic respiration, that increases the pCO_2 (Neal et al. 1998; Abril et al. 2000). By contrast, boreal headwaters in Scottish peatlands show much lower pCO_2 in comparison with a very high DOC. This is due to the soil origin of the CO_2 and its rapid evasion to the atmosphere in these fast flowing headwaters (Hope et al. 2001; Billett et al. 2004). In lakes, DOC is negatively correlated with water residence time, showing the predominant role of microbial and photochemical oxidation (Tranvik 2005). In some temperate eutrophic rivers, a seasonal and sometimes annual uptake of atmospheric CO_2 is observed (Fig. 16.6). In such case, atmospheric carbon is fixed by aquatic primary production and further transported laterally as organic carbon downstream to the estuary. The Loire River, for instance, transports large quantities of algal carbon during summer, but these are totally mineralized in the estuarine turbidity maximum (Meybeck et al. 1988), leading to hypoxia and high CO_2 degassing in the estuary (Abril et al. 2004). Along the river basin continuum, estuaries, particularly when macrotidal, behave as "hotspots" for CO_2 degassing, owing to the quantity of organic carbon they receive from the land as well as to the long residence time of waters and suspended sediments (Frankignoulle et al. 1998; Abril et al. 2002; Abril and Borges 2004). Gas transfer velocities in continental waters also vary in a wide range, and are controlled by the wind stress (lakes, estuaries, to a lesser extend rivers) and by the land topography, water flow, and bottom roughness (streams, rivers, estuaries) (Wanninkhof et al. 1990; Borges et al. 2004). The relative scarcity of CO_2 data in continental waters, relative to the high spatial and temporal variations, renders a bottom-up estimate at the European scale rather uncertain. In addition, surface areas of some ecosystems are also accompanied by significant uncertainties and the highest CO_2 flux rates occur in ecosystems with the

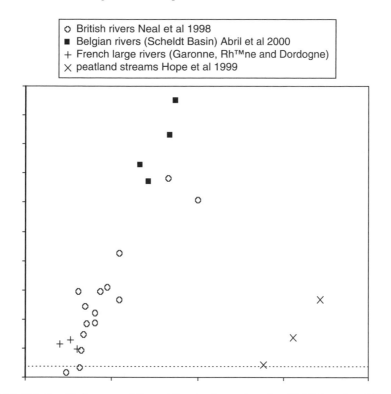

Fig. 16.6 Relationship between pCO_2 and dissolved organic carbon (DOC) in selected European river systems. British lowland rivers from Neal et al. (1998); Belgian lowland rivers (Scheldt watershed) from Abril et al. (2000); large French rivers from Abril and Commarieu (Unpublished) (Garonne and Dordogne) and from Aucour et al. (1999) (Rhône); and Scottish upland peat streams from Hope et al. (2001).

smallest surface areas, such as estuaries and rivers. For DOC, POC, and atmospheric-derived DIC, the lateral transport of carbon to the coastal zone by rivers has been compiled using the main European rivers database (Meybeck and Ragu 1996) and then extrapolated for the European seas catchment (8.16 Mkm²) on the basis of runoff, landcover, and rock types similarities. Southern, Central, and Northern Europe can be differentiated (Table 16.3) showing marked diversity in export rates and carbon species. Estuarine filters are included in this estimate (Abril and Meybeck, in preparation) taking into account the carbon species fate.

Table 16.2 summarizes carbon fluxes through the European river system on an order-of-magnitude basis and compares river carbon lateral transport with CO_2 outgassing. CO_2 fluxes from each subecosystems (peatland streams, lakes, rivers, and estuaries) are based on a compilation of published pCO_2 distributions, on typical gas transfer velocities, and on available information on surface areas of subecosystems (Abril and Meybeck, in preparation). Because the river transport as summarized in Table 16.3 is based on nontidal river sampling and is calculated for the entrance of

Table 16.4 Sensitivity of lateral carbon fluxes to global change

| | River atmospheric carbon[a] | | | |
| | Natural C fluxes | Anthropogenic C | C storage | CO₂ evasion from |
Global change	to ocean	fluxes to ocean	on land	water mirror
CO$_2$ increase	x			x
Warming	x			x
Sea level rise			x	
Runoff increase	xx			
Direct human impacts				
Waste management		xx		xx
Eutrophication		x		x
Damming			xxx	xx
Irrigation			xxx	
channelization			x	
Land use change	x to xx			

[a] From headwaters to estuary included

estuaries, we have distinguished the CO$_2$ degassing in freshwaters from those in estuaries. Although European rivers transport 53 TgC year^{-1} to estuaries, they emit 90 TgC year^{-1} of CO$_2$ to the atmosphere. A majority of this CO$_2$ outgassing is occurring at northern latitudes. Owing to their large surface area (183,000 km^2) and despite their lower CO$_2$ flux density, lakes contribute up to 35% of the total CO$_2$ outgassing from European freshwaters (lakes, reservoirs, rivers, and their flood-plains but excluding wetlands and estuaries). The CO$_2$ degassing from European estuaries has been previously estimated to 30–60 TgC year^{-1} by Frankignoulle et al. (1998). This range is probably an overestimate for two reasons: (1) the surface area of European estuaries (112 × 10^3 km^2) was uncertain and much higher than indicated by recent estimates of 36 × 10^3 km^2 compiled in the Global Lakes and Wetlands Database (GLWD) (2) the investigated estuaries are in the majority macrotidal, where net heterotrophy and CO$_2$ degassing are favored (Abril and Borges 2004; Borges et al. 2006). Little or no CO$_2$ data are available in fjords, fjärds, deltas, and coastal lagoons. Applying the available CO$_2$ fluxes to the surface area of coastal wetlands from the GLWD results in an estuarine CO$_2$ source to the atmosphere of 10–20 TgC year^{-1}, which is much closer to the organic carbon transport by European rivers before the estuarine filter of 20 TgC year^{-1} reported in Table 16.3. We estimated that 90% of the DIC transported by rivers is of atmospheric origin, that is the DIC transport is matched by a CO$_2$ sink in ecosystems. The net carbon balance of the continent is given in Table 16.4.

$$\text{Net carbon budget} = \text{Ecosystem CO}_2 \text{ sink forming river} \qquad (16.3)$$
$$\text{DIC} - \text{Net flux to estuaries} +$$
$$\text{CO}_2 \text{ outgassing from estuaries} +$$
$$\text{CO}_2 \text{ outgassing by freshwaters}$$

16.5   Carbon Dioxide Fluxes in Coastal Seas

Coastal seas not only receive nutrient and organic matter inputs from estuaries but also exchange water and matter with the open ocean across marginal slopes. In European coastal seas, the gross water fluxes across marginal slopes are about 250–2,000 times more intense than that of the respective freshwater discharge; water flux exchanges between open ocean and adjacent coastal seas are 80–520 times higher than that of the respective freshwater discharge (Huthnance 2006). The overall flow of carbon will also depend on the content that is much higher in estuaries than in the adjacent open ocean. Nevertheless, the inputs of carbon from the open ocean to the coastal seas are significant because of the much higher water fluxes involved. In the North Sea, the inputs of DOC and DIC through the northern boundary from the North Atlantic Ocean are 45 and 140 times higher, respectively, than the inputs of the same quantities from estuaries. The input of the same species from the Baltic Sea is roughly equivalent to those from estuaries, but the inputs of DOC and DIC from the English Channel are 3 and 13 times higher, respectively, than the inputs of the same species from estuaries (Thomas et al. 2005).

Unlike macrotidal estuaries that emit CO_2 to the atmosphere throughout the year, coastal seas usually exhibit a distinct seasonal cycle of air–sea CO_2 exchange, shifting from a source to the atmosphere to a sink from the atmosphere. This shift to a large extent is related to the variation in biological activity. The direction and the intensity of the air–sea fluxes of CO_2 are predominantly controlled by the net ecosystem production (NEP), as exemplified in Fig. 16.6 for the Southern Bight of the North Sea (SBNS). The SBNS acts as a sink of atmospheric CO_2 in April and May during the phytoplankton blooms, and as a source of CO_2 to the atmosphere during the rest of the year due to the degradation of organic matter. Overall, the SBNS is a sink of atmospheric CO_2 because of the seasonal decoupling of organic matter production and degradation, with a probable export of organic matter to the adjacent areas in relation to the short exchange time of the water mass in this area (on average 70 days).

Note that besides NEP, air–sea CO_2 fluxes in coastal seas are also modulated by other factors such as additional biogeochemical processes (e.g., $CaCO_3$ precipitation/dissolution), purely thermodynamic effects (temperature change, salinity change, Revelle factor), exchange of water with adjacent aquatic systems, the residence time of the water mass within the system, and the decoupling of organic carbon production and degradation across the water column in relation to the physical conditions of the system (e.g., Borges 2005; Borges et al. 2005, 2006).

Figure 16.7 shows the annually integrated air–sea CO_2 fluxes for various European coastal seas. A significant spatial heterogeneity is clearly apparent. Seasonal patterns also differ from one coastal sea to another (not shown in the figure) as discussed in detail by Borges (2005) and Borges et al. (2005, 2006).

Overall, the European coastal seas are a sink of atmospheric CO_2 of about 68 TgC year^{-1}. The air–sea CO_2 fluxes over European coastal seas are also significantly

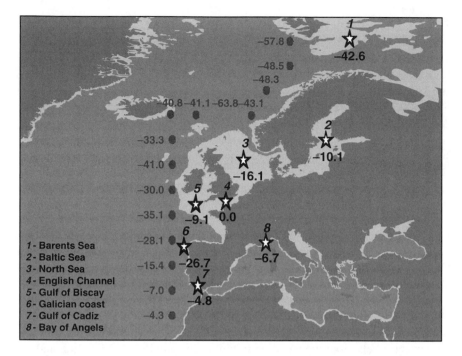

Fig. 16.7 Compilation of annually integrated air–sea CO_2 (gC m^{-2} $year^{-1}$) fluxes in European coastal seas (stars and black numbers) (adapted from Borges et al. 2006) and adjacent open ocean grid nodes from the Takahashi et al. (2002) climatology (red circles and numbers) (*See Color Plate* 12).

different from those in the adjacent open ocean grid nodes in the Takahashi et al. (2002) climatology (Fig. 16.7). The latter is commonly used in atmospheric CO_2 inversion models which probably will lead to a significant but yet unquantified bias in the flux estimated derived by these models. In particular, large biases are expected for terrestrial regions adjacent to extensive coastal seas, such as, the Gulf of Biscay and the North Sea.

16.6   Conclusions and Outlook

In summary, the carbon budget of a continent is more complex than just the sum of photosynthesis and respiration. A number of processes pump carbon away from ecosystems after its fixation by photosynthesis. The carbon displaced by lateral pumps can travel over long distances, but eventually gets oxidized and released back to the atmosphere in the form of CO_2, thus closing the cycle. Lateral transport of carbon occurs within Europe, resulting in local and regional imbalances between CO_2 sources and CO_2 sinks. It also occurs across the borders of the European continent, resulting in a CO_2 sink inside Europe which is balanced by a source outside the continental boundaries. This makes inverse modeling more difficult, since ignoring lateral CO_2 fluxes may bias inversion results. Consequently, the comparison of approaches measuring CO_2 fluxes (inversion modeling and eddy covariance) with those measuring carbon fluxes (soil and forest biomass inventories) must account for lateral fluxes.

At the local scale, the imbalance in the CO_2 flux, created by carbon lateral transport, can be very large, especially in cultivated areas and harvested forests. However, when considered at the continental level, food and wood trade result only in a very small net source of CO_2 to the atmosphere. We also found that a large source of carbon is released to the atmosphere in the form of RCC (183 TgC year^{-1}), but its impact on the net carbon balance is complex, involving the rapid oxidation of RCC into CO_2 within the boundary layer itself and the deposition of RCC at the surface. Overall, we estimated that the cycle of RCC causes a net carbon loss of 76 TgC year^{-1}.

Rivers altogether transport a large carbon of atmospheric origin, that is an ecosystem CO_2 sink of 160 TgC year^{-1}. En route to the oceans, part of this river carbon is outgassed to the atmosphere; part is sequestered in lakes, dams, and estuarine sediments, with the rest being delivered to inner estuaries where a large fraction is outgassed. In total, the river geologic carbon cycle causes a net carbon sink of 113 TgC year^{-1} at the continental level.

Coastal seas were considered because they are subject to important lateral transport (their CO_2 budget largely depends on lateral fluxes), and because coarse resolution inversions' results encompass coastal seas in their estimates. The sink for atmospheric CO_2 in coastal seas of 68 TgC year^{-1} is comparable in magnitude to the terrestrial carbon fluxes. The air–sea CO_2 fluxes in European coastal seas are significantly different from those in the adjacent open waters. The latter are commonly used as a prior in atmospheric CO_2 inversion models, leading to a yet unquantified bias in the output from these models.

In conclusion, the fluxes transported by wood and food trade, RCC cycle, and river geological cycle are significant. Altogether, the displaced carbon pump amounts to roughly 20% of the NPP of the European continent. The contribution of these fluxes to the continental carbon balance is not negligible either, amounting to a CO_2 sink of 165–155 TgC year^{-1} and to a net carbon sink of 81 TgC year^{-1}.

River lateral carbon fluxes have not been considered on the environmental agenda of European countries. For improving estimates of these fluxes, what is needed is regional statistics of wood and food harvest and consumption. It is also needed measurements of RCC fluxes from representative ecosystems. For rivers, data sources are limited to bicarbonates, as a component of ionic analysis, and recently total organic carbon, as an indicator of "organic pollution" in some rivers. It is now necessary to create a European network for river carbon sources fluxes and sinks from headwaters to the continental shelf including DOC, POC, DIC, and CO_2. So far, most of the carbon budgets have been realized in macrotidal estuaries of Western Europe. Other estuarine types must be considered in order to improve our knowledge of net carbon fluxes to the European continental shelf.

References

Abril G. and Borges A.V. (2004). Carbon dioxide and methane emissions from estuaries. In Greenhouse Gas Emissions: Fluxes and Processes. Hydroelectric Reservoirs and Natural Environments. A. Tremblay, L. Varfalvy, C. Roehm and M. Garneau (Eds.), Environmental Science Series, Springer-Verlag, Berlin, Heidelberg, New York. 187–207.

Abril G., Etcheber H., Borges A.V., and Frankignoulle M. (2000). Excess atmospheric carbon dioxide transported by rivers into the Scheldt Estuary. Comptes Rendus de l'Académie des Sciences. Série IIA 330: 761–768.

Abril G., Nogueira M., Etcheber H., Cabeçadas G., Lemaire E., and Brogueira M.J. (2002). Behaviour of organic carbon in nine contrasting European estuaries. Estuarine, Coastal and Shelf Science. 54: 241–262.

Anthoni P., Freibauer A., Kolle O., and Schulze E.-D. (2004). Winter wheat carbon exchange in Thuringia, Germany. Agricultural and Forest Meteorology. 121: 55–67.

Aucour M.-A., Sheppard S.M.F., Guyomar O., and Wattelet J. (1999). Use of ^{13}C to trace origin and cycling of inorganic carbon in the Rhône river system. Chemical Geology. 159: 87–105.

Aumont O., Orr J.C., Monfray P., Ludwig W., Amiotte-Suchet P., and Probst J.L. (2001). Riverine-driven interhemispheric transport of carbon. Global Biogeochemical Cycle. 15: 393–405.

Billett M.F., Palmer S.M., Hope D., Deacon C., Storeton-West R., Hargreaves K.J., Flechard C., and Fowler D. (2004). Linking land-atmosphere-stream carbon fluxes. Global Biogeochemical Cycles. 18(1): 1–12.

Borges A.V., Delille B., Schiettecatte L.S., Gazeau F., Abril A., and Frankignoulle M. (2004). Gas transfer velocities of CO_2 in three European estuaries (Randers Fjord, Scheldt and Thames). Limnology and Oceanography. 49(5): 1630–1641.

Borges A.V. (2005). Do we have enough pieces of the jigsaw to integrate CO_2 fluxes in the Coastal Ocean ? Estuaries. 28(1): 3–27.

Borges A.V., Delille B., and Frankignoulle M. (2005). Budgeting sinks and sources of CO_2 in the coastal ocean: Diversity of ecosystems counts. Geophysical Research Letters. 32: L14601 (doi:10.1029/2005GL023053).

Borges A.V., Schiettecatte L.S., Abril G., Delille B., and Gazeau F. (2006). Carbon dioxide in European coastal waters. Estuarine Coastal and Shelf Science. 70: 375–387.

Ciais P., Bousquet P., Freibauer A., and Naegler T. (2006). On the horizontal displacement of carbon associated to agriculture and how it impacts atmospheric CO_2 gradients. Global Biogeochemical Cycles, in revisions.

Cole J.J. and Caraco N.F. (2001). Carbon in catchments: Connecting terrestrial carbon losses with aquatic metabolism. Marine and Freshwater Research. 52: 101–110.

Cole J.J., Caraco N.F., Kling G.W., and Kratz T.W. (1994). Carbon dioxide supersaturation in the surface waters of lakes. Science. 265: 1568–1570.

CORINE Land cover (2000). EEA online publications: http://reports.eea.eu.int/COR0- landcover/en., edited.

Enting, I. G., and Mansbridge J. V.: Latitudinal Distribution of Sources and Sinks of CO2 - Results of an Inversion Study, Tellus, 43B, 156–170, 1991.

FAO (2004). Food and Agriculture Organization database, in http://faostat.fao.org/faostat/collections?subset = agriculture, edited.

Folberth G., Hauglustaine D., Ciais P., and Lathière J. (2005). On the role of atmospheric chemistry in the global CO_2 budget. Geophysical Research Letters. doi:10.1029.

Frankignoulle M., Abril G., Borges A., Bourge I., Canon C., Delille B., Libert E., and Théate J.M. (1998). Carbon dioxide emission from European estuaries. Science. 282: 434–436.

Gazeau F., Gattuso J.-P., Middelburg J.J., Brion N., Schiettecatte L.-S., Frankignoulle M., and Borges A.V. (2005). Planktonic and whole system metabolism in a nutrient-rich estuary (the Scheldt Estuary). Estuaries. 28(6): 868–883.

Graneli W., Lindell M., and Tranvik L. (1996). Photo-oxidative production of dissolved inorganic carbon in lakes of different humic content. Limnology and Oceanography. 41: 698–706.

Grosbois C., Négrel P., Fouillac C., and Grimaud D. (2000). Dissolved load of the Loire River: Chemical and isotopic characterization. Chemical Geology. 170: 179–201.

Hauglustaine D.A., Hourdin F., Jourdain L., Filiberti M.A., Walters S., Lamarque J.F., and Holland E.A. (2004). Interactive chemistry in the laboratoire de météorologie dynamique general circulation model: Description and background tropospheric chemistry evaluation. Journal of Geophysical Research. 109.

Hope D., Palmer S., Billet M., and Dawson J.J.C. (2001). Carbon dioxide and methane evasion from a temperate peatland stream. Limnology and Oceanography. 46: 847–857.

Huthnance J.M. (2006). North-East Atlantic margins. In Carbon and Nutrient Fluxes in Global Continental Margins. Atkinson L., Liu K.K., Quinones R., and Talaue-McManus L. (Eds.), Springer, New York.

Imhoff M.L., Bounoua L., Ricketts T., Loucks C., Harriss R., and Lawrence W.T. (2004). Global patterns in human consumption of net primary production. Nature. 429: 870–873.

Janssens I.A., et al. (2003). Europe's terrestrial biosphere absorbs 7 to 12% of European Anthropogenic emissions. Science. 300: 1538–1542.

Janssens I.A., et al. (2005). The carbon budget of terrestrial ecosystems at country-scale—A European case study. Biogeosciences. 2: 15–26.

Jones J.B. and Mulholland P.J. (1998). Carbon dioxide variation in a hardwood forest stream: An integrative measure of whole catchment soil respiration. Ecosystems. 1: 183–196.

Kempe S. (1982). Long term record of CO_2 pressure fluctuations in freshwaters. Mitteilungen aus dem Geologish-Paläontologishen Institut der Universität Hamburg. 52: 91–332.

Kempe S. (1984). Sinks of the anthropogenically enhanced carbon cycle in surface fresh waters. Journal of Geophysical Research. 89: 4657–4676.

Kesselmeier J. (2005). Volatile organic carbon compound emissions in relation to plant carbon fixation and the terrestrial carbon budget, Global Biogeochemical Cycles. 16: 11.

Lafont S., Kergoat L., Dedieu G., Chevillard A., Karstens U., and Kolle O. (2002). Spatial and temporal variability of land CO_2 fluxes estimated with remote sensing and analysis data over western Eurasia. Tellus. 54B: 820–833.

Ludwig W., Meybeck M., and Abousamra F. (2003). Riverine transport of water, sediments and pollutants to the Mediterranean Sea. Medit. Action Technical Report Series #141, UNEP/MAP Athens, 111 pp.

Meybeck M. (1993). Riverine transport of atmospheric carbon: Sources, global typology and budget. Water, Air Soil Pollution, 70: 443–464.

Meybeck M. (2005). Global distribution and behaviour of carbon species in world rivers. In Soil Erosion and Carbon dynamics. Roose E., Lal R., Feller C., Barthès B., Stewart B.A. (Eds.), Advances in Soil Science Series, CRC Boca Raton, FL, 209–238.

Meybeck M. and Ragu A. (1996). River discharges to the oceans. An assessment of suspended solids, major ions, and nutrients. Environment Information and Assessment Report. UNEP, Nairobi, 250 pp.

Meybeck M., Cauwet G., Dessery S., Somville M., Gouleau D., and Billen G. (1988). Nutrients (Organic C, P, N, Si) in the eutrophic river Loire and its estuary. East Coast Shelf Science. 27: 595–624.

Myneni R., Dong J., Tucker C., Kaufmann R.K., Kauppi P.E., Uski J., Zhou L., Alexeyev V., and Hughes M.K. (2001). A large carbon sink in the woody biomass of Northern forests. PNAS. 9: 14784–14789.

Neal C., House W.A., Jarvie H.P., and Eatherall A. (1998). The significance of dissolved carbon dioxide in major rivers entering the North Sea. The Science of the total Environment. 210/211: 187–203.

Pacala S. W., et al. (2001). Consistent land- and atmosphere-based U.S. carbon sink estimates. Science. 292: 2316–2320.

Peylin P., Bousquet P., LeQuéré C., Sitch S., Friedlingstein P., McKinley G.A., Gruber N., Rayner P., and Ciais P. (2005). Multiple constraints on regional CO_2 fluxes variations over land and oceans. Global Biogeochemical Cycles. 19: GB1011, doi:10.1029/2003GB002214.

Ramankutty N. and Foley J. (1998). Characterizing patterns of global land use: An analysis of global croplands data. Global Biogeochemical Cycles. 12: 667–685.

Sarmiento J.L. and Sundquist E.T. (1992). Revised budget for the oceanic uptake of anthropogenic carbon dioxide. Nature. 356: 589–593.

Schiettecatte, L.-S., Thomas, H., Bozec, Y., and Borges, A.V. (2007). High temporal coverage of carbon dioxide measurements in the Southern Bight of the North Sea., Marine Chemistry, submitted.

Sobek S., Algesten G., Bergstrom A.-K., Jansson M., Tranvik L.J. (2003). The catchment and climate regulation of pCO2 in boreal lakes. Global Change Biology, 9: 630–41.

Takahashi T., Sutherland S.C., Sweeney C., Poisson A., Metzl N., Tilbrook B., Bates N.R., Wanninkhof R., Feely R.A., Sabine C., Olafsson J., and Nojiri Y. (2002). Global sea-air CO_2 flux based on climatological surface ocean pCO_2, and seasonal biological and temperature effects. Deep-Sea Research. II 49(9–10): 1601–1622.

Tans P.P., Fung I.Y., and Enting I.G. (1995). Storage versus flux budgets: The terrestrial uptake of CO_2 during the 1980s. In Biotic Feedbacks in the Global System. Woodwell G.M. and Mackensie F.T. (Eds.), Oxford University Press, New York. 351–366.

Thomas H., Bozec Y., De Baar H.J.W., Elkalay K., Frankignoulle M., Schiettecatte L.-S., and Borges A.V. (2005). The carbon budget of the North Sea. Biogeosciences. 2(1): 87–96.

Tranvik L. (2005). Terrestrial dissolved organic matter—A huge but not unlimited subsidy to aquatic ecosystems. ASLO summer meeting, Santiago de Compostella. 19–24 June 2005.

Vörösmarty C.J., Meybeck M., Fekete B., Sharma K., Green P., and Syvitski J. (2003). Anthropogenic sediment retention: Major global-scale impact from the population of registered impoundments. Global Planetary Changes. 39: 169–190.

Wanninkhof R., Mulholland P.J., and Elwood J.W. (1990). Gas exchange rates for a first order stream determined with deliberate and natural tracers. Water Resources Research 26: 1621–1630.

Chapter 17
Multiple Constraint Estimates of the European Carbon Balance

Martin Heimann, Christian Rödenbeck, and Galina Churkina

17.1 Introduction

The multiple-constraint approach has become a paradigm of carbon cycle research, in particular in assessments of regional carbon balances and its temporal evolution. In principle, trace gas budgets can be estimated by two complementary approaches: in the bottom-up method, local point-wise information (e.g., flux measurements or inventory data in representative locations) is scaled up to the region of interest using a combination of geographical information system (GIS) and remote sensing data. For the upscaling, various extrapolation procedures, diagnostic or prognostic models have to be used. In contrast, the top-down approach is based on atmospheric concentration measurements of the trace gas under consideration. In this case, the atmosphere is used as a natural integrator of the fluxes from the heterogeneous region of interest. Since the sources and sinks of the trace gas are reflected in spatial and temporal atmospheric concentration variations, observations of the latter can be used in an inverse model of atmospheric transport in order to determine the surface sources and sinks.

In the following two sections, we briefly describe the principles of the top-down and the bottom-up approach, illustrate them with an exemplary result, and discuss their inherent limitations. Section 17.4 addresses the problem of consistency between the top-down and bottom-up approach, which is complicated by additional "minor" carbon fluxes that need to be assessed from additional information. In Sect. 17.5, we briefly outline the problem of merging the different carbon cycle observation data streams into a single data assimilation modeling framework.

This chapter is focused on the European carbon balance. In principle, the multiple-constraint approach is also applicable to other regions of the globe and to other species such as CH_4 (see Chap. 14). Although we believe that similar methodical limitations prevail in other areas, the relative importance may be different because of different environmental and historical conditions as well as different density and quality of the observations.

17.2 The Top-Down Inversion Approach

17.2.1 Method

The consistent inference of the distribution and temporal variations of CO_2 and other long-lived greenhouse gases from atmospheric concentration observations constitutes a problem of considerable complexity. It has to be addressed with the help of a numerical atmospheric transport model, which relates the surface fluxes to the atmospheric concentrations at the observing sites. Such a model essentially numerically solves the continuity equation for an atmospheric trace species given the three-dimensional, time-varying meteorological fields describing the state of the atmosphere. These are either obtained from analyses of numerical weather forecast models or from atmospheric general circulation models running in climate mode. Currently employed global atmospheric transport models have horizontal resolutions of 2–4° latitude and longitude and up to 50 layers in the vertical dimension. In order to represent a particular region more closely, a nested higher resolution grid or a nonuniform zoom region may be employed. Ultimately, the resolution of atmospheric transport models is limited by the resolution of the parent model providing the meteorological fields. This holds also for the temporal resolution, which typically is 6 h as determined by the availability of the meteorological analyses.

Since the atmospheric transport for a conservative trace gas (e.g., CO_2) is linear, the inversion problem can be seen as a linear regression problem. The domain of interest is expanded into a finite series of spatio-temporal source patterns. Using the atmospheric transport model, a "base function" is computed for each of these source patterns, defined as the atmospheric response at the observation locations and times, subsequent to a unit input emitted from the source pattern. Finally, a linear combination of the base functions is determined that optimally matches the measurements at the observation points. The determined weights of the base functions then constitute the source strengths of the different source patterns.

Formally, the atmospheric inversion problem constitutes an ill-posed mathematical problem, which is usually regularized by means of a priori information on spatial and temporal variability of sources and sinks including a priori estimates of their uncertainty. This prior information may be obtained from flux field estimates derived from the bottom-up method or it may be specified on ad hoc plausibility arguments. The prior information is used to define a smaller number of spatio-temporally more complex base functions in order to solve for fewer unknowns compared to the number of observations (Enting et al. 1995; Rayner et al. 1999; Fan et al. 1998; Bousquet et al. 2000; Gurney et al. 2002). Alternatively, one may perform a high-resolution inversion by using the prior information in a Bayesian approach to regularize the problem (Kaminski et al. 1999; Rödenbeck et al. 2003).

A more recent development involves replacing the base function representation of the surface flux with a flux model with unknown parameters (Kaminski et al. 2002; Rayner et al. 2005). The parameters of the combined surface flux—atmospheric transport model—are then determined by minimizing the mismatch between the simulated and observed concentrations. With this approach, the inversion problem generally becomes nonlinear, and, depending on the complexity of the surface flux model, computationally expensive. On the contrary, this approach typically reduces the number of unknowns and thus regularizes the inversion problem essentially using a priori process information as represented in the surface flux model. This parameter optimization approach represents a first step toward a more comprehensive "carbon cycle data assimilation" system (Rayner et al. 2005; see also Sect. 17.4 below).

17.2.2   An Illustrative Example

As an illustrative example, we show in Fig. 17.1 a map of the anomalous surface–atmosphere CO_2 flux for the growing season (May–September) of the year 2003 over Europe as determined by the atmospheric inversion modeling system developed at the Max-Planck-Institute for Biogeochemistry (Rödenbeck et al. 2003; Rödenbeck 2005). The anomalies are computed with respect to a 5-year average 1998–2002. The map shows only the European region, cut out from a global inversion.

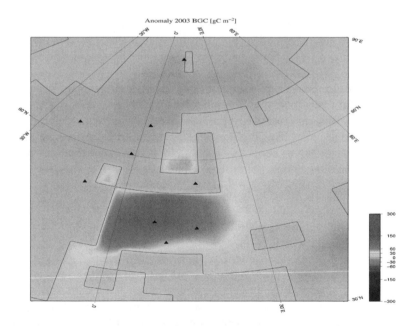

Fig. 17.1 Example of a top-down inversion-based surface–atmosphere CO_2 flux estimate for the European domain. Displayed is the anomalous flux in 2003 (in gC m^{-2}) with respect to the reference period 1998–2002 integrated over the growing season May–September. The black triangles denote the location of the stations whose atmospheric concentration observations were included in the inversion (*See Color Plate* 12).

The coarse grid outline of Europe indicates the resolution of the inversion model (4° latitude × 5° longitude). The atmospheric stations in the European domain that were included in the inversion are indicated by the black triangles.

No prior information on the interannual variability was used in the inversion. Still, the inversion is able to deduce the large-scale pattern of the anomalous surface flux from the atmospheric observations during the hot summer of 2003 (Ciais et al. 2005).

17.2.3   Limitations

There exist a series of important limitations with the inversion problem:

1. Current atmospheric transport models, in particular the numerical representation of subgrid scale mixing (e.g., surface and planetary boundary layer dynamics, vertical transport in convective clouds), are far from perfect.
2. The observational network is very sparse, that is, there exist only a small number (≈ 100) of monitoring stations worldwide, a small number when compared to the heterogeneity of the terrestrial or oceanic carbon sources. Furthermore, at some stations, the sampling frequency is low, and there are often gaps in the observation records.
3. Technically, the "inversion" of the atmospheric transport model is not trivial and requires much larger computing resources than running the model in the forward mode. This limits the spatial and temporal resolution that can be achieved in the inversion.
4. Individual measurements are often not representative of the appropriate temporal and spatial scale of the transport model.
5. A minor limitation may also arise because individual concentration observations are of limited accuracy and precision, and observations from different monitoring networks may not be perfectly comparable, because of differences in measurement techniques and uses of different standards.

The most serious limitation of the top-down approach follows from the limited number of observations and the need to adequately represent sources and sinks with a relatively high spatial and temporal resolution. In general, there exist a very large number of possible surface source-sink configurations that are, in principle, consistent with the atmospheric observations. An example is provided by Kaminski and Heimann (2001), which demonstrate that an unrealistic sink of $2\,\mathrm{PgC}\ a^{-1}$ over Europe may be perfectly compatible with the entire observations from the global monitoring station network. The current atmospheric observations alone are thus not sufficient to uniquely determine the sources at the surface of the Earth.

The mathematical framework of the inversion problem allows to define a formal error of the inferred flux estimates. However, because of the limitations outlined above, this error most likely is too optimistic. Experience with the top-down inverse method has shown an increasing uncertainty with decreasing spatial scale, which is due on the one hand to the low density of the observing stations, and on the other hand to increasing transport errors of the relatively coarse grid numerical models, in particular their subgrid scale mixing process representation, for example, the planetary boundary layer development in regions of strong source-sink heterogeneities. Because of this, current inversion estimates are meaningful only for continental and semicontinental regions. Conversely, in the time domain, the temporal variability (seasonal and interannual) can be better quantified from atmospheric measurements than the longer-term decadal average flux patterns. This is principally caused by the representation error induced by the mismatch of the relatively coarse grid atmospheric transport model (>100-km mesh size) and the point measurement observations. These representation errors are not random, and thus longer-term (e.g., decadal) averaged flux fields determined by atmospheric inversions suffer from systematic biases, which are very difficult to quantify. Because shorter-term flux variations (diurnal, synoptic seasonal) are typically substantially larger than the long-term mean flux, these representation errors tend to be less important on shorter timescales.

The discussion so far addressed the general atmospheric top-down inversion problem. A further complication arises if the carbon budget of a heavy industrialized region, such as Europe, is considered. In Europe, the CO_2 emissions from fossil fuels are a major component of the total surface–atmosphere flux, at least on annual and longer timescales. In order to determine the carbon budget of European ecosystems with the top-down approach, the fossil fuel contribution is subtracted. While the annual emissions, at least on a national basis, are known within an accuracy of 5–10%, the higher spatial resolution of these fluxes, as well as their intraannual temporal variations, are very poorly known and hence introduce an important further uncertainty into top-down estimates of the carbon balance of European ecosystems.

Given the limitations as outlined above, the question arises of how to quantify realistic uncertainties associated with top-down method-based flux estimates. At this point in time, this can be addressed only by means of model intercomparison activities. This is attempted in a systematic way for the global scale in the TRANSCOM experiments (Gurney et al. 2002). For the European region, a similar activity, albeit with fewer models, is currently being performed in CarboEurope-IP.

17.3 The Bottom-Up Approach

17.3.1 Method

The term "bottom-up" approach refers to methods in which local surface flux or surface stock change observations are scaled up to the region of interest using a gridded surface flux model of various complexity and drivers. The fundamental carbon cycle data streams that can be used for estimating the carbon budget of ecosystems in a region such as Europe consists essentially of

1. Surface–atmosphere flux measurements from the network of eddy covariance flux towers in representative ecosystems, and
2. Carbon stock and stock change inventories, primarily for forests.

For the upscaling, a host of additional data streams are available:

1. Gridded meteorological data (temperature, precipitation, humidity, radiation, wind speed, etc.) from analyses of the weather forecast services and from climatologies;
2. Satellite-based remotely sensed ecosystem parameters such as fraction of absorbed photosynthetically active radiation (FAPAR) and above-ground biomass;
3. Various gridded GIS layers of surface properties such as distribution of ecosystem classes, land use, soil types, soil hydraulic properties, nitrogen deposition, etc.

Two fundamental types of bottom-up modeling approaches can be distinguished, which differ in the way these primary data streams are used:

Process-based prognostic terrestrial ecosystem models. These are models that attempt to simulate carbon cycling in terrestrial ecosystems on the basis of fundamental theoretical and empirical process information. They may be termed "prognostic" since they simulate the carbon pool dynamics essentially ab initio, given external drivers such as weather/climate and other environmental conditions. In the bottom-up regional carbon budget estimation problem, these models are run on a spatial grid covering the region of interest and driven by the meteorological data stream and other environmental data (e.g., soil map, hydrology, land use, and N-deposition). A careful initialization procedure is needed in order to eliminate spurious flux trends because of the long turnover times of some of the major carbon pools, such as wood or soils. Typically, it is assumed that prior to the industrial revolution, terrestrial ecosystems were approximately in equilibrium with the prevailing climate. Hence, the models are spun up to equilibrium for this time period. Subsequently, a transient simulation until the present time is performed using available reconstructed historical external drivers such as the rising atmospheric CO_2 concentration, weather or climate, land use history, and/or nitrogen deposition (McGuire et al. 2001). The current direct carbon cycle data streams (flux tower measurements and forest inventories) are used only in the model development phase (Knorr and Kattge 2005) or in the model evaluation (Sitch et al. 2003).

Diagnostic data-driven models. Models in this class use besides the environmental external drivers also direct observations of the carbon/ecosystem itself, primarily remote sensing data such as FAPAR. Examples include light-use efficiency-based models (Reichstein et al. 2003) or a neural network-based upscaling scheme, which is trained at individual flux tower observations and then upscaled to whole regional ecosystems (Papale and Valentini 2003). Forest inventory-derived estimates of regional forest carbon balances are also intrinsically based on a diagnostic bookkeeping approach, which takes into account repeated observations of the age class and biomass distributions and propagates them forward in time (Caspersen et al. 2000).

In practice, the two types of bottom-up modeling approaches outlined above are not exclusive but span a spectrum from purely data driven to fully prognostic ecosystem models.

17.3.2 A Bottom-Up Example and a Comparison with the Corresponding Top-Down Flux Estimates

In Fig. 17.2, we show again the growing season integrated anomalous surface flux over the European domain as computed by the prognostic process-based Lund-Potsdam-Jena (LPJ) dynamical vegetation model (Sitch et al. 2003). For this simulation, the LPJ model was driven by the high-resolution (0.25° × 0.25°) meteorological fields from the REMO regional circulation model, which was run in a

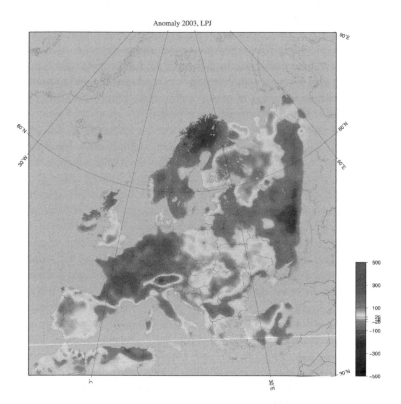

Fig. 17.2 Example of a bottom-up process-based model simulation of the surface–atmosphere CO_2 flux for the European domain. Displayed is the anomalous flux (in gC m^{-2}) in 2003 with respect to the reference period 1998–2002 integrated over the growing season May–September as computed by the LPJ model (Sitch et al. 2003) (*See Color Plate* 13).

data assimilation mode (Feser et al. 2001). Details of the simulation protocol and the driver data as well as a detailed comparison with several other bottom-up models are discussed in the reference Vetter et al. (2007).

The anomalous flux field shown in Fig. 17.2 corresponds to the flux field displayed above in Fig. 17.1 as derived by a top-down inversion. The different spatial resolution of the two approaches is evident. Nevertheless, the large-scale pattern of the summer 2003 anomaly in South-Central and Western Europe is clearly captured by both approaches.

17.3.3   Limitations of the Bottom-Up Approach

Evidently, the major limitation of the bottom-up approach concerns the limited process understanding in ecosystem functioning. Contrary to physical models, for example, of the atmosphere, a fundamental theory based on first principles to describe the cycling of carbon in terrestrial ecosystems does not yet exist. Hence, many of the basic processes have to be modeled by means of empirical relationships obtained from in situ or laboratory manipulative experiments, from comparative studies along, for example, a climate gradient, or simply from heuristic reasoning. Since these observations inevitably do not span the full range and combination of environmental conditions, any application of the models in an extrapolation mode, that is applying it, for example, for an entire continent, will result in flux estimates with uncertainties that are very difficult to estimate.

A second fundamental limitation in the bottom-up approach is inconsistent, inadequate, and/or missing driver data. Examples include dramatic differences in land cover depending on the employed database (Jung et al. 2006), which, when used to drive the models, result in substantially different flux estimates (Jung et al. 2007). A critical factor for bottom-up simulations is the meteorological driver data. In general, the ecosystem models are developed and evaluated at sites with detailed local meteorological observations, such as well-established and documented eddy covariance flux tower sites. For the bottom-up extrapolation, however, gridded meteorological data for the whole domain are used from weather analyses or from climate data. These are found to be inconsistent with the local in situ weather data and also among themselves. Using different meteorological driver data sets yields different flux estimates and thus severely biases the model output (Chen et al. in preparation).

Finally, depending on the region of interest, some of the most important driver data may not be available. For example, the carbon balance of Europe depends to a significant extent on the past history. Historical weather and atmospheric CO_2 concentration over the last 100 years is available (albeit with some uncertainties). Also, the land cover history can be assessed from national statistical reports (Ramankutti et al. 2007) albeit with problematic biases due to differences in the national reporting. However, a consistent, continent-wide history of land management is not available at present. This definitely constitutes a serious limitation for the quality of the bottom-up extrapolated continental carbon balances, which, certainly on a decadal average, is substantially determined by this driving factor. Because of this, as in the top-down approach, also the decadal average bottom-up carbon balance estimates are expected to be less reliable than shorter-term variations (e.g., seasonal cycle, interannual variability).

As in the top-down approach, the classic uncertainty assessment approach is to systematically compare and evaluate against independent observation a suite of models using as much as possible the same driver data and model simulation protocol. Various international bottom-up model intercomparison studies of this kind have been performed, for the global domain (Cramer et al. 2001; McGuire et al. 2001), as well as for individual continents such as Europe (e.g., Vetter et al. 2007).

17.4 Consistency

Given the various limitations of the top-down and bottom-up approaches as described in the previous section, it is natural to ask if consistency between the two approaches can be achieved, and if so, whether the uncertainty of the estimated continental carbon balance can be reduced. For the United States and for Europe, a combined assessment of the top-down and bottom-up carbon decadal average balance estimates have been performed (Pacala et al. 2001; Janssens et al. 2003) and are discussed in Chap. 18 of this volume.

Following up on the examples given in the previous sections, we provide in Fig. 17.3 a quantitative comparison between the top-down and the bottom-up flux estimates. The left panels display the seasonal cycle of the land-atmosphere CO_2 flux integrated over the "western region" (indicated by the letter "W" in the upper right inset), for each month of the year, both for the 5-year average 1998–2002 (red lines) as well as the anomalous year 2003 (light blue dashed lines). It is seen that the bottom-up simulation with the LPJ model exhibits a substantially (60%) larger seasonal cycle amplitude than the top-down inversion. However, the anomaly of 2003 in that region is picked up similarly by both models (lower right panel), even though with not the exact timing of the onset in spring and the decay in fall.

The relatively satisfactory agreement between the flux estimates in this case may be fortuitous and needs further corroboration. A more comprehensive comparison of the top-down and bottom-up approach on a range of spatial and temporal scales is currently being performed as part of the Continental Integration component activity of CarboEurope-IP.

The direct comparison of the modeled bottom-up and the estimated top-down surface–atmosphere fluxes is further complicated by the fact that the atmosphere integrates the terrestrial ecosystem CO_2 exchange fluxes driven by photosynthesis and respiration. In addition to the already mentioned CO_2 fluxes from fossil fuel combustion, there are also a series of additional poorly known "minor" carbon fluxes, which have to be estimated and taken into account. A recent assessment of these fluxes for Europe has been provided by Ciais et al. (2006).

These minor fluxes include the following:

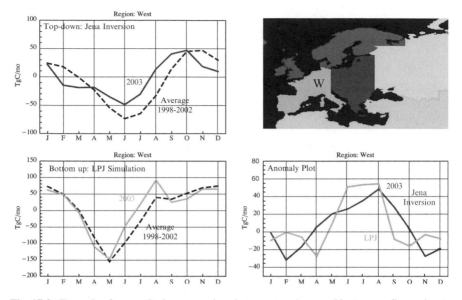

Fig. 17.3 Example of a quantitative comparison between top-down and bottom-up flux estimates over Europe. Left hand panels: time series of the seasonal cycle of the net surface–atmosphere CO_2 flux integrated over the "western region" (green region indicated by the letter "W" in the upper right panel). Black dashed lines: 5 year average (1998–2002) monthly fluxes, colored lines: monthly fluxes in 2003. Top-down approach (upper left panel): fluxes determined by the Jena Inversion modeling system. Bottom-up approach (lower left panel): simulation by the LPJ process-based model. Lower right panel: time series of the monthly anomalous fluxes in 2003, blue: top-down inversion, green: bottom-up simulation (*See Color Plate* 13).

1. Carbon fluxes into and out of the region of interest due to carbon-containing goods (primarily agriculture and wood products)
2. Emissions, transformation, deposition, and atmospheric export of non–CO_2-reduced carbon compounds (e.g., CO and VOC)
3. Carbon flows due to weathering and river transport, including burial in lake and river sediments and estuaries
4. Carbon fluxes into and out of the coastal seas (which cannot be distinguished separately in the top-down inversion)

For Europe, estimates of these fluxes analyzed as a net effect on the European terrestrial ecosystem (vegetation + soils) carbon pool balance yields a small loss of ~0.05 PgCa^{-1}. Analyzed as a contribution to the net surface–atmosphere CO_2 fluxes, their contribution is larger reflecting a sink of about 0.1 PgCa^{-1} (Ciais et al. 2006). (The difference between these estimates arises because there are atmosphere–surface fluxes which do not show up as an accumulation or loss in the vegetation + soil carbon pools.) These flux estimates are longer-term climatological averages and some of these minor fluxes are expected to vary seasonally as well as from year to year.

The magnitude of these numbers is highly uncertain. Unless they are better quantified, they impose a lower limit on the agreement between top-down and bottom-up estimates that ultimately may be achievable, in addition to the errors in the individual budget estimates themselves.

17.5   Toward Carbon Data Assimilation Systems

Conceptually, the multiple-constraint method to estimate a regional carbon balance can be schematically depicted as shown in Fig. 17.4. In the left hand column, the major data streams are listed that are then used by the various models and methods to infer flux maps for the region of interest. By comparing these for consistency, it is hoped that more realistic uncertainty estimates can be derived. Essentially, this scheme represents a separate data analysis and carbon budget assessment method for each primary carbon cycle data stream (concentration, flux, and inventory data).

It is tantalizing to envisage a data assimilation system in which the different input data streams can be jointly inserted in a direct way into a suitable modeling framework. A simplified schematic of such a framework is shown in Fig. 17.5. It also includes an ocean component which has to be included in a comprehensive carbon cycle data assimilation system.

While this clearly must be a goal for the future, we believe that there are still substantial obstacles that have to be overcome before a credible assimilation scheme of this kind may become feasible. On one hand, the major limitations of the top-down and bottom-up approaches, as outlined in the previous sections, will also

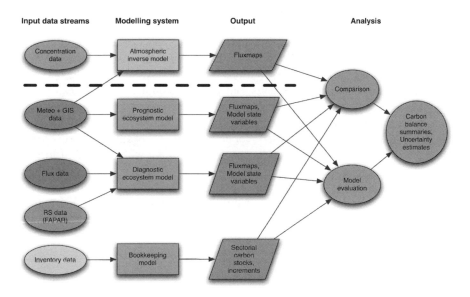

Fig. 17.4 Schematic of the single carbon cycle data stream analysis as currently performed for estimating regional carbon budgets. The dashed black line separates the top-down from the bottom-up approaches (*See Color Plate* 14).

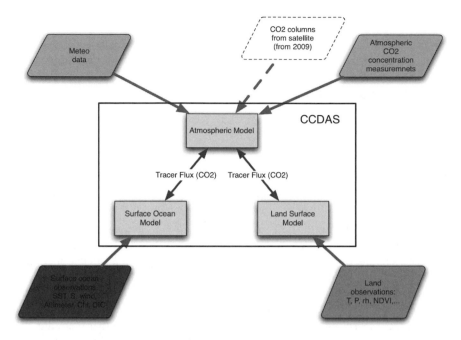

Fig. 17.5 Simplified schematic of a future comprehensive carbon cycle data assimilation system, including atmospheric, oceanic, and land components (*See* Color Plate 14).

pertain to a more comprehensive data assimilation system. Furthermore, the inclusion of carbon cycle data streams of very different temporal and spatial characteristics might introduce substantial mathematical and conceptual difficulties. For example, it is not without good reason that the weather forecast data assimilation community does not assimilate observations of meteorological fluxes, such as rain rate or evaporation flux measurements, but prefers to adjust only the major state variables of the modeling system (e.g., temperature, pressure, wind, humidity, soil water content). A conceptual problem will be the representativity of the data streams with respect to the grid-based model: how can, for example, the flux tower measurements obtained for a particular specific ecosystem with its local history be incorporated in a consistent way into a gridded model with finite (30–100 km) grid size? A further problem will consist of the relative weighting of the different data streams in the optimization procedure. All these problems are not insurmountable, but will necessitate substantial scientific research.

17.6   Conclusions

Despite the current limitations, as outlined in this chapter, the multiple-constraint approach must reside at the core of a regional/continental carbon cycle monitoring strategy—there is no fundamental alternative. A successful implementation, however, depends very much on the monitoring goals to be achieved, as well as the region of interest. There are essentially two objectives:

1. A carbon cycle observing system must provide the necessary scientific information on how the carbon cycle changes due to direct human activities as well as changes in climate those are already underway. This information is indispensable for diagnosing the Earth system as well as for improving comprehensive Earth system models, which are needed for the construction of credible mitigation and adaptation scenarios for the next 100 years. The system must also be designed to capture early enough potential surprises, such as a rapid decay of vulnerable carbon pools (permafrost, wetlands)
2. A carbon cycle observing system has also to serve as a verification tool for documenting and corroborating regional and national trace gas emission reductions

In "hot spots" of the global carbon cycle (e.g., Siberia, Amazonia, the North Atlantic, or the Southern Ocean), clearly objective (1) prevails. In a global perspective, Europe is a major greenhouse gas emitter, but its natural carbon sources and sinks as well as its natural carbon pools are relatively minor, hence a carbon cycle observing system must be designed to address objective (2). This is difficult, because, as discussed in the earlier sections, neither the top-down nor the bottom-up approach allow at present to quantify in a credible way the longer term (pentadal or decadal average) surface–atmosphere carbon fluxes with the necessary accuracy, for example, to detect an emission decrease of 10% over 10 years (corresponding for entire Europe and European Russia to about $150\,\mathrm{TgCa^{-1}}$).

In order to achieve this goal for Europe, several critical improvements to the present system have to be made:

1. Obviously, a more dense, high-quality, and homogeneous long-term observing station network is needed in order to significantly reduce the uncertainties.
2. In terms of data analysis, the most important Achilles heel of the present modeling frameworks is the representativity problem, that is, how a single point measurement can be related to the regional and global models with a finite grid size. This problem affects both the top-down and bottom-up approach and limits to a large extent their validity.
3. Model improvements are also indispensable: for the atmosphere, this means higher spatial and temporal resolution, and also better parameterizations of sub-grid scale transport processes. For the bottom-up models, this involves the incorporation of processes that are currently mostly neglected, such as forest age structure, land management, nitrogen cycling, or adverse effects from ozone and other pollutants.
4. Improved driver and auxiliary data sets have to be compiled. Foremost, this addresses the need for improved, Europe-wide homogeneous fields of fossil fuel CO_2 emissions with higher spatial and temporal resolution than currently available. This also includes longer and continued meteorological reanalyses for homogeneous transport simulations, boundary conditions for nested mesoscale models, and drivers of bottom-up models. Furthermore, a reassessment and recompilation of the historical data sets is needed (in particular, the land management history).

5. Finally, as discussed in Sect. 17.5, the merging of observational data streams of very different spatial and temporal characteristics into a more comprehensive data assimilation modeling framework has to be pursued.

References

Bousquet, P., Peylin, P., Ciais, P., Le Quéré, C., Friedlingstein, P. and Tans, P.P., 2000. Regional changes in carbon dioxide fluxes of land and oceans since 1980. Science, 290(5495): 1342–1346.

Caspersen, J.P., Pacala, S.W., Jenkins, J.C., Hurtt, G.C., Moorcroft, P.R. and Birdsey, R.A., 2000. Contributions of land-use history to carbon accumulation in U.S. forests. Science, 290: 1148–1151.

Ciais, P., Reichstein, M., Viovy, N., Granier, A., Ogée, J., Allard, V., Aubinet, M., Buchmann, N., Bernhofer, C., Carrara, A., Chevallier, F., De Noblet, N., Friend, A.D., Friedlingstein, P., Grünwald, T., Heinesch, B., Keronen, P., Knohl, A., Krinner, G., Loustau, D., Manca, G., Matteucci, G., Miglietta, F., Ourcival, J.M., Papale, D., Pilegaard, K., Rambal, S., Seufert, G., Soussana, J.F., Sanz, M.J., Schulze, E.D., Vesala, T. and Valentini, R., 2005. Europe-wide reduction in primary productivity caused by the heat and drought in 2003. Nature, 437(7058): 529–533.

Ciais, P., Borges, A.V., Abril, G., Meybeck, M., Folberth, G., Hauglustaine, D. and Janssens, I.A., 2006. The impact of lateral carbon fluxes on the European carbon balance. Biogeosciences Discussions, 3: 1529–1559.

Cramer, W., Bondeau, A., Woodward, F.I., Prentice, C., Betts, R.A., Brovkin, V., Cox, P.M., Fisher, V., Foley, J.A., Friend, A.D., Kucharik, C., Lomas, M.R., Ramankutty, N., Sitch, S., Smith, B., White, A. and Young-Molling, C., 2001. Global response of terrestrial ecosystem structure and function to CO_2 and climate change: Results from six dynamic global vegetation models. Global Change Biology, 7(4): 357–373.

Enting, I.G., Trudinger, C.M., and Francey, R.J., 1995. A synthesis inversion of the concentration and 13C of atmospheric CO_2. Tellus, 47B: 35–52.

Fan, S., Gloor, M., Mahlman, J., Pacala, S., Sarmiento, J., Takahashi, T. and Tans, P., 1998. A large terrestrial carbon sink in North America implied by atmospheric and oceanic carbon dioxide data and models. Science, 282(5388): 442–446.

Feser, F., Weisse, R. and von Storch, H., 2001. Multi-decadal atmospheric modeling for Europe yields multi-purpose data. EOS Transactions, 82: 305–310.

Gurney, K.R., Law, R.M., Denning, A.S., Rayner, P.J., Baker, D., Bousquet, P., Bruhwiler, L., Chen, Y.-H., Ciais, P., Fan, S., Fung, I.Y., Gloor, M., Heimann, M., Higuchi, K., John, J., Maki, T., Maksyutov, S., Masarie, K., Peylin, P., Prather, M., Pak, B.C., Randerson, J., Sarmiento, J., Taguchi, S., Takahashi, T. and Yuen, C.-W., 2002. Towards robust regional estimates of CO_2 sources and sinks using atmospheric transport models. Nature, 415(6872): 626–630.

Janssens, I.A., Freibauer, A., Ciais, P., Smith, P., Nabuurs, G.-J., Folberth, G., Schlamadinger, B., Hutjes, R.W.A., Ceulemans, R., Schulze, E.-D., Valentini, R. and Dolman, A.J., 2003. Europe's terrestrial biosphere absorbs 7 to 12% of European anthropogenic CO_2 emissions. Science, 300: 1538–1542.

Jung, M., Vetter, M., Herold, M., Churkina, G., Reichstein, M., Zaehle, S., Ciais, P., Viovy, N., Bondeau, A., Chen, Y., Trusilova, K., Feser, F. and Heimann, M., 2007. Uncertainties of modelling GPP over Europe: A systematic study on the effects of using different drivers and terrestrial biosphere models. Global Biogeochemical Cycles, 21: GB4021, doi:10.1029/2006GB002915.

Jung, M., Henkel, K., Herold, M. and Churkina, G., 2006. Exploiting synergies of global land cover products for carbon cycle modeling. Remote Sensing of Environment, 101: 534–553.

Kaminski, T. and Heimann, M., 2001. Inverse modeling of atmospheric carbon dioxide fluxes. Science, 294(5541): 259a–259a.

Kaminski, T., Heimann, M., and Giering, R., 1999. A coarse grid three-dimensional global inverse model of the atmospheric transport. 2. Inversion of the transport of CO_2 in the 1980s. Journal of Geophysical Research–Atmospheres, 104(D15): 18555–18581.

Kaminski, T., Knorr, W., Rayner, P.J. and Heimann, M., 2002. Assimilating atmospheric data into a terrestrial biosphere model: A case study of the seasonal cycle. Global Biogeochemical Cycles, 16(4): 1066, doi:10.1029/2001GB001463.

Knorr, W. and Kattge, J., 2005. Inversion of terrestrial ecosystem model parameter values against eddy covariance measurements by Monte Carlo sampling. Global Change Biology, 11: 1333–1351.

McGuire, A.D., Sitch, S., Clein, J.S., Dargaville, R., Esser, G., Foley, J., Heimann, M., Joos, F., Kaplan, J., Kicklighter, D.W., Meier, R.A., Melillo, J.M., Moore, III B., Prentice, I.C., Ramankutty, N., Reichenau, T., Schloss, A., Tian, H., Williams, L.J. and Wittenberg, U., 2001. Carbon balance of the terrestrial biosphere in the twentieth century: Analyses of CO_2, climate and land use effects with four process-based ecosystem models. Global Biogeochemical Cycles, 15(1): 183–206.

Pacala, S.W., Hurtt, G.C., Baker, D., Peylin, P., Houghton, R.A., Birdsey, R.A., Heath, L., Sundquist, E.T., Stallard, R.F., Ciais, P., Moorcroft, P., Caspersen, J.P., Shevliakova, E., Moore, B., Kohlmaier, G., Holland, E., Gloor, M., Harmon, M.E., Fan, S.-M., Sarmiento, J.L., Goodale, C.L., Schimel, D. and Field, C.B., 2001. Consistent land- and atmosphere-based U.S. carbon sink estimates. Science, 292(5525): 2316–2320.

Papale, D. and Valentini, A., 2003. A new assessment of European forests carbon exchanges by eddy fluxes and artificial neural network spatialization. Global Change Biology, 9(4): 525–535.

Ramankutti, N. et al., 2007. Global land-cover change: Recent progress, remaining challenges, In Land Use and Land Cover Change: Local Processes, Global Impacts, E. Lambin and H. Geist (eds.), Springer Verlag, New York.

Rayner, P.J., Enting, I.G., Francey, R.J., and Langenfelds, R., 1999. Reconstructing the recent carbon cycle from atmospheric CO_2, delta C-13 and O_2/N_2 observations. Tellus, 51B: 213–232.

Rayner, P.J., Scholze, M., Knorr, W., Kaminski, T., Giering, R. and Widmann, H., 2005. Two decades of terrestrial carbon fluxes from a carbon cycle data assimilation system (CCDAS). Global Biogeochemical Cycles, 19(2): GB2026, doi:10.1029/2004GB002254.

Reichstein, M., Rey, A., Freibauer, A., Tenhunen, J., Valentini, R., Banza, J., Casals, P., Cheng, Y.F., Grünzweig, J.M., Irvine, J., Joffre, R., Law, B.E., Loustau, D., Miglietta, F., Oechel, W., Ourcival, J.-M., Pereira, J.S., Peressotti, A., Ponti, F., Qi, Y., Rambal, S., Rayment, M., Romanya, J., Rossi, F., Tedeschi, V., Tirone, G., Xu, M. and Yakir, D., 2003. Modeling temporal and large-scale spatial variability of soil respiration from soil water availability, temperature and vegetation productivity indices. Global Biogeochemical Cycles, 17(4): 1104, doi:10.1029/2003GB002035.

Rödenbeck, C., Houweling, S., Gloor, M. and Heimann, M., 2003. CO_2 flux history 1982–2001 inferred from atmospheric data using a global inversion of atmospheric transport. Atmospheric Chemistry and Physics, 3: 1919–1964.

Rödenbeck, C. 2005. Estimating CO_2 sources and sinks from atmospheric mixing ratio measurements using a global inversion of atmospheric transport. Tech. Rep. #6, Max-Planck-Institute for Biogeochemistry, Jena, 61 pp.

Sitch, S., Smith, B., Prentice, I.C., Arneth, A., Bondeau, A., Cramer, W., Kaplan, J.O., Levis, S., Lucht, W., Sykes, M.T., Thonicke, K. and Venevsky, S., 2003. Evaluation of ecosystem dynamics, plant geography and terrestrial carbon cycling in the LPJ dynamic global vegetation model. Global Change Biology, 9(2): 161–185.

Vetter, M., Churkina, G., Jung, M., Reichstein, M., Zaehle, S., Bondeau, A., Chen, Y., Ciais, P., Feser, F., Freibauer, A., Geyer, R., Jones, C., Papale, D., Tenhunen, J., Tomelleri, E., Trusilova, K., Viovy, N. and Heimann, M., 2007. Analyzing the causes and spatial pattern of the European 2003 carbon flux anomaly in Europe using seven models. Biogeosciences Discussions, 4: 1–40.

Chapter 18
A Roadmap for a Continental-Scale Greenhouse Gas Observing System in Europe

A. Johannes Dolman, Philippe Ciais, Riccardo Valentini,
Ernst-Detlef Schulze, Martin Heimann, and Annette Freibauer

18.1 Introduction

The global growth rate of CO_2 emissions from fossil fuel burning and other industrial processes has been accelerating from 1.1% year^{-1} in 1990–1999 to more than 3% year^{-1} in the period 2000–2004 (Raupach et al. 2007). This increase has contributed to an observed increase in the atmospheric growth rate of CO_2; during 1990–1999, the atmospheric CO_2 growth rate was estimated at 3.1 Pg C year^{-1}; for the period of 2000–2005, this is estimated at 4.2 Pg C year^{-1} (Denman et al. 2007). The increase in emissions has put the growth rates of atmospheric carbon in recent years at the top end of the SRES (Special Report Emission Scenarios) prepared by the Intergovernmental Panel on Climate Change (IPCC) (Nakicenovic et al. 2000). These scenarios are the most important forcing agents used in current climate models to predict the effects of greenhouse gases and aerosols on the Earth's climate. The dominant factor causing the imbalance appears to be a faster increase in fossil fuel emissions relative to the sink capacity of the land and oceans.

The analysis of Raupach et al. (2007) and the various chapters of this book point to the clear need to continue observing the carbon cycle as a means of both to verify and to fingerprint the source of emissions. Raupach et al. furthermore suggest that increased understanding of where the emissions are accelerating, and where decarbonisation of energy use takes place, has important implications for global equity and burden sharing in the costs of global change. This makes understanding the global carbon balance an issue of direct and fundamental relevance to society.

Under the Kyoto protocol, the majority of industrial countries have agreed to a reduction of emissions and take other measures to reduce the amount of greenhouse gases in the atmosphere. The effectiveness of greenhouse gas emission reductions needs to be monitored through observations. Although countries regularly report to the United Nations Framework Convention on Climate Change (UNFCCC), this reporting is partial, not always verifiable and not in any way related to the atmospheric increase which it is aimed at to halt. This book also shows how the current greenhouse gas monitoring can be improved upon by new techniques and methods.

Denman et al. (2007) noted for the IPCC's Fourth Assessment Report that the budget of anthropogenic CO_2 can now be calculated with improved accuracy. On land, they suggest that there is a better understanding of the contribution to the build up of CO_2 in the atmosphere since 1750 associated with land use and of how the land surface and the terrestrial biosphere interact with a changing climate. At the global scale, inverse modelling techniques (Chaps. 3, 7, 17) used to infer the magnitude and location of major fluxes in the global carbon cycle have continued to mature, reflecting both the refinement of the techniques and the availability of new observations over the interior of the continents.

18.2   Monitoring the Carbon Cycle: The Multiple Constraint Method

Monitoring the carbon cycle and understanding the underlying carbon sequestration and emission mechanisms thus serves essentially two goals (e.g. Chap. 2): increased understanding of the carbon cycle in the context of the Earth system and its future evolution and provision of data and analysis required to support policy decisions about management of the global carbon cycle. The chapters of this book show how precise and carefully executed observations combining atmospheric and ecosystem networks can improve regional-scale flux estimates and the attribution of flux and flux changes to the underlying processes. However, it must also be said that our current scientific and technological capacity is not yet fully up to the task of providing accurate regional budgets of greenhouse gas sources and sinks integrated across each nation's border. This makes it highly complicated for instance to determine the response of eco- and agro-systems to climate and management change. This uncertainty hinders future CO_2 and climate change projections and verification of approaches to curb emissions.

The distribution of regional fluxes over land and oceans can be retrieved using observations of atmospheric CO_2 and related tracers within models of atmospheric transport. This is feasible because the atmosphere mixes and integrates surface fluxes that vary both spatially and temporally very efficiently. Atmospheric inversions are based on this approach and determine an optimal set of fluxes that minimise the mismatch between modelled and observed concentrations, while accounting for measurement and model errors (Chap. 17). Chapters 3, 7 and 17 provide examples of this approach for both CO_2 and CH_4 at the continental to regional scale, although the latter is still very much in development (Chap. 14). This is what is generally called the 'top–down' approach to estimating fluxes.

Top–down inversion results based upon the current set of atmospheric observation sites and global, relatively coarse-scale transport models (Chaps. 3, 17) strongly suggest the existence of a significant carbon sink over northern ecosystems. As mentioned, the atmospheric network however is still too sparse over most of the continents to fully use the potential of the technique. The current sparseness of the atmospheric network severely limits adding more spatial resolution to the continental or even latitudinal result. However, in the CARBOEUROPE project, a new set of eight tall towers and additional surface stations fill in this gap, enabling to determine fluxes at higher resolutions using mesoscale transport models. The implementation of such a denser atmospheric network is still at early stages, but the towers are very useful in that they also provide information on other trace gases such as CO, CH_4, N_2O, SF_6, H_2, N_2/O_2 ratio and several isotopic variants. At the regional-scale level, more dense observation networks are required to achieve the desired improved spatial and temporal resolution (e.g. Chap. 14).

In addition to the inversion estimates, regional time-integrated fluxes can also be determined by measuring carbon stock changes at repeated intervals, from which the time-integrated fluxes can be deduced, or by direct observations of the fluxes at the ecosystem–atmosphere interface. This is known as the 'bottom–up' approach. Ideally, the top–down and bottom–up estimates should agree within error bounds. It is worth considering two recent papers that tried to compare various bottom–up and top–down estimates of the carbon balance of the European and North American continents (Janssens et al. 2003; Pacala et al. 2001). Table 18.1 compares the bottom–up estimates of the two papers.

Broadly, the bottom–up estimates are within the range of the inversions. This is largely due to the large uncertainty surrounding the inversion estimates that suffer from poor resolution capability and performance in longitude. When biomass inventories (e.g. Chap. 10) are used to derive estimates of the sink strength of forest

Table 18.1 Comparison of bottom–up estimates of European and North American carbon balance based on Pacala et al. (2001) and Janssens et al. (2003)

	Area USA (Mha)	Area EU (Mha)	NBP USA	NBP EU
Emissions			1.3	1.8 (including eastern Europe)
Forest trees (g C m^{-2} year^{-1})	247	339	−44.5, −60.7	−107
Other wooded land (g C m^{-2} year^{-1})	247	50	−12.1, −60.7	−28.0
Croplands (g C m^{-2} year^{-1})	184	326	21.5	92
Non-forest, non-cropland (including grassland and woody encroachment) (g C m^{-2} year^{-1})	335	151	−35.8, −38.8	−67
Total area (Mha) and total land sink (Pg C year^{-1})	1,013	866	0.3, −0.58	−0.14, 0.21
Inversion estimates (Pg C year^{-1})	Original studies in Pacala et al. and Janssens et al.		0, −1.7	−0.08, −0.56 (mean 0.29)
Denman et al. 2007 (Pg C year^{-1})	IPCC 4AR range		−0.6 to −1.1	−0.9 to −0.2

A negative sign indicates a sink (uptake)

IPCC 4AR IPCC Fourth Assessment Report

from measured tree growth increments at repeated intervals, lower values of uptake Net Biome Production Net Ecosystem Production, Net Ecosystem Exchange (NBP) are found than those obtained by inverse models. However, Table 18.1 also highlights important differences between the inventories of both continents. These become apparent when we look at the rates of sequestration used in the two studies, rather than their overall total sink estimates. The first difference is the rate of forest C sequestration. Pacala et al. using forest inventory data calculate a value that is only half the value used by Janssens et al. (2003), which is based on (direct) eddy correlation flux measurements. The latter are reduced using a fixed Net Ecosystem Production (NEP)/NBP ratio of 0.47, taken from forest biomass inventory estimates. Pacala et al. (2001), in contrast, use the estimates from the forest inventory (Chap. 10) and model estimates to account for losses of litter, slash, woody debris, fires and harvest. Rates of carbon uptake by eddy correlation Net Ecosystem Exchange (NEE) in the USA and EU derived from eddy correlation tend to be similar, even higher in the USA (Chap. 11). There is thus still considerable uncertainty associated with these numbers and their correct use and interpretation in carbon balance estimates. The most obvious difference between the two continental-scale analyses is the estimate for croplands. In the EU these are estimated to be sources of carbon, whereas in the USA they are estimated as sinks. There is, as Janssens et al. (2003) argue, considerable uncertainty surrounding this number, as it is based on a single model study that can hardly be validated with existing empirical data. Chapter 13 shows how progress in this area can be achieved by modelling and observation of the full greenhouse gas balance at the farm and regional level. Importantly, data on land-use management are required to achieve improved realism in these estimates.

Eddy covariance techniques allow continuous monitoring of CO_2, H_2O, heat fluxes over vegetation canopies (Chap. 11). They serves as important data providers for development and testing of ecological models, although considerable uncertainty remains in splitting up the net measurements into respiration and assimilation components. Also the methodology is not completely robust enough to deal adequately with problems of night-time fluxes and advection (Chap. 11). They are increasingly used to measure the fluxes of the greenhouse gases of CH_4 and N_2O. The temporal resolution of a day or so for eddy flux towers is sufficient to capture the variability in terrestrial fluxes as it is driven by changing weather patterns (e.g. the effect of frost or drought on forests). There is a need to transform these largely research-based networks of flux towers into a routinely operating system that also provide long-term records of key ecological and soil variables that directly influence the biospheric uptake or emissions. However, terrestrial ecosystem carbon fluxes are so heterogeneous and variable that it will be virtually impossible to measure them over all kinds of ecosystems continually over Europe and adjacent regions (e.g. Chaps. 11–13). As it is impossible to sample all the possible permutations of land use (history) and climate with eddy covariance towers, a limited number of ecosystem observation sites may need to be selected where continuous and intense measurements of the CO_2, H_2O and other greenhouse gases as well as heat fluxes are made. Flux data at these sites can then be complemented by carbon and nitrogen pool and biophysical parameters measurements to gain full understanding of the processes that control the greenhouse balance of ecosystems. These long-term ecosystem observation sites can be the basis for upscaling, using advanced models which can integrate the terrestrial greenhouse gas fluxes.

In general, our analysis shows that forests are relatively well sampled in eddy covariance networks (Chap. 11), but we note that this sampling density is rather different from those of forest inventories that have a much higher spatial sampling density (e.g. Chap. 10). However, uncertainties in both the CH_4 and the N_2O (Chaps. 7, 8) budgets would require an increased focus on peatland (Chap. 12) and grassland ecosystems (Chap. 13). Redesigning the networks to move from a carbon observing system to a full greenhouse gas observing system may require major relocations and improvements to the current networks, including precise regional atmospheric observations of these greenhouse gases, also in more polluted areas.

A full greenhouse monitoring network is ideally be planned as a long-term network designed from the very beginning to measure both the atmosphere and the representative ecosystems in an integrated manner (e.g. Chaps. 11, 17). There is a clear need to improve our capability to observe the temporal and spatial distribution of regional fossil fuel emissions. The most reliable approach towards a better quantification of high-resolution fossil fuel CO_2 over Europe is (Chap. 3) an observation-based approach that combines for instance weekly integrated high-precision $^{14}CO_2$ measurements with precise continuous CO observations that can be used as tracer (Chap. 4).

As Chap. 16 shows, there is also a need to determine the precise magnitude and variability of the lateral fluxes in trade and river transport that affect continental-scale-averaged budgets. These lateral fluxes over Europe imply a CO_2 uptake in ecosystems of equivalent magnitude as the carbon accumulation in forests. They have an even larger impact on regional carbon budget estimates.

The chapters in this book show that approaches to establish greenhouse budgets so far have relied on a combination of various methods to assess the individual budget components and that this is likely to remain the case in the near future. This book has attempted to show both the 'top–down' and the 'bottom–up' approaches, and in particular how these need to be merged into a single multiple constraint estimate to provide realistic estimates of the continental-scale greenhouse gas budget. We have deliberately chosen to show that a combination of process understanding and data integration is required to achieve the ambitious goal of a multiple constraint methodology (e.g. Chap. 17).

18.3   An Integrated Carbon Observing System

Regional, integrated carbon measurement networks already exist in the USA (for the atmosphere), and in Japan and Australia. Yet, a unified European long-term carbon observing network is still lacking. The state of the art is that diverse groups working in the framework of the CARBOEUROPE project (2004–2008) are currently bearing the burden of running the flux towers and the atmospheric observatories (e.g. Chaps. 3, 11, 17). Although, measuring sites are numerous in Europe (see Figs. 18.1–18.3), the funding is short term and depends strongly on national funds. Despite the continuing CARBOEUROPE efforts, some sites have closed recently, and instrumentation is not yet perfectly traceable on uniform scale. Such measurement variations make it difficult to detect the changes in trends that reflect the efficiency of management strategies or reveal surprises. The CARBOEUROPE series of projects essentially led to a research network that explored the feasibility and the methodologies of the integrated top–down and bottom–up process, but this is not an operational observation network. Nevertheless, a lot of work still needs to be done, also concerning the extension of the networks towards other greenhouse gases such as N_2O.

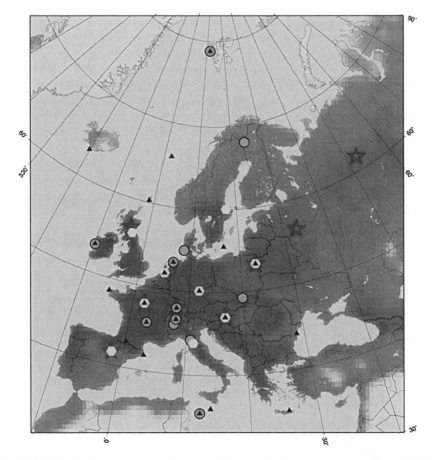

Fig. 18.1 Existing European research network of atmospheric concentration among which the Integrated Carbon Observing System (ICOS) main sites and associated sites will be selected and essential new sites implemented. Stars are regular lower troposphere profiling by aircraft, diamonds the continuous CO_2, CH_4, CO, SF_6, N_2O measurements on tall towers. Diamonds in grey: continuous measurements on intermediate size towers. Circles: continuous surface stations. Triangles: flask sampling sites (*See Color Plate* 15).

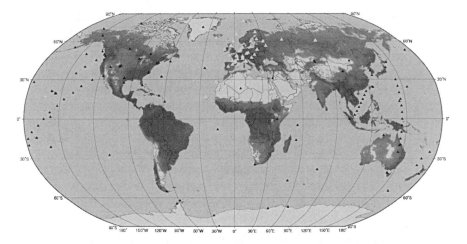

Fig. 18.2 Existing global atmospheric concentration network (ref.) (*See Color Plate* 15).

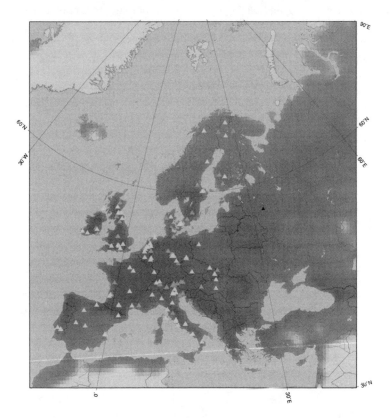

Fig. 18.3 Existing European research network of ecosystem observation sites among which the Integrated Carbon Observing System (ICOS) main sites and associated sites will be selected and essential new sites implemented. White triangles are CARBOEUROPE flux sites and black triangles flux sites from other projects (*See Color Plate* 16).

Operational networks ideally should adhere to the monitoring principles of the Global Climate and Terrestrial Observing Systems (GCOS and GTOS). These consist of an established set of principles (http://www.wmo.int/pages/prog/gcos/Publications/GCOS_Climate_Monitoring_Principles.pdf) that are far from trivial in their practical execution. They concern detailed measurement protocols, quality control and data management plans for secure long-term operation. For Europe, plans for an Integrated Carbon Observing System (ICOS; not yet a full greenhouse gases observing system) exist. They involve the establishment of about 30 atmospheric main sites including remote sites and tall towers with co-located aircraft observation, and 30 ecosystem main sites (e.g. Chap. 11) over Europe. A permanent quality control system would be established to ensure the long-term integrity of these main sites. The main sites, once selected in this project using optimal network design modelling (Chap. 3) and logistical adequacy criteria, would form a backbone observing network, guaranteed to become the future operational network. Determining accurately the time and space distribution of greenhouse gas fluxes at a high resolution requires, however, a much larger number of associated regional sites (see also Figs. 18.1 and 18.4). This implies that strategic choice needs to be made of having either a few main sites, at which the largest number of core parameters will be measured operationally with the highest precision or a larger number of associated regional sites, at which measuring the full list of parameters at the highest precision will not be mandatory. The associated sites would then form the basis of the Regional Observing Network. The ICOS could support the Regional Observing Network through common data management systems, powerful outreach tools, and access to cheap and low-maintenance sensors and standards to ensure that these sites can be maintained with a minimum baseline funding.

The impending launch of satellites specially targeted at monitoring components of the global carbon cycle, especially the atmospheric reservoir, could provide a quan-

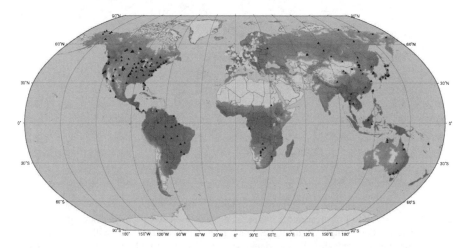

Fig. 18.4 Existing global research network of ecosystem observation sites (ref.) (*See Color Plate* 16).

tum leap in our ability to diagnose the behaviour of this cycle at global and regional scales. The two currently proposed missions, Orbiting Carbon Observatory (OCO) and Greenhouse Gases Observing Satellite (GOSAT), are experimental missions designed to test the technology. These tests will require the closest possible integration with ground-based remote sensing and in situ measurements of biomass and other biophysical variables. This requires both sets of data to be 'integration ready'.

As is the case for ground-based flask data, satellite measurements of atmospheric greenhouse gas concentrations provide no direct information on carbon exchange processes, but they can constrain atmospheric transport models whose boundary conditions are the surface fluxes we wish to know. In contrast, other satellite measurements can provide information on relevant surface properties Leaf Area Index (albedo, LAI) and processes fraction of Photosynthetically Active Radiation (fPAR), but the connection to carbon exchange estimates is normally indirect, and can only be achieved by interfacing them with an ecosystem model. Effective use of the data requires that this interface be well founded. Chapter 15 explores how this can be achieved through increased use of data assimilation techniques, but Chap. 17 also highlights some of the inherent difficulties.

The successful deployment and long-term operational viability of such an ICOS also require central facilities for data acquisition and quality control and dedicated analysis. Integration with other data sources such as those from forest inventory, soil carbon inventories, agricultural data, land-use history and accurate regional fossil fuel estimates is necessary to make such a system work to its full capacity. Again, this is far from trivial. The data streams of the various components, which show great temporal and spatial variability, need to be merged in a single or multiple coherent data assimilation network. So far, only prototypes of such systems exist (Chaps 3, 15, 17, e.g. Peters et al. 2005). Nevertheless, we are confident that a preparatory phase of ICOS will be established in Europe.

18.4 Conclusions

An integrated global carbon observation system is a key contribution to the strategy of the Group on Earth Observations (GEO) and the Global Earth Observation System of Systems (GEOSS), as called for in recent Earth Observation Summits. The development of a global system of carbon cycle observations is the goal of the Integrated Carbon Observing System (IGCO). The IGCO Implementation Plan described a large number of actions (over 100 for all reservoirs, including oceans, land and atmosphere) that are needed to expand the current observing system in such a way that a fully integrated observation system of core variables is achieved. As this book shows, some actions are already being carried out, while others are still waiting to be addressed. Completing the plan will have to involve scientists, agency representatives, technicians and policymakers together. Because of the connection between the carbon cycle and other components of the Earth System, an ICOS not only supports carbon cycle studies and assessments but also provides key information for managing and understanding the water cycle, improving weather prediction, monitoring and managing terrestrial ecosystems, and supporting agricultural sustainability.

A fully integrated carbon observing system that operates at global scale is, however, still a long way away. The expansion of a uniform monitoring system across the globe requires putting together several, independently funded, continental-scale observational programmes. This requires, among others, harmonizing of acquisition and data access and quality control protocols and procedures of the carbon cycle variables across international partners. Improvement of the interoperability of data sets that are used in global-scale carbon cycle studies is crucial to an improved assessment of regional sources and sinks distributions. Only then can we narrow down uncertainties and decrease differences in emerging global data sets that are aimed at providing constraints on the vulnerability of the global carbon cycle.

References

Denman, K.L., Brasseur, G., Chidthaisong, A., Ciais, P., Cox, P.M., Dickinson, R.E., Hauglustaine, D., Heinze, C., Holland, E., Jacob, D., Lohmann, U., Ramachandran, S, da Silva Dias, P.L., Wofsy, S.C. and Zhang, X. 2007. Couplings Between Changes in the Climate System and Biogeochemistry. In: *Climate Change 2007: The Physical Science Basis. Contribution of Working Group I to the Fourth Assessment Report of the Intergovernmental Panel on Climate Change* [Solomon, S., D. Qin, M. Manning, Z. Chen, M. Marquis, K.B. Averyt, M.Tignor and H.L. Miller (eds.)]. Cambridge University Press, Cambridge, United Kingdom and New York, NY, USA.

Janssens, I.A., Freibauer, A., Ciais, P., Smith, P., Nabuurs, G.-J., Folberth, G., Schlamadinger, B., Hutjes, R.W.A., Ceulemans, R., Schulze, E.-D., Valentini, R., Dolman, A.J. 2003. Europe's biosphere absorbs 7–12% of anthrogogenic carbon emissions. Science 300, 1538–1542.

Nakicenovic, Nebojsa Joseph Alcamo, Gerald Davis, Bert de Vries, Joergen Fenhann, Stuart Gaffin, Kenneth Gregory, Arnulf Grübler, Tae Yong Jung, Tom Kram, Emilio Lebre La Rovere, Laurie Michaelis, Shunsuke Mori, Tsuneyuki Morita, William Pepper, Hugh Pitcher, Lynn Price, Keywan Riahi, Alexander Roehrl, Hans-Holger Rogner, Alexei Sankovski, Michael Schlesinger, Priyadarshi Shukla, Steven Smith, Robert Swart, Sascha van Rooijen, Nadejda Victor, Zhou Dadi 2000. *IPCC Special Report on Emission Scenarios*. Cambridge University Press, UK.

Pacala, J., Hurtt, S.W., Baker, G.C., Peylin, P., Houghton, R.A., Birdsey, R.A., et al., 2001: Consistent land- and atmosphere-based U.S. carbon sink estimates, Science, 292, 2316–2320.

Peters, W. et al. 2005. An ensemble data assimilation system to estimate CO surface fluxes from atmospheric trace gas observations. J. Geophys. Res., 110(D24304): doi:10.1029/2005JD006157.

Raupach et al. 2007. Global and regional drivers of accelerating CO_2 emissions. PNAS, doi:10.1073/pnas.0700609104.

Index

Ecological Studies

Current List of Titles in Ecological Studies Series:

Printed in The United States of America